Climatic Effects on Individual, Social, and Economic Behavior

Recent Title in
Bibliographies and Indexes in Geography

A Bibliography of Geographic Thought
Catherine L. Brown and James O. Wheeler, compilers

Climatic Effects on Individual, Social, and Economic Behavior

A Physioeconomic Review of Research Across Disciplines

Philip M. Parker

Bibliographies and Indexes in Geography, Number 2

GREENWOOD PRESS
Westport, Connecticut • London

Library of Congress Cataloging-in-Publication Data

Parker, Philip, 1960-
Climatic effects on individual, social, and economic behavior : a physioeconomic review of research across disciplines / Philip M. Parker.
p. cm.—(Bibliographies and indexes in geography, ISSN 1044–8349 ; no. 2)
Includes bibliographical references and indexes.
ISBN 0–313–29400–3 (alk. paper)
1. Climatology—Bibliography. 2. Climatology—Economic aspects—Bibliography. 3. Man—Influence of climate—Bibliography.
I. Series
Z6683.C5P37 1995
[QC981.8.C5]
016.3042'5—dc20 94–41521

British Library Cataloguing in Publication Data is available.

Library of Congress Catalog Card Number: 94–41521
ISBN: 0–313–29400–3
ISSN: 1044–8349

First published in 1995

Greenwood Press, 88 Post Road West, Westport, CT 06881
An imprint of Greenwood Publishing Group, Inc.

Printed in the United States of America

The paper used in this book complies with the Permanent Paper Standard issued by the National Information Standards Organization (Z39.48–1984).

10 9 8 7 6 5 4 3 2 1

Contents

Preface

The volume of published works across various fields complicates the researcher's task in understanding our climate's effects on individual and social behavior. This book represents an attempt to summarize these interrelated works in a single volume. Chapter 1 reviews the literature using a *physioeconomic* framework which closely follows the views of Montesquieu (*The Spirit of the Laws*, 1748). The ideas presented in Chapter 1 originate, however, from conversations I had with an econometrician from San Miguel, a large food company and brewery based in the Philippines with operations throughout Asia. The company had been frustrated in developing international demand models which were applicable to various geographic regions around the world, noting that large regional variations in income per capita, or similar measures of economic wealth or standards of living, often failed to explain variances in consumption. This observation led me to consider a class of models based on climatic and other natural resource variables (*physioeconomic models*). I have relied on these as a general framework from which the literature from both the physical and social sciences can be positioned. Chapter 2 provides a number of key climatic variables for over 230 countries and economic regions which are original to this research and are not readily available from commercial or public sources.

The remaining chapters present select bibliographies in the order suggested by the physioeconomic framework introduced in Chapter 1. Chapter 3 focuses on physiological effects, the most important of these being acclimation (acclimatization), allergies, disease, hormonal secretion, metabolism, and thermoregulation (heat exchange). Chapter 4 covers psychological effects of climate including affective disorders, aggression, personality, and mental illness (including schizophrenia). Chapter 5 considers social or cross-cultural effects, including accidents, crime, vital statistics (fertility, mortality) and suicide. Chapter 6 focuses on economic aspects which some authors feel result from individual and social behaviors generated by climate. These include our consumption behaviors, or preferences for various goods, and production behaviors which largely depend on climatic factors (e.g: agriculture, tourism, and energy). Finally, the literature exploring macroeconomic and public policies is presented, including recent publications on the impact of global climatic change on economic institutions.

Every attempt has been made to include references from both traditional sciences (e.g. medicine, geology, climatology) and the social sciences (e.g. sociology, economics). Provided that the topic of discussion is the effect of climate on individual or social behavior, references readily qualified for inclusion. Four criteria were used, however, during the selection process and should be seen as caveats to the completeness of the bibliographies. The first was that the publication have an English title or abstract (original or translated). The second is that the publication be published in book, dissertation, report, or article form; less formal works which are not readily available to researchers were excluded. To this criteria was added the condition that articles be of an academic nature, meaning that references to the existing literature are generally included in the work, or that the article was subject to a peer review process. Popular press articles, therefore, are generally excluded from the bibliography. Third, only publications which are expressly concerned with climatic effects are included. For example, in numerous books on suicide there are frequent discussions on the role of climate or the seasons. These discussions are made, however, only in passing and in reference to other works. The bibliographies only include publications whose primary or secondary objective is to explore the effects of climate. Works meeting this criteria are typically those cited in more general discussions of the topics in question. Had this criteria not been imposed, well over 2000 references per subject could have been generated. Likewise, articles concerned with non-human behavior or purely engineering issues were also excluded. Fourth, there is an implicit boundary on the term "climate." Climatologists are quick to point out the multifaceted nature of climate. These begin with absolute latitude, or solar climate; the term *climate* being derived from the Greek *klima* meaning latitude from the equator, or one's angle to the sun. Climates are further affected or characterized by the earth's topology, or physical climate, ocean currents, wind conditions or interactions amongst these which generate prevailing weather patterns (short-term variations in temperature, humidity, rainfall, cloud cover, barometric pressure, etc.). These factors in turn are generally held responsible for variances in the world's marine biologic, vegetation and zoological resources. Every attempt has been made to incorporate all of these geophysical aspects, provided climate is discussed in the work. Certain effects of climate, however, are not likely to be included as these either represent marginal areas in the literature, or have not as yet been treated in the sociology or economics literature (e.g. the effects of ozone depletion on skin cancer). This is especially true for Chapter 3 which is concerned with physiological effects; the number of articles dedicated to seasonal allergies (e.g. hay fever) largely exceed the size of the four bibliographies combined. To this extent, the bibliographies should be seen as broad, as opposed to deep, in scope.

Finally, although not a strict criteria, the bibliographies are biased toward more recent publications. This is the true for two reasons. The first is the extensive use of computer databases which tend to index the more recent literature streams. The second is that more recent publications have frequent citations to the important works pre-dating systematic indexing by computers. Nevertheless, a special effort was made to include the more cited references published from the mid-18th century to the mid-20th century, and recent publications not readily available from databases. The usual disclaimer applies, however, as certain works may have been regrettably overlooked. Every effort has been made to avoid a particular regional bias by the extensive use of international databases and libraries, with their networks, in North America, Asia, and Europe. Furthermore, no reporting bias is introduced in the bibliographies to the extent that the results reported were not used as a criteria for inclusion. Citations from various schools of thought, including those showing a lack of climatic effect on a particular behavior, are represented.

There are over 3000 references in the four bibliographies. In addition to original works, a number of databases were consulted including general online systems (DIALOG, Internet connections to libraries, MELVYL) and specific databases (MEDLINE, NTIS, Dissertation Abstracts, SOICOFILE, Social Science Index, Social Science Citation Index, General Science Index, General Science Citation Index, Applied Science and Technology Index, GEOREF, Books in Print, AGRICOLA, PSYCHLIT, ECONLIT, Life Sciences Index and BIO Abstracts, among others). Due to space limitations only first authors are cited in the references and, in many cases, abbreviated journal titles are used. All of the journal abbreviations are those used by the National Library of Medicine (MEDLINE, MEDLARS onLINE) and provide sufficient information for complete identification. These abbreviations are indexed in "List of Serial Index for Online Services" (267 pages), available from the National Technical Information Service (NTIS), U.S. Department of Commerce, 5285 Port Royal Road, Springfield, VA 22161 (tel.: 703-487-4650, 4640); $28.00. For further information on cited dissertations, the reader is referred to Dissertation Abstracts; dissertations available from UMI are noted. For further information on reports, the reader is referred to NTIS. Readers will also note that virtually all of the citations include a brief run-in annotation of subjects discussed in the reference. These have been drawn from the keywords, text, abstracts, or descriptors of the works in question and put in standard format for cross-reference. For more information on a particular citation, the reader should consult further with online services or databases.

Acknowledgements

A number of libraries in North America (the United States Library of Congress, the University of California libraries), Asia (Hong Kong University of Science and Technology - HKUST), and Europe (The British Lending Library, the European Institute of Business Administration - INSEAD) were consulted to compile the bibliographies. Thanks are owed to the librarians in these institutions and especially to Thomas Lundstrom (HKUST) and Christophe Cartay (INSEAD). I am also indebted to Patrick Poon (HKUST), Adam Wong (HKUST), Marie-Louise Berry (INSEAD), and William Fisk (INSEAD), who have assisted me during various phases of this project. Greatest thanks are owed to Eva Szekeres (INSEAD) who helped me in editing and proofing the citations. She and Julian Glaive also helped generate some of the climatic data in Chapter 2; all errors in these are, however, my responsibility.

Certain aspects of Chapter 1 were originally presented at a joint seminar held in mid-1993 at the Hautes Etudes Commerciales (HEC, Paris), and later at the INSEAD Business Economics and Finance seminar in the Fall of 1993. The feedback from participants of those seminars is gratefully acknowledged, as are the comments received on this research from colleagues and students at INSEAD and HKUST. I am also indebted to Miklos Sarvary, Doctoral Candidate (INSEAD), who has been a constant sounding board for the ideas presented in Chapter 1.

Finally, thanks are owed to the Research and Development Committee of INSEAD which has financed portions of this research, and to my family who has let me invest their time in this project.

1

Review of the Literature: A Physioeconomic Framework

This chapter reviews and integrates into a simple framework the extant literature concerning climatic effects on human behavior. While Hippocrates (*Airs, Waters and Places* 460 B.C.) described how climate is responsible or has therapeutic value for certain illnesses, Montesquieu (1748) provides one of the earliest discussions on how climate affects human physiology, sociology, and economics. In *The Spirit of the Laws* (Book 14), Montesquieu attributes a number of behaviors and national characteristics to variations in climate, including culture, work ethic, suicide, alcoholism, aggressivity, religious beliefs (superstition), mortality, fertility, stature, obesity, sexism, combativeness, and industrial development (agriculture, trade, and commerce). Several contemporary works, representing over 3000 citations indexed in Chapters 3 to 7, make similar observations or extend these to additional areas.

In this review, a physioeconomic framework is introduced which integrates this literature. The term *physioeconomics* reflects the wedding of physiological behavior with economics. The key assumption behind this framework is that humans act within exogenous constraints imposed by nature or natural endowments; all natural endowments find their origins in climate. As described later, climatic-driven constraints result in physical, physiological and psychological mechanisms which affect our utilities for various goods and services, our budget constraints (assets and wealth), as well as the relative prices of alternative choices we face, especially when these are considered at the national level. From this perspective, the literature presented in Chapters 3 to 6 can be positioned. For some, but not all behaviors, the principal conclusions drawn from this framework are (1) societies having *absolute* differences in exogenous endowments may have *divergent* long-run *absolute* equilibrium behaviors (i.e. aggregate measures of behavior across countries), (2) societies may *converge* over time to similar levels of behavior *relative* to their endowments, (3) as time progresses, natural endowments may explain cross-society variances in certain social and economic behavior to a greater extent (e.g. vis-a-vis public policy, culture, or other variables), and (4) certain cross-national or cross-cultural studies which fail to control for these physioeconomic constraints can be systematically flawed; this is especially true for studies examining a large number of countries. The last conclusion is based on two arguments. The first is purely statistical in nature and reflects problems associated with positive spatial autocorrelation and panel heterogeneity. The second is more fundamental: both dependent and independent variables in many comparative economic and social science studies are endogenous to or are dependent on climatic and resulting natural resource constraints. The failure to account for these can lead to biased estimation of tested relationships (e.g. the effect of political economy on income per capita). Before the framework is introduced, however, it is useful to define *climate* and briefly discuss methodologies used to estimate its effects.

DEFINITIONS AND METHODOLOGIES

Definitions

The literature generally defines *climate* (e.g. tropical, polar, temperate, high-altitude, etc.) as average meteorological conditions specific to a geographic region over a period of several years or decades (e.g. 30 years). *Weather* is typically defined as short term fluctuations in meteorology, including changes in temperature, humidity, cloudiness, rainfall, and barometric pressure. General weather patterns are associated with particular climatic conditions (e.g. strong seasonality is highly associated with temperate climates). Climatologists have long recognized a hierarchy in climate formation. The earth's climates originate from solar radiation which varies in relation to one's absolute latitude from the equator (the square of latitude provides an approximation of radiation received; the lower the latitude the more the radiation). If the earth had no land masses, the resulting climate would be identical at all points equally distant above and below the equator. Deviations from the solar climate are generated by land areas and topology (including elevation) which result in physical climates and ocean currents. The interactions amongst these generate prevailing weather patterns and microclimates.

Methodologies

Studies of climatic effects in sociology and economics typically consider two forms of effects using both climatic (latitude, elevation) and meteorological (temperature fluctuations) variables: (1) temporal effects using time series data of meteorological patterns, and (2) cross-sectional effects across various climatic regions, including cross-cultural studies. Especially in studies of physiology and psychology, data for both of these approaches have been collected using experimental or clinical (case study) methodologies. In exceptional cases, climatic effects are investigated over time and cross-sectionally. A third approach involves longitudinal studies of climatic effects on current and previous civilizations; these studies are mostly authored by anthropologists, archaeologists, geographers and historians (indexed under *Civilization, Culture & History* in Chapter 5). These works explore dramatic changes in local climates (e.g. prolonged drought, or Ice Ages) as factors affecting migration, housing, agriculture, food consumption patterns, and general customs. Often these studies rely on single case-study observations or civilization/climate combinations. Among the three approaches, time series and cross-sectional studies have greatest frequency in the social sciences and economics literature. In these disciplines, authors frequently test models which independently control for climatic effects in order to study other relationships of interest.

Time Series Studies

Using time series data collected on a daily, monthly, or quarterly basis, numerous studies indexed in Chapters 3 to 6 investigate the role of seasonality on human behavior: the demand for certain ethical drugs, suicide, mood, conception, fertility, mortality, consumer sales, and monetary policy, among others. Seasonality in data, however, does not imply a theoretical link between climate and behavior. It is useful to think of the term "seasonality," as commonly used in the literature, in three separate cases:

Case 1: climatic seasonality
Case 2: non-climatic periodicity
Case 3: non-climatic seasonality

In the first case, exogenous climatic conditions can be held partly responsible for temporal variation in behavior. Housing starts in Wisconsin, for example, follow a seasonal trend for which the severe weather of the winter is generally held to be a recurring cause; large amounts of snowfall, in this case, shift cost (and perhaps demand, from a search point of view) curves in such a way to produce non-economic fluctuations in housing starts. The modeller often filters the dependent behaviors for exogenous seasonal patterns in order to study other causal relationships (e.g. the effects of short-term interest rates on housing starts). When climatic seasonal patterns are controlled for, the author is modelling the tilt of the earth on its axis and its position to the sun. Though this description of seasonal filtering is rarely given, theoretical and empirical models manifestly assume that the earth's tilt is partly responsible for certain behavior over time (within a given climatic area); see *Methodologies* indexed in Chapter 6.

In the second case, seasonality is manifestly independent of climatic conditions. For example, the demand for certain ethical drugs shows a systematic increase every year in the United States around April 15. Without developing a theory for why this occurs (e.g. tax filing tends to cause individual stress), one might control for this periodicity using some seasonal filtering procedure. It is the systematic variation in demand, and not a climatic justification, that makes one control for seasonality in this second case. Though one school of thought (*environmental determinism*) might argue that the tax filing date is itself a climate-driven date, seasonality reflects non-climatic periodicity. In the third case, economic or social behavior seems to follow a regular seasonal pattern which might be mistaken for climatic effects. For example, gift giving appears to be highest during colder periods of the year in certain countries. One might conclude that cold weather causes this generous behavior. To determine if sales are high or low during the winter season, one deseasonalizes the data first for exogenous (non-economic) periodicity. In doing so, we obtain a relative measure of sales (e.g. given the "season"). In this case, the variables used to "deseasonalize" the dependent data may not model the tilt of the sun (though they are correlated with this), but measure the impact of non-climatic cultural rituals (Chinese New Year, or Christian gift giving traditions). In the three cases illustrated, identical modelling approaches might be used to estimate *seasonality*. The research question of interest is whether the phenomenon observed is generated by climatic factors. Cross-sectional studies have supplemented our knowledge in this respect.

Cross-Sectional Studies

There are two prevalent forms of cross-sectional studies in the literature. The first typically compares average behavior across a number of geographic regions with differing climates; climate chambers have been used in experimental settings. While appealing, this method may be limited to descriptive conclusions by failing to identify the mechanisms driving observed differences, especially when behavior is measured at the level of an entire society or economy. The second, more insightful, approach involves meta-analyses or data collected from multiple regions which are simultaneously observed for seasonal behavior. As discussed above, seasonal variation does not imply that behavior is driven or caused by climatic effects. The lack of seasonality in one region and strong seasonality in another does not, however, imply the absence of climatic effects. In fact, the opposite may be true for some behaviors since an observed seasonality may provide evidence of climatic effects in a cross-sectional sense. For example, in temperate climates one sees a dramatic change in dressing behavior over time in co-variance with the seasons (similar observations can be made for most forms of food consumption including protein, fat, and caloric intake). In the city of Washington D.C., for example, we observe that people tend to dress in light garments in the sultry summer and wear heavy coats in the winter. Based on this seasonal observance, one might hypothesize that equatorial climates lead to the behavior of dressing in light garments versus polar climates leading to the behavior of dressing in heavier garments. At the equator, there should be little seasonal variation in dressing behavior as the temperature is relatively invariant over time.

Figure 1 indicates examples of possible outcomes from cross-sectional analyses of temporal data. Each figure represents the time variation of behavior from two distinct geographic areas (e.g. having differing solar climates or absolute latitudes). The top figure depicts, for example, the typical effects noted in the literature for suicidal behavior (see *Suicide* indexed in Chapter 5): countries closer the equator have lower suicide rates which show far less seasonal variation than countries located nearer the poles which have strong seasonal variation. The literature has hypothesized that hormonal differences generated by natural sunlight and/or solar radiation result in these variances within and across climates. Both the cross-sectional and seasonal variations support this conjecture. In other domains, the magnitudes can be reversed, with strong seasonal patterns in, for example, rainfall seasons observed in the lower latitudes not generating equivalent behavior in upper latitudes (e.g. agricultural labor). The middle figure illustrates cases when two climates generate seasonal patterns of a similar nature, though the mean level is divergent at all points. Combinations of these effects with differing leads or lags may also be observed. The bottom figure indicates weak climatic effects. In this case, effects are seasonal in one climate and constant in another; both climates generate, however, similar aggregate behaviors when averaged over time.

In addition to studies which pool time series data are studies which consider aggregate (typically mean) behavior across countries having substantially different climates (e.g. average mortality rates by type of death for countries having temperate versus equatorial climates). While generally accepted as descriptive, these studies have been criticized for not eliminating the possibility of non-climatic factors (e.g. culture, economic policies) generating observed patterns. Studies which begin to overcome this criticism are conducted by researchers who consider the behavior of military personnel and their dependants having the same nationality, socioeconomic status or cultural background (from the armed forces of the U.S.A., Britain, France or the former Soviet Union) stationed at locations having different climates (e.g. absolutes latitudes or elevations). Combined, cross-sectional and time series studies have found climatic-dependent behavioral differences in terms of food consumption (e.g. caloric intake), nutrition and housing requirements, illness, clothing, mood, and physical/cognitive performance.

Figure 1

Seasonality and Climatic Effects on Behavior by Absolute Latitude (Solar Climate)

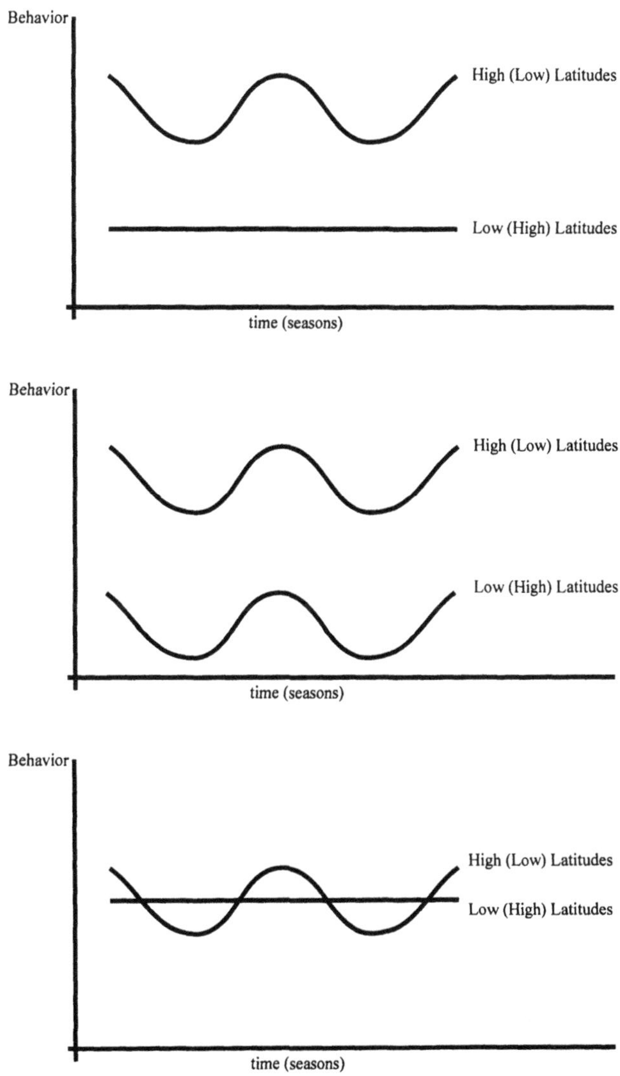

Problems

The lingering problem in most studies of climate is attribution. During the early 20th century, certain geographers, among others, were labelled as holding a determinist view of climatic effects; climate not only correlates with cross-cultural differences in behavior, but also determined these (e.g. the works of E. Huntington). The lack of strong theoretical underpinnings and certain extremes in early writings, especially as directed toward individual behavior including generalizations concerning racial differences, lead to a general discredit of climate-based theories of behavior in geography, sociology and economics. Though some authors have recently argued that these early criticisms were themselves suspect of certain extremes or lacked rigor (see, for example, J. Rotton), the most important problem in attribution is the identification of causal mechanisms.

Recent work in physiology and psychology, however, fills certain gaps in identifying mechanisms lacking in earlier debates. Together, these would tend to suggest that climate, and climate-related natural resources should be considered as "contributing factors" affecting behavior (rather than factors with determining roles). While some debate has occurred as to whether these effects are direct, or are mediated by social factors, the preponderance of evidence would suggest that both temporal and cross-sectional (cross-cultural) studies of behavior ignoring climatic effects may suffer from substantial specification errors. The following physioeconomic framework is presented to illustrate this point.

A PHYSIOECONOMIC FRAMEWORK

Assumptions

The basic premise of this framework is the assumption that all human behaviors are inherently constrained to time-invariant laws of physiology, among other physical laws, which prevail irrespective of cultural, sociological or economic institutions. While the biological and medical sciences indicate that these are numerous (e.g. humans require oxygen to mentally and physically function), three physiological constraints in particular are noteworthy. The first results from humans being homeothermal (or homothermal) mammals (homeotherms). Humans, being warm-blooded, must constantly maintain an internal core body temperature of 37 degrees Celsius, irrespective of where they live or the environments that they face (homoiothermy or homeostatic equilibrium). Heat-exchange behavior balances incoming and outgoing heat by means of physiological processes of thermoregulation in reaction to external factors of radiation, ambient temperature, moisture/humidity, and air movement (see citations indexed under *Thermoregulation* in Chapter 3)[1]. Homeostatic equilibrium, which is controlled by the hypothalamus via nervous and hormonal mechanisms, requires that the human body balance energy absorption and production. In the short-run this process may result in

---

1 This heat transfer relationship can be expressed as:

$$0 = M +- CD +- CV +- R - E$$

where M is the heat generated from metabolism (food consumption, exercise), CD is heat gained/lost by conduction, CV the heat gained/lost by convection, R the heat gained/lost by radiation, and E the heat loss due to evaporation; see Stanier, M.W. et al., *Energy Balance and Temperature Regulation*, Cambridge University Press: Cambridge, 1984.

shivering or sweating, or, the long run, adjustments to metabolic functions (see *Acclimation & Physical Comfort* and *Metabolism* indexed in Chapter 3). In all cases, core or deep body temperature must remain constant irrespective of the environment encountered. Reflecting the need to produce energy via metabolism, physiologists note that some 80 percent of food consumption is required for thermoregulatory needs (the rest being attributed to variety seeking and other behaviors). Failure to maintain energy balance, especially in the extremes, results in various illnesses or death (hypothermia, and heat stroke in the extremes); see Chapter 3, *Thermoregulation*, and *Accidents & Injuries* indexed in Chapter 5. Bioclimatology and biometeorology attributes a number of consumption behaviors to this constraint: the use of tile floors in hot climates (affecting conduction; see *Architecture & Housing* indexed in Chapter 6), an inverse relationship between caloric intake or digestive energy and ambient temperature (the hotter the climate, the less one eats, especially of foods which require high energies to digest such as meat; see *Diet & Nutrition* indexed in Chapter 3), the colder the climate, the greater the need to install artificial heating systems and/or wear heavy/layered clothing (see *Clothing* and *Energy* indexed in Chapter 6) and preferences for alcoholic and non-alcoholic beverages (see *Alcohol & Other Beverages* indexed in Chapter 6). This climatic effect has further been held to explain differences in physiological utility for physical work or exertion; an equivalent level of physical work has greater utility in temperate or polar regions than in equatorial or hot arid regions (see *Labor & Unemployment* indexed in Chapter 6).

The second mechanism, related to the first, affects physical comfort which is dependent on changes in and long-run levels of ambient temperature and humidity. Acclimatization (or acclimation) involves an internal physiological adjustment process, also controlled by the hypothalamus (see *Acclimation & Physical Comfort* indexed in Chapter 3 and *Stress* indexed in Chapter 4). For each climate there exists a temperature-humidity threshold (TH) which, if passed, generates physical discomfort or illness for those acclimated to a particular climate; in hotter climates, human physiology acclimates to allow for higher TH levels before discomfort is generated[2]. Persons acclimated to colder climates have lower TH levels (i.e. are able to feel comfortable at lower TH levels). Persons living in the tropics, for example, have less discomfort than tourists visiting the tropics (as Scandinavian's have less discomfort than tourists coming from the tropics to Scandinavia). Thresholds are generated by both mean climatic conditions and seasonal variations. Persons living in climates with strong seasonality (very cold winters and very hot summers) have highest levels of discomfort all year round, and will have greatest physiological utility for simultaneously owning both indoor heating and summer air conditioning (non-climatic factors being held constant). Acclimation has also been used to explain variances in labor productivity for a variety of industries including mining where the underground environments differ substantially from surface climates (see *Mining*, indexed in Chapter 6).

Based on these two physiological mechanisms, certain climatologists and geographers argue that climate is the single most important factor determining variances in numerous human behaviors, including dietary patterns (caloric/protein intake), housing requirements

---

2 This level is typically expressed as follows:

$$TH = f(Td, Tw, Z)$$

where TH is a temperature humidity index, Td is the dry-bulb temperature, and Tw is the simultaneous wet-bulb temperature; and Z is a vector of constants, which vary by climate.

(heating, insolation, architecture), and clothing requirements (light versus heavy). Since these climate-dependent activities alone represent from 30 to 50 percent of total household consumption in developed economies and up to 90 percent in lesser developed economies, climate, it is argued, cannot but have a substantial impact on aggregate measures of behavior which incorporate these items (e.g. cross-cultural measures of aggregate consumption per capita).

The third mechanism is the subject of recent studies in medicine and physiology which have demonstrated the role of climate in hormonal secretions, some of which result from thermoregulation and acclimatization (e.g. see *Hormones* indexed in Chapter 3). Variances in hormone levels in individuals across climates are used to explain, for example, strong seasonal affective disorders in temperate or polar climates, yet weak affective disorders in equatorial climates (see *Affective Disorders* indexed in Chapter 4); such disorders are commonly treated using light or phototherapy (see also *Mood & Personality* indexed in Chapter 4). Evidence of hormonal influences can also be seen in the seasonality of demand for mental health services (see *Health Care* indexed in Chapter 6). Similar tendencies have been observed for suicide and suicide ideation (see *Suicide* indexed in Chapter 5); persons residing closer to the poles are more likely to experience seasonal affective disorders, depression, and suicide ideation. Suicide and mood disorders rise in the winter months in temperate or polar climates and reach their peaks toward the end of winter/early spring when cumulative imbalances are greatest.

Care must be taken not to confuse these mechanisms with direct causes of specific behaviors at the individual level. In the case of suicide, for example, sociologists (including E. Durkheim in the late 19th century) discarded climate as a factor affecting cross-national suicide rates as climate is rarely, if ever, directly linked to a particular act of suicide. It goes without question that individual acts of suicide are frequently brought on by specific tragedy (as opposed to variations in the weather). This observance does not eliminate the possibility that climate affects one's propensity to commit suicide given a specific personal tragedy (i.e. several persons facing similar tragedies may have differing propensities to commit suicide given their respective prevailing climates -- polar versus equatorial). In some cases, climatic effects may be observed through non-meteorological factors (e.g. alcoholism) which themselves are affected by climate (polar countries generally having higher levels of alcoholism than equatorial climates).

To the mechanisms discussed above can be added a variety of physical constraints discussed at length by human and economic geographers. These reflect necessary physical conditions to various behaviors at the individual or societal level. Variances in these necessary conditions, it is argued, affect equilibrium behavior. Certain sports, for example, can only be practised in geographic areas having appropriate climates (e.g. ice fishing, coral photography, alpine skiing). Certain diseases are limited in nature to areas which have not experienced frosts over the last 10,000 years (e.g. tropical diseases, including malaria; see *Disease & Illness* indexed in Chapter 3). Virtually all animal and plant species are geographically bounded by prevailing climatic conditions (e.g. oak forests and their animal inhabitants), the proximity of which affects the economics of human consumption and transformation. Various industrial activities are also directly bounded by the severity of climate: e.g. the economics of mineral extraction varies by climatic conditions (see *Mining* indexed in Chapter 6); elevation and/or solar radiation prevents the cultivation of certain crops and directly affects soil fertility (see *Agriculture* indexed in Chapter 6). These constraints, including the availability of climate-dependent fresh water, are held to affect the supply and demand of various goods and services, certain cultural traits of populations, economic development patterns, and vital statistics.

Figure 2 graphically summarizes the various effects discussed by identifying a simply hierarchy. Using this framework, one can better understand the incremental role of both exogenous climatic factors and endogenous factors which affect behavior. In this case, "exogenous" factors are meant to include all "inherited" or non-human natural resources. Endogenous factors include any of human origin, including population, culture, religion, language, economic policies, and social systems. Within exogenous and endogenous factors, various sub-hierarchies can be considered. Figure 2 notes the existence of human-generated climate (e.g. in-door air conditioning, heating, pollution and human-induced greenhouse effects) as endogenous to natural endowments which are mostly time invariant over the last ten centuries (e.g. the absolute latitude of a region, the natural existence of mineral deposits, etc.).

Estimating Effects

Estimating climatic effects may prove problematic given the complexity of factors affecting behavior. Climate affects water resources and temperature which affect natural vegetation, which in turn affects dietary preferences and consumption patterns. In the case of cross-national comparisons, for example, we might consider the interrelationships between suicide, alcoholism, income per capita, political systems, and various demographic variables (e.g. mortality levels). Each of these variables have been shown to be dependent on climate; many of these variables may be mediating factors to others. Short of applying numerous structural variable approaches (LISREL), one can filter each variable for climatic effects in the same vein that we deseasonalize time series data. In this case we are not interested in directly estimating all mediating effects (e.g. the effects of alcoholism, net of climate, on suicide), but rather the net effects of climate on suicide and other variables (these in turn can be modelled in some dependence structure).

Figure 3 shows a filtering procedure based on the physioeconomic framework. Natural historians have recognized a hierarchy of exogeneity whereby certain natural phenomenon are necessary conditions (and simultaneously historical precedents) to others. Climate, for example, is a necessary condition for marine and animal/human life, which is in turn a necessary condition for culture or economic behavior. Each level of the hierarchy has sub-hierarchies. Solar climate is the most exogenous of factors in natural history. Physical climate (continental formation) follows, resulting in land and ocean topology; again, without solar radiation, physical-induced climates would not exist. Following the hierarchy of natural history, climatic forces are generally held responsible for variances in the world's natural marine biology, vegetation and zoological resources. Again, each level in the hierarchy is a necessary condition for the next: terrestrial life depending on prior existence of marine life, which itself depends on climatic formation. Minerals are considered after these (though this might be debated) since many are formed based on the prior existence of topology (physical climate) and animal or plant life; their observance, measurement or extraction also necessarily follows, or is endogenous to, animal life. Beyond these purely exogenous factors, difficulties arise in establishing hierarchies in human-generated factors. In contrast to the natural endowments, it may be difficult to accept that these factors are necessary conditions by order of exogeneity (e.g. religion may not be a necessary condition to language). For the sake of illustration, Figure 3 follows a hierarchy generally accepted by historians with respect to current social systems. Modern religions, for example, are generally more exogenous than modern

Figure 2

Physioeconomic Framework

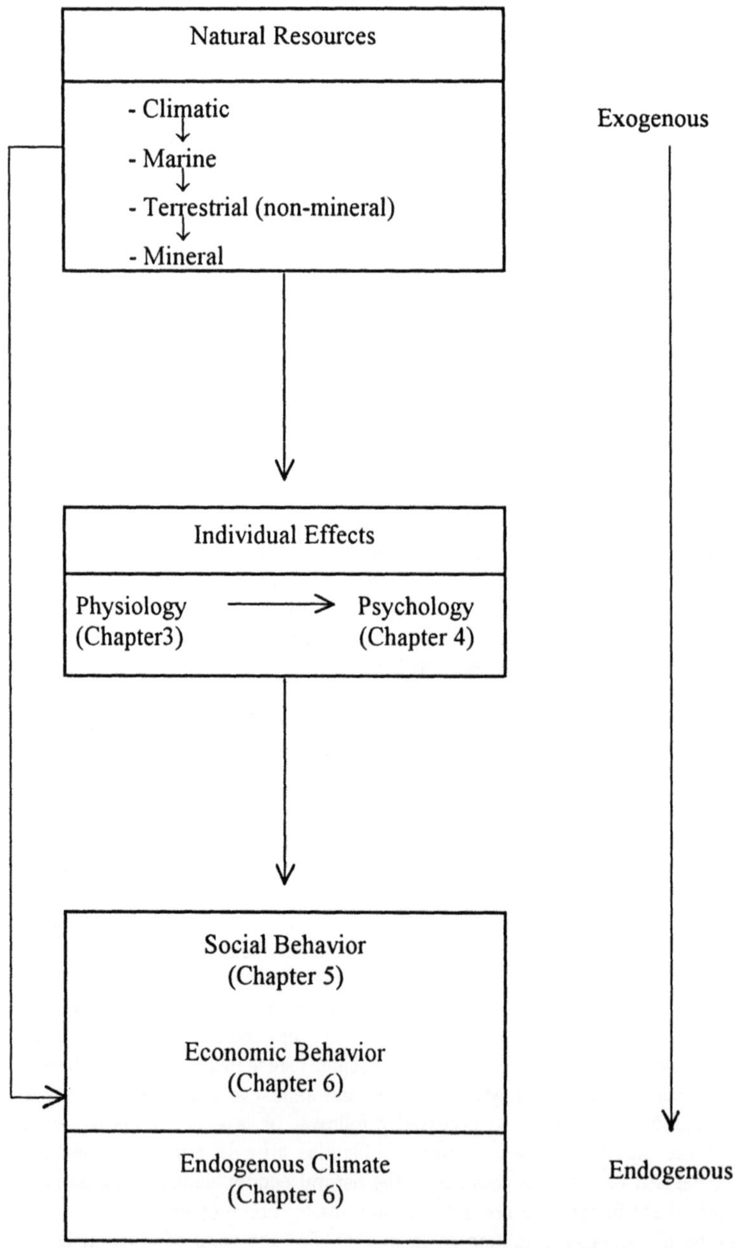

Figure 3

Filtering Sequence
(typical incremental explained variance in parentheses)

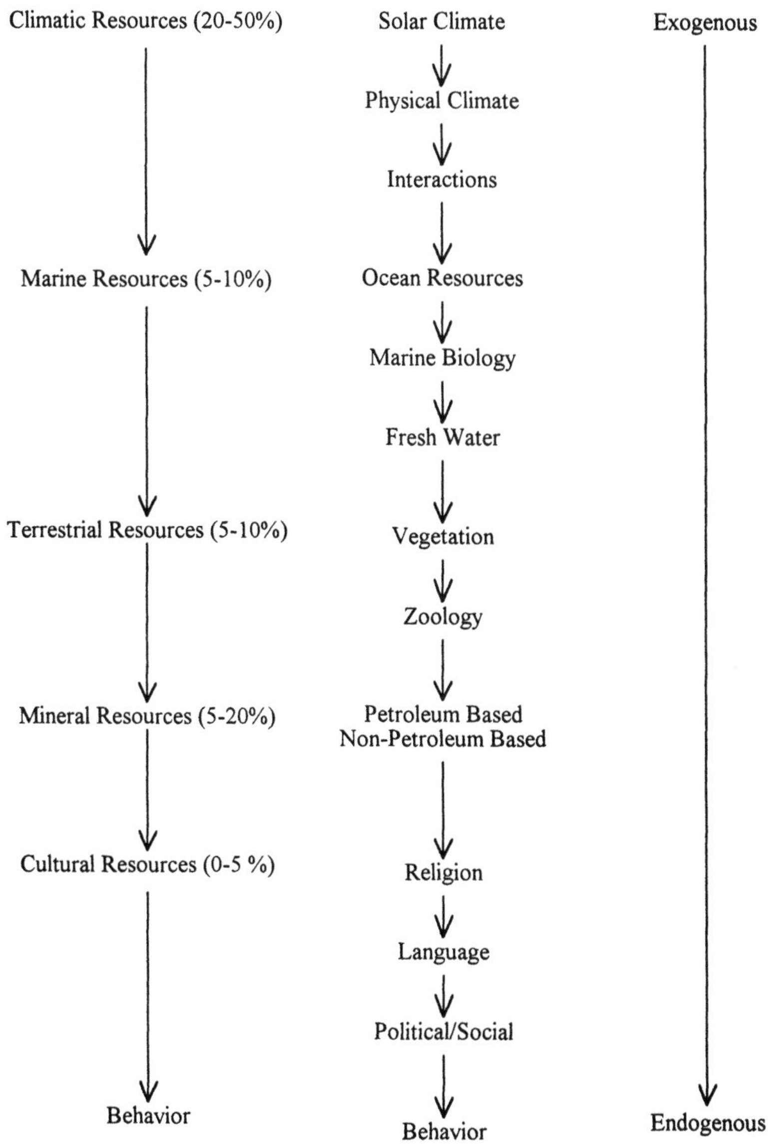

languages which are in turn more exogenous than current economic policies, political systems or social institutions.

This framework can be operationalized for any given human behavior via sequential filtering. The dependent variable (e.g. caloric consumption) is first filtered for solar climate. The resulting residuals become the transformed dependent variable which is in turn filtered for physical climate, and so forth through the hierarchy. This process is analogous to that commonly used in time series studies which attempt to measure the short-run affects of policy on economic behavior which varies in a seasonal pattern (see *Methodologies* indexed in Chapter 6). An alternative approach is to simultaneously filter the original dependent variable for all exogenous factors (which may have been previously factor analyzed). The primary advantage of the sequential filtering process is the insight gained by having a direct measure of the incremental explanatory power of each element along the filtering hierarchy. For example, we can estimate the extent to which Protestantism (or a Protestant work ethic) explains labor productivity beyond variances explained by exogenous elements (i.e. "work ethic" may itself be explained by climate). Similarly, we can test whether religious affiliation (Catholic versus Protestant) affects suicide rates after we control for exogenous geographic effects (given that Catholic countries are more concentrated toward the equator, compared to Protestant countries concentrated toward the poles).

Three additional reasons supporting this approach are worth noting. The first is that it overcomes a fundamental problem observed in many cross-cultural (or cross-national) studies: positive spatial autocorrelation. In a study of income per capita across members of the United Nations, for example, we might expect that the development of the Benelux or Scandinavian countries do not represent independent observations as these tend to have similar patterns due to proximity. The "theoretical" degrees of freedom in unfiltered data are far less than the number of statistical observations (countries) used in the study (often leading to an exaggeration of measured effects). Each filter in sequence, therefore, increases the theoretical degrees of freedom available to the researcher while simultaneously eliminating positive spatial autocorrelation (e.g. behaviors across absolute latitudes, or solar climates, being similar). The second reason is that cross-cultural studies assume that countries are independent panel members whose behaviors do not systematically vary via an omitted exogenous factor; this assumption is inherently violated when studies include a large number of dissimilar countries (e.g. not all countries have similar climatic or mineral resources). Filtering controls for this violation.

Finally, the examination of the filtered data can provide insight not otherwise evident or measurable from unfiltered data (e.g. measures of economic performance or social behavior relative to each country's exogenous endowment). This insight is especially critical to social planners interested in measuring the incremental effect of a specific policy on economic development or progress. By proceeding with a filtering procedure, we can compare the effects of economic policies on countries like Chad, which have no mineral resources, no coastlines, no-inland waterways, desert soils, and long periods of drought, versus countries rich in marine and mineral resources with more generous climates, like the United States. In doing so, we can ascertain the extent to which countries like Chad are "poor" due to their exogenous endowment versus their current economic policies. Equivalently, we can determine if Chad has higher "real" economic performance, given its endowment, than endowment rich countries having alternative policies.

For each level of the filtering hierarchy, Figure 3 indicates in parentheses the general range of incremental percent explained variation, adjusted for degrees of freedom, across some thirty cross-national measures frequently used in demography (e.g. mortality, fertility, population, population growth rates), sociology (e.g. crime, suicide) and economics (e.g. manufacturing, mineral extraction, goods consumption, caloric intake, aggregate output, savings)[3]. These ranges should be seen as typical to many, but not general to all behaviors. Some behaviors are left completely unexplained by the hierarchy (e.g. short-term inflation rates across countries). Across the various levels in Figure 3, solar climate generally explains the largest variances in behavior; frequently over 70% of the variance observed across the behaviors mentioned are explained by the entire hierarchy, including income per capita. Cultural variables typically explain little variance beyond that already explained by natural endowments (e.g. controlling for climatic effects, religion does not have an impact on labor productivity or "work ethics"). It is interesting to note that solar climate alone (absolute latitude squared) has statistically equal or greater explanatory power than income per capita for a variety of behaviors which are typically thought of as being "development driven" or, if not development driven, "inferior" behaviors (e.g. when the consumption of a good is not positively related to income). Behaviors which are uncorrelated with income per capita but explained by solar climate include (statistically significant Pearson correlation coefficients are given in parentheses): milk consumption per capita (.53), and cereal consumption per capita (.57). Behaviors which are better explained by solar climate than income per capita include: butter consumption per capita (.59), flour consumption per capita (.35), potato consumption per capita (.66), sugar consumption per capita (.46), caloric intake per capita (.74), coffee consumption per capita (.75), cigarette consumption per capita (.33), surface cleaner consumption per capita (.52), percent of households owning a microwave oven (.76), percent of households owning a personal stereo (e.g. Sony Walkman, .84), percent of households owning a personal computer (.59), suicide rates (.62), and homicide rates (-.35). Behaviors which are equally well explained by solar climate and income per capita include: average population growth rates (-.50), number of major diseases present (-.52), literacy rates (.46), the percent of homes with piped water (.66), the percent of homes with flush toilets (.73), daily newspaper circulation per capita (.76), telephones per capita (.65), beef consumption per capita (.36), pork consumption per capita (.64), egg consumption per capita (.54), beer consumption per capita (.56), steel consumption per capita (.65), toilet paper consumption per capita (.72), the percent of homes owning a video camera (.66), and the percent of homes owning cable television, (.48)[4]. In general, the correlations are higher when the data are limited to European countries, or industrial countries located in very different climatic regions (e.g. desert,

---

3 Solar climate is measured as absolute latitude squared (to account for the earth's curvature). Measures of physical climate include elevation, and total land area which simultaneously captures other terrestrial resources. Interactions are measured using climatic classifications. Marine resources are measured using coastline length. Similar measures are used for land, and mineral resources. Cultural measures include measures of the five largest world religions, numbers of ethnic groups present, and the use of major colonial languages. Political and economic measures include dummy variables for alternative regimes (e.g. communism, dictatorship, democracy).

4 Data on these behaviors are taken from sources listed in Chapter 2, from various Euromonitor publications (Europen Marketing Data and Statistics; International Marketing Data and Statistics), and from various United Nations statistical reports. Consumer electronics data are limited to European countries.

polar, temperate, tropical and Mediterranean).

Validation and Dynamics

A number of validation procedures have been used in the literature. These include studies conducted on military personnel, controlled (laboratory) experiments, and the combined use of time series and cross-sectional methodologies which may generate confirming (or conflicting) results. Approaches used to validate estimated climatic effects, when applying the filtering approach or similar methodologies reported in the literature, include replication using subsets of the data which simultaneously offer sufficient variation in climate and degrees of freedom (i.e. country observations). Clusters can include countries having non-democratic political systems, certain dominating religions, or colonial languages. One can also group the data by exogenous controls including the hemispheres, time zones, form of topology, or wealth in mineral endowments. In both cases, these validity tests must allow for sufficient variation in climate (or climate-dependent variables). Within-country tests often do not provide this variation unless these have substantial variances in latitude, elevation, or other climatic factors.

While the percentages reported in Figure 3 are applicable to the validity tests discussed above, the results are static (as is common for cross-cultural studies of climate). Two questions arise: (1) are the results applicable over time, and (2) in which direction are they likely to change in the future. The first question is best evaluated using data collected over extended time periods using static samples (i.e. the same countries must be represented over time); the variables used in the exogenous filters themselves never change over time. Changes in coefficients and explained variances can address the first question. The second question is best answered by examining the dynamics of the residuals (corrected dependent variables) after the final filter of interest. Strong deviations from the mean (of zero) may foretell future dynamics reflecting a convergence or regression to the mean. In the case of zero mean shifts over time (an empirical question), those countries far below the mean have higher probability of increasing their behaviors toward the mean, whereas those substantially above the mean will eventually fall to the mean. Alternatively (e.g. in cases of positive mean shifts in behavior over time, driven by innovation or other growth mechanisms), those countries substantially above (or below) the mean may be on a boundary to which other countries converge (e.g. the saturation point on a diffusion curve); this boundary may shift in time. Empirical outcomes are likely to vary from one behavior to another.

Some Implications

The implementation of physioeconomic models may be revealing to researchers and policy makers. Consider, for example, the historical study of communism and its effect on income per capita across countries. Without controlling for climatic endowments, the average income per capita of communist countries can be compared against those of non-communist countries. Clearly, one should give more weight to comparisons such as North Korea to South Korea, North Vietnam to South Vietnam, Yugoslavia to Italy, East Germany to West Germany, Eastern Europe to Western Europe, Benin to Togo, Cuba to Jamaica, Nicaragua to Honduras, or the former Soviet Union to the United States. Across these pairs we have implicitly controlled for various exogenous natural endowments: locality-driven climatic conditions, physical size of country, mineral and marine resources. Proximity in location is not as critical as proximity in environmental "starting position" (e.g. we can reasonably compare the U.S.A. to the former Soviet Union which are centered on opposite sides of the northern hemisphere). Using a sample drawn from United Nations members, traditional comparisons indicate that communism had no effect on income per capita, whereas filtered measures of income which control for relative starting positions in natural endowments indicate that, on average, countries governed by communist regimes suffered an average reduction in income per capita by over $3600 per year (1985 dollars).

The filtering of income or consumption per capita also reveals that approximately 70 percent of variances observed internationally are explained by variances in natural endowments (e.g. certain countries may be underdeveloped more due to climatic effects and less due to current social policies). Furthermore, based on cross-sectional data which adjust for purchasing power parity and which are available back to the 1950s, this explained variance has been increasing over time. In the early 1990s, solar climate alone explained some 50 percent of the variance in income per capita across countries; the gradual increases in explained variance has occurred while average consumption levels have substantially increased over time. Furthermore, filtered measures of development are often not correlated with unfiltered measures. When comparing each country's level of *relative* development, or its level of economic performance given it exogenous endowments, one finds that resulting income differences are relatively small compared to absolute levels (e.g. within a few thousand dollars, versus tens-of-thousands of dollars). By focusing on these filtered measures, one may be able to identify *relatively* successful countries; the policies of these countries may prove useful in other countries with relatively low performance. Tropical countries, for example, may benefit from studying social and economic policies in Costa Rica or the Cayman Islands which have managed to obtain relatively high economic performance given their natural endowments.

The concept of "relative performance" raises an interesting question concerning cross-national measures of development. From the previous discussion and the literature indexed in Chapters 3 to 6, there appear to be fundamental cross-country differences in both physiological and psychological utility ("basal utility") for a number of human activities including food consumption, dressing behavior, housing requirements, energy consumption, and the use of in-door entertainment. Cross-country data, including those limited to industrialized countries, seem to indicate that some portion of the observed variances in measures of development can be explained by climatic factors which my affect basal utility. For certain measures of development incorporating these behaviors (e.g. aggregate consumption per capita), holding all countries (or climatic regions) to a unique or absolute standard, regardless of where that standard is established, can be problematic. Considering the case of caloric intake, European data spanning several decades indicates substantially lower caloric consumption in warmer countries, than colder countries (consistent with thermoregulation). These levels have remained relatively constant over time, despite large increases in real income per capita. Similar observations can be made with respect to the consumption of various climate-affected behaviors (e.g. clothing, the use of internal heating, insolation, etc.). If we are interested in measuring development from an absolute utility perspective, differences in basal utilities should be taken into account (i.e. two divergent levels of absolute consumption may generate similar levels of utility, the difference in consumption being explained by differences in thermoregulatory functions). For example, Belgians consume some 50% more calories per capita than persons living in Bermuda (3850 versus 2545, respectively); it is difficult to imagine that all of this difference is due to economic development or that

Belgians receive that much more absolute utility from their diets. While adjustments might be made to economic development measures to account for basal utility differences across countries, these adjustments run the risk of overestimating development for those countries having the lowest absolute values on the original scale. An alternative to adjusting current measures is to place more emphasis on alternative measures not subject to basal utility shifts, or indirect physiological or psychological mechanisms indexed in Chapters 3 and 4. Alternative scales which, in part, reflect the direct or life threatening impacts of climate may prove more useful (e.g. the prevalence of certain diseases, etc.). These scales should indicate relative levels of development which cannot be confounded with differences in basal utility (a lower level on the scale should always indicate lower performance and, preferably, in a linear manner).

Since climate substantially contributes to human illness (e.g. malaria), and economic poverty (e.g. due to droughts), efforts might be better focused on approaches to directly combat its deleterious effects (as opposed to emphasizing or, worse, blaming confounding factors such as culture). Innovation and discovery, in the past, has demonstrated that the "parameters of the model" can and do change over time. Care must be taken not to interpret strong climatic effects (high explained variances) as indicating that countries will remain underdeveloped. Absolute levels of behavior have shifted over time for all countries despite climate playing a role in affecting variances. While current policies may be ineffective in counteracting certain climatic effects, the literature indexed in Chapters 3 to 6 reveals that human initiative has been able to overcome climatic effects when these are properly diagnosed and understood. In order to facilitate research which stands to benefit from the incorporation of climatic effects, Chapter 2 provides estimates for some relevant cross-country variables not readily available from secondary sources. The hope is that these will serve to foster meaningful and objective research in this area.

2

Cross-Country Climatic Data

This chapter provides ten climatic variables for over 230 countries or geographic regions: absolute latitude (degrees from the equator), elevation (in meters), temperature (in degrees Celsius), rainfall (average per month in millimeters), humidity (average daily percentage), rain days (average number of days of rainfall per month), coastline length (in kilometers), total land area (in square kilometers), July barometric pressure, and January barometric pressure. All numbers should be considered as gross estimates/averages which have undergone rounding and certain adjustments. For example, estimates of latitude and elevation reflect averages calculated across major population centers (e.g. cities having populations greater than 100,000 inhabitants) and are not geographic averages of a country's topology; this adjustment is especially important for large and/or mountainous countries; in many smaller countries, the latitude/elevation reported is that of the capital city. Meteorological estimates are averages recorded over a 30 year or similar period for major population centers. In some cases, estimates are generated from adjacent or neighboring countries with similar climates when data were lacking (mostly small islands, or recently established nations). For all estimates, care should be exercised in interpreting estimates for populous countries or those with varying topologies.

The statistics in this chapter were generated using either original estimates or averages calculated from multiple sources. Should one wish to refine the estimates given, the following references are recommended: E. A. Pearce and Gordon Smith, *World Weather Guide*, New York: Times Books, Randomhouse, 1990; Willy Rudolph, *World Climates: with Tables of Climatic Data and Practical Suggestions*, Stuttgart: Weissenschaftliche, 1981; The Central Intelligence Agency, *The World Factbook*, Washington D.C.: The Directorate of Intelligence of the United States Central Intelligence Agency, 1994; Rand McNally, *Rand McNally World Atlas & Review*, New York: Rand McNally and Company, 1993; and *The World Almanac and Book of Facts*, Mahwah, New Jersey: Funk and Wagnalls, and World Almanac Books, 1994. Cross-country climatic databases in software form include *PCGlobe The New World Order* (version 5.0), Broderbund, 500 Redwood Boulevard, Novato, CA 94948-6121; *PCGlobe Maps-N-Facts*, Broderbund, P.O. Box 6125, Navato, California, 94948-6125; *World Atlas: The Software Toolworks*, CD version 3.2.1., 60 Leveroni Court, Novato, CA. 94949; *World Fact Book*, CD version, Bureau Development Inc., 141 New Road, Parsippany, NJ 07054.

Cross-Country Climatic Data

Country Name	Absolute Latitude	Elevation (metres)	Temp. (celsius)	Rain (mm/month)	Humidity (percentage)	Rain Days (per month)	Coastline (length km)	Land Area (km squared)	July Pressure	January Pressure
Afghanistan	33	1820	10	51	56	4	0	647500	999	1020
Albania	41	120	16	122	65	10	362	28750	1011	1017
Algeria	36	50	21	69	66	6	1183	2381740	1013	1020
American Samoa	14	5	27	268	77	15	116	199	1014	1008
Andorra	42	1080	9	70	70	10	0	450	1017	1020
Angola	8	40	26	59	78	4	1600	1246700	1017	1011
Anguilla	18	40	26	141	65	12	61	91	1017	1017
Antigua & Barbuda	17	6	27	141	65	12	153	440	1014	1017
Argentina	39	20	17	83	78	8	4989	2766890	1006	1003
Armenia	40	990	8	31	66	7	0	29800	1008	1020
Aruba	12	10	29	66	73	10	69	193	1014	1014
Australia	28	20	17	49	60	9	25760	7686850	1020	1011
Austria	47	200	8	62	71	13	0	83850	1014	1020
Azerbaijan	40	990	14	31	66	7	0	86600	1008	1020
Bahamas	25	5	25	104	75	10	3542	13940	1017	1020
Bahrain	26	1	25	9	74	1	161	620	999	1017
Bangladesh	23	8	25	220	71	8	580	144000	1002	1017
Barbados	13	50	26	117	72	13	97	430	1015	1015
Belarus	54	150	4	62	68	18	0	207600	1011	1017
Belgium	51	70	10	74	83	17	64	30510	1017	1014
Belize	17	5	26	181	90	10	386	22960	1014	1017
Benin	6	10	27	190	80	7	121	112620	1011	1014
Bermuda	32	40	22	126	76	12	103	50	1020	1020
Bhutan	27	2420	17	188	74	11	0	47000	1002	1017
Bolivia	18	2410	12	61	70	12	0	1098580	1017	1011
Botswana	24	1010	20	54	65	4	0	600370	1020	1014
Brazil	16	990	25	89	78	11	7491	8511970	1020	1011
British Indian Ocean	8	5	26	228	76	11	120	60	1011	1014

Cross-Country Climatic Data

Country Name	Absolute Latitude	Elevation (metres)	Temp. (celsius)	Rain (mm/month)	Humidity (percentage)	Rain Days (per month)	Coastline (length km)	Land Area (km squared)	July Pressure	January Pressure
British Virgin Islds	18	40	26	141	65	12	80	150	1017	1017
Brunei	5	3	27	182	80	12	161	5770	1011	1011
Bulgaria	43	560	9	58	71	10	354	110910	1011	1020
Burkina Faso	12	300	29	139	51	7	0	274200	1014	1014
Burma	19	6	28	293	75	13	3060	676550	1005	1017
Burundi	16	780	24	65	69	10	0	27830	1014	1011
Cambodia	11	10	28	132	78	10	443	181040	1008	1014
Cameroon	4	380	25	159	83	14	402	475440	1014	1011
Canada	53	80	4	73	60	12	58808	9976140	1011	1020
Cape Verde	15	20	25	57	65	4	965	4030	1014	1014
Cayman Islands	19	30	26	98	75	6	160	260	1014	1017
Central African Rep	4	380	27	116	77	11	0	622980	1014	1011
Chad	12	300	28	160	52	11	0	1284000	1011	1011
Chile	36	260	17	44	66	3	6435	756950	1014	1011
China, People's Rep	35	30	11	123	60	8	14500	9596960	1005	1026
Christmas Island	1	20	27	205	71	10	54	135	1011	1011
Colombia	6	1550	14	106	71	13	2414	1138910	1014	1014
Comoros	11	50	26	228	76	11	340	2170	1017	1011
Congo	4	510	25	146	73	5	169	342000	1017	1011
Cook Island	15	2	26	268	77	15	120	240	1014	1011
Costa Rica	10	1170	21	155	76	13	1290	50900	1014	1014
Cote D'Ivoire	6	7	26	268	86	11	515	322460	1014	1014
Croatia	45	130	12	71	70	11	1778	56538	1011	1017
Cuba	22	40	25	110	75	8	3735	110860	1017	1017
Cyprus	35	150	19	39	59	7	648	9250	1011	1017
Czechoslovakia	49	190	9	43	71	12	0	127870	1014	1017
Denmark	56	5	8	52	76	14	3379	43070	1011	1014
Djibouti	11	6	32	13	65	2	314	22000	1008	1014

Cross-Country Climatic Data

Country Name	Absolute Latitude	Elevation (metres)	Temp. (celsius)	Rain (mm/month)	Humidity (percentage)	Rain Days (per month)	Coastline (length km)	Land Area (km squared)	July Pressure	January Pressure
Dominica	15	8	27	168	68	16	148	750	1017	1017
Dominican Republic	19	10	26	111	76	9	1288	48730	1017	1017
East Germany	52	30	8	53	79	15	901	108330	1014	1014
Ecuador	2	1410	15	98	69	15	2237	283560	1014	1014
Egypt	28	20	24	3	45	1	2450	1001450	1008	1017
El Salvador	13	690	24	167	71	11	307	21040	1014	1017
Equatorial Guinea	3	20	26	154	90	14	296	28050	1014	1017
Estonia	59	40	5	55	77	16	1393	45100	1011	1017
Ethiopia	12	2140	17	153	62	15	1094	1221900	1011	1017
Falkland Islands	51	2	7	55	82	14	1288	12170	1002	999
Faroe Islands	62	20	6	117	83	21	764	14000	1011	1002
Fiji	18	9	25	247	78	17	1129	18270	1014	1008
Finland	62	10	5	55	77	16	1126	337030	1008	1011
France	46	60	13	50	77	15	3427	547030	1014	1017
French Guiana	5	9	27	291	78	15	378	91000	1014	1011
French Polynesia	17	2	26	268	77	15	2625	3941	1017	1011
Gabon	0	3	27	188	82	10	885	267670	1014	1017
Gaza Strip	31	40	20	66	56	6	40	380	1008	1017
Georgia	41	490	13	45	61	10	310	69700	1008	1020
Germany	51	40	8	57	75	15	1488	248580	1014	1014
Ghana	7	60	27	97	83	6	539	238540	1014	1014
Gibraltar	36	3	20	77	71	5	12	7	1017	1023
Greece	38	5	19	39	56	9	13676	131940	1011	1017
Greenland	64	20	-2	58	88	12	44087	2175600	1008	996
Grenada	12	90	28	180	79	17	121	340	1014	1014
Guadeloupe	16	10	23	301	81	23	306	1780	1014	1014
Guam	13	70	27	204	81	17	126	541	1011	1011
Guatemala	14	1500	19	139	74	13	400	108890	1014	1017

Cross-Country Climatic Data

Country Name	Absolute Latitude	Elevation (metres)	Temp. (celsius)	Rain (mm/month)	Humidity (percentage)	Rain Days (per month)	Coastline (length km)	Land Area (km squared)	July Pressure	January Pressure
Guernsey	50	5	12	51	74	13	50	194	1014	1017
Guinea	10	40	27	651	82	15	346	245860	1014	1014
Guinea-Bissau	12	20	28	127	69	7	274	36120	1014	1014
Guyana	6	2	27	189	81	16	459	214970	1014	1014
Haiti	19	40	27	132	62	8	1771	27750	1017	1017
Honduras	15	580	22	91	85	9	820	112090	1014	1017
Hong Kong	22	30	21	213	75	10	733	1040	1005	1020
Hungary	47	110	9	53	72	11	0	93030	1014	1014
Iceland	64	20	5	68	76	18	4988	103000	1011	999
India	21	110	24	92	50	4	7000	3287590	1002	1017
Indonesia	5	7	28	172	80	11	54716	1919440	1011	1011
Iran, I.R. of	33	910	15	25	60	2	3180	1648000	999	1020
Iraq	33	30	21	14	45	3	58	434920	1005	1020
Ireland, Republic of	52	10	9	60	82	12	1448	70280	1014	1011
Isle of Man	54	80	12	51	74	13	113	588	1014	1011
Israel	32	400	20	66	56	6	273	20770	1008	1017
Italy	42	20	17	72	67	6	4996	301230	1014	1017
Jamaica	18	30	26	98	75	6	1022	10990	1014	1017
Japan	37	4	15	141	70	9	13685	372310	1011	1017
Jersey	49	5	12	51	74	13	70	117	1014	1017
Johnston Atoll	18	10	29	155	81	18	10	3	1011	1014
Jordan	32	700	19	37	51	4	26	91880	1008	1017
Kazakhstan	43	60	5	10	66	4	0	2717300	1008	1029
Kenya	2	880	18	113	66	11	536	582650	1014	1011
Kingman Reef	6	20	27	205	71	10	3	1	1011	1011
Kiribati	1	2	28	205	71	10	1143	717	1011	1008
Kuwait	29	5	25	14	57	2	499	17820	999	1020
Kyrgyzstan	43	770	2	63	66	9	0	198500	1002	1020

Cross-Country Climatic Data

Country Name	Absolute Latitude	Elevation (metres)	Temp. (celsius)	Rain (mm/month)	Humidity (percentage)	Rain Days (per month)	Coastline (length km)	Land Area (km squared)	July Pressure	January Pressure
Laos	18	5	26	153	79	10	0	236800	1008	1014
Latvia	57	40	4	55	77	16	531	64100	1011	1017
Lebanon	34	10	21	96	66	8	225	10400	1008	1017
Lesotho	29	1570	16	50	48	5	0	30350	1017	1014
Liberia	6	20	25	514	86	16	579	111370	1014	1014
Libya	32	20	24	47	63	6	1770	1759540	1011	1017
Liechtenstein	47	460	8	100	75	14	0	160	1014	1014
Lithuania	54	40	5	55	77	16	108	65200	1011	1017
Luxembourg	49	300	9	63	82	17	0	2586	1014	1017
Macau	22	20	23	213	75	10	40	16	1005	1020
Madagascar	19	1380	18	154	75	14	4828	587040	1023	1011
Malawi	15	1010	19	109	58	10	0	118480	1017	1011
Malaysia	4	20	28	196	79	16	4675	329750	1011	1011
Maldives	4	1	28	157	77	9	644	300	1011	1011
Mali	14	310	29	174	55	9	0	1240000	1011	1014
Malta	36	70	23	55	69	7	140	320	1014	1017
Martinique	14	4	26	168	83	21	290	1100	1014	1017
Mauritania	18	5	27	52	60	2	754	1030700	1014	1017
Mauritius	20	50	23	10	75	4	177	1860	1020	1014
Mayotte	13	5	26	228	76	11	165	375	1017	1011
Mexico	24	1310	16	88	58	16	9330	1972550	1014	1017
Micronesia, Fed.	5	10	29	155	81	18	6112	702	1011	1008
Midway Islands	28	9	19	186	80	11	15	5	1011	1014
Moldova	47	170	10	61	74	11	0	33700	1011	1020
Monaco	43	50	17	72	72	4	4	2	1014	1017
Mongolia	48	1300	-1	38	67	6	0	1565000	1005	1032
Montserrat	17	40	26	141	65	12	40	100	1017	1017
Morocco	33	50	24	43	78	5	1835	446550	1014	1020

Cross-Country Climatic Data

Country Name	Absolute Latitude	Elevation (metres)	Temp. (celsius)	Rain (mm/month)	Humidity (percentage)	Rain Days (per month)	Coastline (length km)	Land Area (km squared)	July Pressure	January Pressure
Mozambique	23	30	27	72	66	6	2470	801590	1020	1011
Namibia	24	1650	18	40	34	4	1489	824290	1020	1011
Nauru	0	20	28	205	71	10	30	20	1011	1008
Navassa Island	18	30	25	98	75	6	9	5	1014	1017
Nepal	27	1330	17	188	74	11	0	140800	1002	1020
Netherlands	52	3	10	56	79	16	451	37310	1014	1017
Netherlands Antilles	12	20	27	66	73	10	364	960	1011	1014
New Caledonia	22	3	24	98	71	11	2254	19060	1017	1011
New Zealand	41	60	15	109	75	14	15134	268680	1011	1014
Nicaragua	12	50	27	149	83	11	910	129494	1014	1014
Niger	13	220	29	94	49	6	0	1267000	1008	1014
Nigeria	9	3	28	243	78	11	853	923770	1011	1011
Niue	19	2	26	268	77	15	64	260	1014	1011
Norfolk Island	29	9	24	98	71	11	32	40	1017	1011
North Korea	39	20	9	124	72	8	2495	120540	1008	1026
Norway	64	90	5	61	73	12	21925	324220	1011	1008
Oman	23	6	29	14	70	1	2092	212460	1002	1017
Pakistan	29	270	22	135	37	7	1046	803940	999	1020
Panama	9	20	27	135	86	10	2490	78200	1014	1014
Papua New Guinea	7	10	28	106	74	5	5152	461690	1011	1008
Paraguay	25	60	22	98	68	6	0	406750	1017	1008
Peru	10	150	22	4	83	1	2414	1285220	1017	1011
Philippines	11	10	27	223	77	14	22540	300000	1011	1017
Pitcairn Islands	25	2	26	268	77	15	51	47	1014	1011
Poland	52	100	7	64	79	14	491	312680	1011	1017
Portugal	39	90	20	57	67	9	1793	92080	1017	1020
Puerto Rico	18	10	26	115	77	18	501	9104	1017	1017
Qatar	25	10	26	18	67	1	563	11000	999	1017

Cross-Country Climatic Data

Country Name	Absolute Latitude	Elevation (metres)	Temp. (celsius)	Rain (mm/month)	Humidity (percentage)	Rain Days (per month)	Coastline (length km)	Land Area (km squared)	July Pressure	January Pressure
Reunion	21	10	24	10	75	4	201	2510	1023	1014
Romania	45	80	8	74	74	9	225	237500	1011	1020
Russia	55	150	3	62	68	18	37653	7075200	1008	1020
Rwanda	2	1540	20	95	71	12	0	26340	1014	1017
Sahara, Western	27	40	26	18	65	4	1110	266000	1014	1020
San Marino	44	10	17	72	67	6	0	60	1014	1017
Sao Tome E Prin.	0	10	25	75	76	5	209	960	1017	1014
Saudi Arabia	24	290	23	13	40	2	2510	2149690	1005	1017
Senegal	14	20	28	127	69	7	531	196190	1014	1014
Seychelles	4	3	27	228	76	11	491	455	1017	1014
Sierra Leone	8	20	27	453	80	14	402	71740	1014	1014
Singapore	1	10	28	214	77	15	193	580	1011	1011
Slovenia	46	130	9	71	70	11	32	20296	1011	1017
Solomon Islands	9	50	27	250	79	14	5313	28450	1011	1008
Somalia	5	10	28	49	80	10	3025	637660	1011	1014
South Africa	30	10	19	49	71	6	2881	1221040	1008	1011
South Korea	36	80	11	198	70	11	2413	98480	1008	1023
Spain	40	440	15	32	62	6	4964	504750	1017	1020
Sri Lanka	8	7	27	220	74	13	1340	65610	1008	1014
St. Helena	15	10	24	10	69	4	60	310	1020	1014
St. Kitts and Nevis	17	40	27	141	65	12	135	360	1017	1017
St. Lucia	14	20	26	180	79	17	158	620	1017	1017
St. Pierre and Miqu.	47	40	9	60	82	12	120	242	1014	1014
St. Vincent & the Gr.	13	20	26	180	79	17	84	340	1017	1017
Sudan	12	230	29	36	34	3	853	2505810	1011	1014
Suriname	6	4	27	193	81	16	386	163270	1014	1014
Swaziland	26	1160	17	137	70	10	0	17360	1023	1011
Sweden	59	30	6	51	74	14	3218	449960	1011	1014

Cross-Country Climatic Data

Country Name	Absolute Latitude	Elevation (metres)	Temp. (celsius)	Rain (mm/month)	Humidity (percentage)	Rain Days (per month)	Coastline (length km)	Land Area (km squared)	July Pressure	January Pressure
Switzerland	47	520	9	100	75	14	0	41290	1014	1017
Syrian Arab Republic	35	390	20	22	51	4	193	185180	1008	1024
Taiwan	24	8	21	186	80	11	1448	35980	1008	1017
Tajikistan	38	470	13	35	59	7	0	143100	999	1023
Tanzania	4	10	22	76	62	6	1424	945090	1017	1011
Thailand	13	8	28	155	77	8	3219	514000	1011	1014
The Gambia	13	2	24	250	66	10	80	11300	1014	1014
Togo	6	20	27	97	83	6	56	56790	1014	1014
Tokelau	9	70	27	250	79	14	101	10	1011	1008
Tonga	21	2	27	268	77	15	419	700	1014	1011
Trinidad & Tobago	10	20	26	144	77	16	362	5130	1014	1014
Tunisia	36	30	24	34	66	8	1148	163610	1014	1020
Turkey	39	460	9	29	59	5	7200	780580	1008	1017
Turkmenistan	38	20	14	14	60	3	0	488100	1005	1023
Turks and Caicos Isl	10	4	20	104	75	10	300	430	1011	1014
Tuvalu	8	2	28	250	79	14	24	26	1011	1008
Uganda	0	1190	24	111	73	14	0	236040	1014	1011
Ukraine	50	170	5	61	74	11	2782	603700	1011	1017
United Arab Emirate	25	5	26	18	67	1	1448	83600	999	1017
United Kingdom	53	40	12	51	74	13	12429	244820	1014	1011
United States	44	70	12	89	63	10	19924	9372610	1014	1020
Uruguay	34	20	20	83	71	6	660	176220	1017	1011
USSR (Former)	53	100	4	62	68	18	42777	2402200	1008	1020
Uzbekistan	41	20	13	14	60	3	0	447400	1005	1023
Vanuatu	17	20	25	247	78	17	2528	14760	1017	1011
Vatican City	42	10	17	72	67	6	0	1	1014	1017
Venezuela	9	830	20	60	72	9	2800	912050	1014	1014
Vietnam	15	10	25	181	75	12	3444	329560	1008	1014

Cross-Country Climatic Data

Country Name	Absolute Latitude	Elevation (metres)	Temp. (celsius)	Rain (mm/month)	Humidity (percentage)	Rain Days (per month)	Coastline (length km)	Land Area (km squared)	July Pressure	January Pressure
Virgin Islands, US	18	5	27	141	65	12	90	352	1017	1017
Wallis and Futuna	14	2	26	268	77	15	129	274	1014	1011
West Bank	32	580	20	66	56	6	0	5860	1008	1017
West Germany	50	100	8	57	75	15	2389	248580	1014	1014
Western Samoa	14	2	26	268	77	15	403	2860	1014	1011
Yemen Arab Rep	15	2210	30	12	66	1	1906	195000	1005	1014
Yemen, P.D. Rep	12	4	29	3	69	1	1383	332970	1005	1014
Yugoslavia	45	130	10	71	70	11	3935	255800	1011	1017
Zaire	7	290	25	113	79	8	37	2345410	1014	1010
Zambia	20	1270	21	116	54	11	0	752610	1017	1008
Zimbabwe	19	1410	18	98	51	9	0	390580	1020	1011

3

Physiological Effects: A Select Bibliography

ACCLIMATION & PHYSICAL COMFORT

1. Abusaleh, A. M. "Physiological responses to heat stress of Arabs and Americans." Ph.D. diss., University of Oregon, 1983. Topics discussed include: heat, physiology, acclimation.

2. "Adjustment to high altitude : proceedings of the International Symposium on Acclimatization, Adaptation, and Tolerance to High Altitude." International Symposium on Acclimatization, Adaptation, and Tolerance to High Altitude (1981 : Bethesda, Md.), and John E. Fogarty International Center for Advanced Study in the Health Sciences. U.S. Dept. of Health and Human Services, Public Health Service, National Institutes of Health, Bethesda, Md., 1983. Topics discussed include: anoxemia, mountain sickness.

3. Allan, J. "The effects of physical training in a temperate and hot climate on the physiological responses to heat stress." *Ergonomics* 8, no. 4 (October 1965): 445-53. Topics discussed include: acclimation, physical effort, stress.

4. Allexenberg, R. S. "St. Louis, Missouri: combating the heat wave of 1980: lessons for the future." *Urban Health* (September 1980): 26-30. Topics discussed include: acclimation, body temperature, mass media, health services, emergencies.

5. Ambler, H. "Thermal comfort in hot and temperate climates." *J Trop Med Hyg* 71, no. 11 (November 1968): 294-6. Topics discussed include: India, timing, adaptation, body temperature, humidity.

6. Armstrong, L. E. et al. *Preparing Alberto Salazar for the Heat of the 1984 Olympic Marathon*, Performer: Army Research Inst. of Environmental Medicine, Natick, MA., January 1985. Topics discussed include: adaptation, acclimation, exercise, blood, dehydration, stress, medicine, biology.

7. Armstrong, L. et al. "Responses to moderate and low sodium diets during exercise heat acclimation." *Int J Sport Nutr* 3, no. 2 (June 1993): 207-21. Topics discussed include: blood, body mass, thermoregulation, effects on heart beat, hemodynamics, exercise physiology, sodium, dosage.

8. Auliciems, A. "Effects of weather on indoor thermal comfort." *Int J Biometeorol* 13, no. 2 (October 1969): 147-62. Topics discussed include: seasonality, gender effects, thermoregulation, temperature sense.

9. ---. *The atmospheric environment; a study of comfort and performance*. 166 pages. University of Toronto. Dept. of Geography. Research publications, 8, Toronto: Published for the University of Toronto,

Dept. of Geography, by the University of Toronto Press, 1972. Topics discussed include: temperature, physiology.

10. Avery, M. "Heat stress and older adults." *Aging* (July 1981): 39-41. Topics discussed include: acclimation, strokes, cardiovascular diseases, self care, environment, heart failure.

11. Avery, W. M. "Hypothermia and heat illness." *Aging* (April 1984): 42-47. Topics discussed include: hypothermia, preventive medicine, body temperature, acclimation.

12. Azhaev, A. et al. "Physiological reaction in humans after flight to a region with a tropical climate." *Voen Med Zh* no. 1 (January 1985): 53-5. Topics discussed include: Asia, seasonality, acclimation, aerospace medicine, blood pressure, body temperature.

13. Babaev, A. "Some problems of acclimtization for construction workers to subtropical climatic conditions." *Gig Sanit* no. 6 (June 1976): 106-8. Topics discussed include: thermoregulation, Tajikistan, temperature, timing, acclimation, occupational medicine.

14. Bartnicki, C. et al. "The effect of seasonal adaptation on changes of some physiological indexes in people exposed to high environmental temperature." *Biul Inst Med Morsk Gdansk* 20, no. 3 (1969): 195-201. Topics discussed include: blood pressure, effects on heart beat, humidity effects, perspiration, acclimation, thermoregulation, heat.

15. Belding, H. "Research on problems of work in heat. 15 years' activity in the Department of Occupational Health, University of Pittsburgh." *Arch Environ Health* 15, no. 5 (November 1967): 660-9. Topics discussed include: body temperature, environmental exposure, effects on heart beat, Pennsylvania, perspiration, acclimation, medicine.

16. Berdyshev, V. "Effect of physical training on body function and work capacity of sailors in the tropics." *Voen Med Zh no.* 3 (March 1979): 55-7. Topics discussed include: acclimation, naval medicine, physical education, tropical climates.

17. Bittel, J. "Influence of thermohygrometric factors on sweat depression during heat acclimatization by controlled hyperthermia procedure in man." *Archives des Sciences Physiologiques* 27, no. 2 (1973): A117-A126.

18. Budd, G. "Effects of cold exposure and exercise in a wet, cold antarctic climate." *J Appl Physiol* 20, no. 3 (May 1965): 417-22. Topics discussed include: blood pressure, body temperature, frostbite, oxygen consumption, physical fitness, shivering, acclimation physiology, cold exertion.

19. Budd, G. et al. "Body temperature, shivering, blood pressure and heart rate during a standard cold stress in Australia and Antarctica." *J Physiol Lond* 186, no. 1 (September 1966): 216-32. Topics discussed include: Antarctic regions, Australia, acclimation, blood pressure, cold climates, effects on heart beat, stress.

20. ---. "Urinary excretion of adrenal steroids, catecholamines and electrolytes in man, before and after acclimatization to cold in Antarctica." *J Physiol Lond* 210, no. 4 (November 1970): 799-806. Topics discussed include: Antarctic regions, creatinine, potassium, sodium, timing, urine physiology, cold climates, hydroxycorticosteroids, ketosteroids.

21. Buguet, A. et al. "Seasonal changes in circadian rhythms of body temperatures in humans living in a dry tropical climate." *Eur J Appl Physiol* 58, no. 3 (1988): 334-9. Topics discussed include: acclimation, humidity effects, Niger.

22. Burch, G. et al. "A perspective on climate and man's health." *JAMA* 221, no. 9 (28 August 1972): 1050-1. Topics discussed include: acclimation, body temperature, cardiovascular system, cold climate effects, homeostasis, humidity effects, seasonality, weather, health.

23. Bytel, H. et al. "A method for measuring body core temperature at work sites exposed to heat stress." *Z Gesamte Hyg* 33, no. 1 (January 1987): 15-7. Topics discussed include: acclimation, exhaustion, occupational diseases, thermometers.

24. Aldo Sir, Castellani, *Climate and acclimatization*, 2d. ed. ed. London, J. Bale, Sons & Curnow, Ltd., 1938. Topics discussed include: climatology, medicine, tropical climates.

25. Chance, N. "Implications of environmental stress. Strategies of developmental change in the North." *Arch Environ Health* 17, no. 4 (October 1968): 571-7. Topics discussed include: acclimation, acculturation, arctic regions, cold climates, crowding, Indians, North America, religion, psychology, social behavior, social welfare.

26. Cohen, J. et al. "Effects of interval training on work heat tolerance of young women." *Med Sci Sports Exerc* 14, no. 1 (1982): 46-52. Topics discussed include: body temperature, physical effort, effects on heart beat, oxygen consumption, perspiration, acclimation, physical education, training.

27. Conu, O. et al. "Adaptability of women to work in high temperature environments." *Viata Med Rev Inf Prof Stiint Cadrelor Medii Sanit* 29, no. 7 (July 1981): 163-5. Topics discussed include: acclimation, heat, work.

28. Dear, R. J. et al. "Indoor climate and thermal comfort in high-rise public housing in an equatorial climate: a field-study in Singapore." *Atmospheric Environment Part B, Urban Atmosphere* 24B, no. 2 (1990): 313-20. Topics discussed include: heating, cooling, psychology, Singapore, energy use.

29. Deriapa, N. R. *Antarctic nature and human acclimatization*. 154 pages. Zapiski geograficheskoe obshchestvo soiuza USSR. Novaia seriia T. 24, Moskva: Nauka, 1965. Topics discussed include: arctic regions, cold climate effects, physiology.

30. Dill, D. et al. "Work capacity in chronic exposures to altitude." *J Appl Physiol* 23, no. 4 (October 1967): 555-60. Topics discussed include: blood pressure physiology, body constitution, desert climate, effects on heart beat, lung physiology, oxygen consumption, gender effects, acclimation.

31. Dolmierski, R. et al. "The effect of work in tropics on psychophysical state of man; preliminary report." *Bull Inst Marit Trop Med Gdynia* 29, no. 3-4 (1978): 229-34. Topics discussed include: work, acclimation, adaptation, physical effort, tropical climates.

32. Drinkwater, B. et al. "Aerobic power as a factor in women's response to work in hot environments." *J Appl Physiol* 41, no. 6 (December 1976): 815-21. Topics discussed include: acclimation, blood pressure, body temperature, cardiac output, environment, effects on heart beat, gender effects, thermoregulation, heat.

33. Duncan, K. "Effects of an artificial acclimatization technique on infantry performance in a hot climate." *Ergonomics* 9, no. 3 (May 1966): 229-43. Topics discussed include: Arabia, United Kingdom, desert climates, military medicine.

34. Edholm, O. et al. "Body temperature during exercise and rest in cold and hot climates." *Arch Sci Physiol Paris* 27, no. 3 (1973): 339-55. Topics discussed include: air movements, body weight, environment, humidity effects, acclimation, thermoregulation, physical effort, heat.

35. Ejsmont, W. et al. "The problem of acclimatization in the tropical zone. VII. Dynamics of some physiological indexes in unadapted persons, under conditions of experimental humid heat." *Biul Inst Med Morsk Gdansk* 19, no. 1 (1968): 97-106. Topics discussed include: body temperature, tropical medicine, tropical climates.

36. ---. "The problem of acclimatization in the tropical zone. IV. Loss of body weight and increase of temperature in persons examined in the high temperature chamber." *Biul Inst Med Morsk Gdansk* 17, no. 4 (1966): 505-10. Topics discussed include: Poland, thermoregulation, body weight, tropical climates.

37. Ellis, F.P. "Heat Illness .3. Acclimatization." *Transactions of the Royal Society pf Tropical Medicine and Hygiene* 70, no. 5-6 (1976): 419-425.

38. Ellis, F. P. "Thermal comfort in warm and humid atmospheres. Observations on Europeans and Asians in Singapore." *J Hyg* 51 (1953): 386-404. Topics discussed include: temperature, environment, biology, behavior, physionomy.

39. Finberg, J. et al. "Plasma renin activity after acute heat exposure in nonacclimatized and naturally acclimatized man." *J Appl Physiol* 36, no. 5 (May 1974): 519-23. Topics discussed include: blood pressure, desert climate, drinking, environmental exposure, effects on heart beat, blood, seasonality, perspiration, water, exertion.

40. Fox, R. et al. "A study of temperature regulation in New Guinea people." *Philos Trans R Soc Lond Biol* 268, no. 893 (1 August 1974): 375-91. Topics discussed include: acclimation, body constitution, chlorides analysis, blood supply, effects on heart beat, tropical climates, thermoregulation, ethnic groups.

41. Frye, A. J. et al. *Sweating Efficiency in Acclimated Men and Women Exercising in Humid and Dry Heat Final rept*, Performer: School of Aerospace Medicine, Brooks AFB, TX., 1983. Topics discussed include: sweat glands, exercise, humidity effects, environment, thermoregulation, medicine.

42. Gagge, A. et al. "Comfort and Thermal Sensations and Associated Physiological Responses at Various Ambient Temperatures." *Environmental Research* 1 (1967): 1-20. Topics discussed include: environment, biology, behavior, comfort.

43. Gertner, A. et al. "Water and electrolyte balance in workers exposed to a hot environment during their work shift." *Ergonomics* 27, no. 2 (February 1984): 125-34. Topics discussed include: acclimation, body temperature, Israel, metallurgy, desert climate, water electrolyte balance, work.

44. Gisolfi C.V. et al. "Relationships among training, heat acclimation, and heat tolerance in men and women - Controversy revisited." *Medicine and Science in Sports and Exercise* 11, no. 1 (1979): 56-59.

45. Givoni, B. et al. "Predicting effects of heat acclimatization on heart rate and rectal temperature." *J Appl Physiol* 35, no. 6 (December 1973): 875-9. Topics discussed include: exertion, timing.

46. Goldman, R. F. *Acclimation to Heat and Suggestions, by Inference, for Microwave Radiation*, Performer: Army Research Inst. of Environmental Medicine, Natick, MA., October 1981. Topics discussed include: microwaves, acclimation, adaptation, stress, medicine, biology radiobiology, medicine.

47. Hall, S. "Heat stress in outdoor manual workers in East Africa." *Ergonomics* 14, no. 1 (January 1971): 91-4. Topics discussed include: acclimation, agriculture, climate, relaxation, perspiration, Tanzania, physical labor, physical effort, occupational medicine.

48. Heath, D. et al. *Man at high altitude : the pathophysiology of acclimatization and adaptation.* 292 pages. Edinburgh; New York: Churchill Livingstone : distributed in the U.S.A. by Longman, 1977. Topics discussed include: adaptation, anoxia physiopathology, physiology, pathological, mountain sickness.

49. Hofer, K. "Climate comfort and health resort." *Wien Med Wochenschr* 117, no. 16 (22 April 1967): 444-6. Topics discussed include: altitude, Austria, humidity effects, respiratory diseases, rheumatic diseases, sunlight, temperature, weather, acclimation, balneology.

50. Hofler, W. "Adaptation in the tropical climate. Climatic chamber field experiments experiences." *Med Klin* 63, no. 25 (21 June 1968): 995-1001. Topics discussed include: physiology, body temperature, environmental exposure, perspiration, acclimation physiology, exertion.

51. ---. "Behavior of sweat secretion during acclimatisation in a naturally tropical climate." *Int Z Angew Physiol* 22, no. 4 (4 August 1966): 317-26. Topics discussed include: thermoregulation, Caucasoid race, Nigeria, relaxation, urination, physical labor, acclimation, perspiration physiology.

52. ---. "The course of heat acclimatization in a natural tropical climate." *Z Tropenmed Parasitol* 17, no. 2 (July 1966): 127-44. Topics discussed include: blood pressure, chlorides urine, diuresis, physical effort, sweat analysis, physiology, thermoregulation.

53. Hofler, W. et al. "Relation between sweat rate, evaporation and body temperature in acclimatization to an artificial tropical climate." *Int Z Angew Physiol* 27, no. 1 (1969): 34-42. Topics discussed include: timing, physical effort, perspiration.

54. Hori, S. et al. "A comparative study on sweating pattern and heat tolerance. A field study on residents of Okinawa in summer." *Int J Biometeorol* 19, no. 3 (November 1975): 184-93. Topics discussed include: Japan, perspiration, acclimation, thermoregulation.

55. Ihzuka, H. et al. "Seasonal variations of physiological responses to heat of subtropical and temperate natives." *Int J Biometeorol* 30, no. 2 (June 1986): 107-13. Topics discussed include: heat, Japan, acclimation, thermoregulation, tropical climates.

56. Iwanaga, K. et al. "Effect of precooling on heat tolerance of resting men in a hot environment: comparison with seasonal effect on it." *Ann Physiol Anthropol* 8, no. 3 (July 1989): 151-4. Topics discussed include: acclimatization physiology, thermoregulation, cold climate effects.

57. Kandror, I. et al. "The thermal state of man as a basis for the sanitary climatological zoning of residential areas." *Gig Sanit* 31, no. 1 (January 1966): 24-30. Topics discussed include: acclimation, environmental exposure, thermoregulation physiology, weather.

58. Kew, M. "Temperature regulation in heatstroke in man." *Isr J Med Sci* 12, no. 8 (August 1976): 759-64. Topics discussed include: acclimation, heat exhaustion complications, kidney tubular necrosis, acute etiology, nervous system diseases, South Africa, thermoregulation.

59. Kolka, M. A. et al. *Heat Exchange Following Atropine Injection Before and After Heat Acclimation*, Performer: Army Research Inst. of Environmental Medicine, Natick, MA., August 1983. Topics discussed include: atropine, acclimation, body temperature, perspiration, exercise physiology, medicine, pharmacology, chemistry, medicine, biology stress.

60. ---. *Effects of Heat Acclimation on Atropine Impaired Thermoregulation*, Performer: Army Research Inst. of Environmental Medicine, Natick, MA., December 1984. Topics discussed include: acclimation, atropine, temperature control, perspiration, exercise physiology, skin anatomy, body temperature, effects on heart beat, physiology.

61. Kozacha, P. et al. "Adaptation times of flight technical personnel to work in subtropical climate conditions." *Kosm Biol Aviakosm Med* 10, no. 1 (January 1976): 54-6. Topics discussed include: seasonality, timing, acclimation, aerospace medicine, tropical climates.

62. Kozyreva, T. et al. "Effect of adaptation to a hot climate on the state of temperature reception." *Hum Physiol* 10, no. 1 (January 1984): 56-8. Topics discussed include: Siberia, Turkmenistan, acclimation, thermoregulation, heat.

63. Krajewski, J. et al. "Scheduling rest for consecutive light and heavy work loads under hot ambient conditions." *Ergonomics* 22, no. 8 (August 1979): 975-87. Topics discussed include: acclimation, body temperature, effects on heart beat, lactates blood, oxygen consumption, skin temperature, perspiration, physical effort, heat, humidity effects, rest.

64. Krivoshchekov, S. "Characteristics of the vascular reaction to cooling in polar explorers in Antarctica depending on the length of winter and occupational activity." *Gig Tr Prof Zabol no.* 11 (November 1983): 58-9. Topics discussed include: Antarctic regions, acclimation, thermoregulation, cold climate effects, blood supply, occupations, skin temperature.

65. Kuhlemeier, K. et al. "Heart rate rectal temperature relationships during prolonged work: males and females compared. pp." In: Folinsbee L.J. et al., ed. *Environmental Stress,* New York, Academic Press, (1978): 71-81. Topics discussed include: acclimation, heat, seasonality, gender effects, stress, physical effort, heart rate.

66. ---. "Pulse rate rectal temperature relationships during prolonged work." *J Appl Physiol* 44, no. 3 (March 1978): 450-4. Topics discussed include: cold, effects on heart beat, heat, industry, meat packing industry, acclimation, physical effort.

67. Lahiri, S. et al. *High altitude and man.* 199 pages. American Physiological Society. Bethesda, Md. Baltimore: American Physiological Society ; Distributed by Williams & Wilkins Co., 1984. Topics discussed include: man, adaptation physiology, acclimation.

68. Lind, A. "Effect of individual variation on upper limit of prescriptive zone of climates." *J Appl Physiol* 28, no. 1 (January 1970): 57-62. Topics discussed include: body temperature, effects on heart beat, heat, acclimation, thermoregulation, exertion.

69. Lind, A. et al. "Influence of age and daily duration of exposure on responses of men to work in heat." *J Appl Physiol* 28, no. 1 (January 1970): 50-6. Topics discussed include: body temperature, body weight, forearm blood supply, effects on heart beat, regional blood flow, acclimation, thermoregulation, physical effort.

70. Matiukhin, V. et al. "The seasonal and the daily changes of oxygen consumption and body temperature among adolescents in the course of studies and adaptation to new climatic conditions." *Gig Sanit no.* 12 (December 1975): 97-9. Topics discussed include: climate, acclimation, oxygen consumption, work.

71. Matyukhin, V. et al. "Seasonal and diurnal dynamics of skin temperature and heat loss in adolescents during adaptation." *Hum Physiol* 4, no. 4 (July 1978): 599-604. Topics discussed include: circadian rhythm, emigration, immigration, microclimate, USSR, acclimation, thermoregulation, skin.

72. McNall, P. et al. "Seasonal Variation in Comfort Conditions for College-Age Persons in the Middle West." *ASHRAE Transactions* 74 (1968): IV.2.1-IV.2.9. Topics discussed include: environment, behavior.

73. Milovidov, E. "Physiological approaches to the climatic treatment of kidney disease." *Ter Arkh* 64, no. 11 (1992): 70-4. Topics discussed include: acclimation, physiology, acid base equilibrium, combined modality therapy, Kazakhstan, physiopathology, blood, reference values, desert climate.

74. Mitchell, D. et al. "Acclimatization in a Hot, Humid Environment - Energy Exchange, Body-Temperature, and Sweating." *Journal of Applied Physiology* 40, no. 5 (1976): 768-778.

75. Nair, C. et al. "The effect of altitude and cold on body temperature during acclimatization of man at 3,300 m." *Int J Biometeorol* 16, no. 1 (January 1972): 79-84. Topics discussed include: anoxia, thermoregulation, cold.

76. ---. "Effect of acclimatization to altitude and cold on basal heart rate, blood pressure, respiration and breath holding in man." *Aerosp Med* 42, no. 8 (August 1971): 851-5. Topics discussed include: aerospace medicine, anoxia physiopathology, arteries, spirometry, blood pressure, cold climates, effects on heart beat, respiration.

77. ---. "Effect of post hypoxic acclimatization on extremity temperature during exposure to acute cold stress in man." *Indian J Med Res* 90 (October 1989): 341-5. Topics discussed include: extremities, physiology, altitude, body physiology, oxygen physiology.

78. Natsume, K. et al. "Preferred ambient temperature for old and young men in summer and winter." *Int J Biometeorol* 36, no. 1 (March 1992): 1-4. Topics discussed include: aging physiology, body regulation physiology, perception physiology, seasonality, acclimation physiology.

79. Nedbaeva, N. "Seasonal dynamics of diurnal rhythms of oxygen consumption and heat production among adolescents in Siberia." *Fiziol Cheloveka* 12, no. 2 (March 1986): 263-8. Topics discussed include: Siberia, acclimation, thermoregulation, circadian rhythm, oxygen consumption.

80. Norberg, K. A. "Physiological adaptation to a hot climate." *Lakartidningen* 72, no. 9 (26 February 1975): 821-2. Topics discussed include: body temperature, dehydration, heat, perspiration, acclimation, desert tropical climates.

81. Pandolf, K. B. et al. *Physiological Differences between Men and Women in Exercise Heat Tolerance and Heat Acclimation*, Performer: Army Research Inst. of Environmental Medicine, Natick, MA., February 1985. Topics discussed include: acclimation, exercise physiology, stress physiology, tolerance.

82. Pandolf, K. B. et al. "Role of physical fitness in heat acclimatization, decay and reinduction." *Ergonomics* 20, no. 4 (1977): 399-408.

83. Pandolf, K. B. et al. *Human performance physiology and environmental medicine at terrestrial extremes*, edited by Kent B. Pandolf, Michael N. Sawka, Richard R. Gonzalez, 637 pages. Indianapolis: Benchmark Press, 1988. Topics discussed include: adaptation, acclimation, exposure.

84. Pepmiller, E. G. "Conserve heat and live." *Aging* (January 1981): 16-17. Topics discussed include: hypothermia, biological aging, acclimation, body temperature, life style, energy conservation, chronic diseases.

85. *Adaptations to climatic changes*, ed P. Pevet. Basel, New York: Karger, 1987. Topics discussed include: acclimation, physiology.

86. "Proceedings of the International Symposium on Acclimatization, Adaptation, and Tolerance to High Altitude." International Symposium on Acclimatization, Adaptation, and Tolerance to High Altitude (1981 : National Institutes of Health), and John E. Fogarty International Center for Advanced Study in the Health Sciences. U.S. Dept. of Health and Human Services, Public Health Service, National Institutes of Health, Washington, D.C., 1983.

87. Purkayastha, S. et al. "Peripheral vascular response to local cold stress of tropical men during sojourn in the Arctic cold region." *Jpn J Physiol* 42, no. 6 (1992): 877-89. Topics discussed include: arctic regions, body temperature, India, Russia, skin temperature, tropical climates, wrist blood supply, acclimation, blood circulation.

88. Raynaud, J. et al. "Body temperatures during rest and exercise in residents and sojourners in hot climate." *Int J Biometeorol* 20, no. 4 (December 1976): 307-17. Topics discussed include: acclimation, Europe, heat, India, perspiration, thermoregulation, physical effort, tropical climates.

89. Rohles, F. H. et al. "Thermal Comfórt in the Elderly." *ASHRAE Transactions* 78 (1972): 131-137. Topics discussed include: temperature, behavior, comfort.

90. Saribekova, T. "Changes in the work and living conditions and the state of health of former inhabitants of the Polar regions, having moved to the moderate climate zone." *Zdravookhr Ross Fed* 0, no. 10 (October 1974): 31-4. Topics discussed include: activities of daily living, adaptation, cold transients, migrants, USSR, work capacity, acclimation, climate.

91. Sen, G. J. et al. "Physiological responses during continuous work in hot dry and hot humid environments in Indians." *Int J Biometeorol* 28, no. 2 (May 1984): 137-46. Topics discussed include: acclimation, effects on heart beat, heat, humidity effects, respiration, thermoregulation, exertion.

92. Senay, L. J. "Changes in plasma volume and protein content during exposures of working men to various temperatures before and after acclimatization to heat: separation of the roles of cutaneous and skeletal muscle circulation." *J Physiol Lond* 224, no. 1 (July 1972): 61-81. Topics discussed include: body temperature, body weight, capillary permeability, cold climate effects, physical effort, timing, water, blood, proteins analysis, muscles, skin blood supply.

93. Shapiro, Y. et al. "Physiological and hematologic responses to summer and winter dry heat acclimation." *J Appl Physiol* 50, no. 4 (April 1981): 792-8. Topics discussed include: blood proteins analysis, body temperature, effects on heart beat, hematocrit, plasma volume, serum albumin analysis, acclimation, seasonality.

94. Shishkin, G. et al. "Lung volume changes during cold air breathing in construction workers in northern regions of Western Siberia." *Fiziol Cheloveka* 18, no. 4 (July 1992): 12-7. Topics discussed include: lung volume measurements, Siberia, acclimation, cold climates, occupational exposure, respiration physiology.

95. Shvartz E et al. "Simple reaction-time during exercise, heat-exposure, and heat acclimation." *Aviation Space and Environmental Medicinė* 47, no. 11 (1976): 1168-1170.

96. ---. "Orthostatism and heat acclimation." *Journal of Applied Physiology* 39, no. 4 (1975): 590-595.

97. Shvartz, E. "Comparision of three methods of acclimatization to dry heat." *Journal of Applied Physiology* 34, no. 2 (1973): 214-219.

98. Shvartz, E. "Effect of neck versus chest cooling on responses to work in heat." *J Appl Physiol* 40, no. 5 (May 1976): 668-72. Topics discussed include: acclimation, skin temperature, perspiration, thermoregulation, cold climate effects, neck physiology, thorax physiology, work.

99. Sil'Chenko, K. "Personal acclimatization to a hot climate." *Voen Med Zh* 5, no. 0 (May 1974): 52-4. Topics discussed include: atmosphere exposure chambers, military personnel, timing, tropical climates.

100. Sundstroem, E. S. "Contributions to tropical physiology with special reference to the adaptation of the white man to the climate of North Queensland." University of California publications in physiology. v. 6, Berkeley, Calif.: University of California Press, 1926. Topics discussed include: acclimation, man.

101. Topchian, Z. et al. "Mechanism of action of climatobalneological factors of Dzhermuk health resort on patients with ischemic heart disease." *Vopr Kurortol Fizioter Lech Fiz Kult no.* 1 (January 1982): 8-11. Topics discussed include: acclimation, altitude, Armenia, coronary physiopathology, balneology, coronary therapy.

102. Torii, M. et al. "Functional differentiation of thermoregulation, with particular references to seasonal variation of whole body sweat rate of exercising man." *Ann Physiol Anthropol* 10, no. 3 (July 1991): 163-70. Topics discussed include: acclimation, temperature, thermoregulation, exercise physiology, perspiration physiology.

103. Tsyganov, E. et al. "Body function in nutrition with a diet made up of canned products during acclimatization to a hot climate." *Vopr Pitan no.* 1 (January 1985): 27-30. Topics discussed include: cardiovascular system physiology, physical effort, higher nervous activity, muscles physiology, nervous system, timing, food preservation.

104. Tuliakova, L. et al. "On the peculiarities of thermoregulation and other responses under conditions of cold climate in man." *Fiziol Zh SSSR* 53, no. 8 (August 1967): 969-73. Topics discussed include: age effects, seasonality, skin, acclimation, thermoregulation, cold climate.

105. Wenzel, H. "Relations between body core temperature and pulse rate of a man performing physical work under warm climatic conditions." *Int Z Angew Physiol* 26, no. 1 (1968): 43-94. Topics discussed include: body regulation, effects on heart beat, humidity effects, acclimation, physical effort, work.

106. Wenzel, H. et al. "Technical experiments on the structure and operation of a climatic chamber for physiologic studies of humans at work." *Int Z Angew Physiol* 25, no. 3 (1968): 235-78. Topics discussed include: acclimation, heat, air, physical effort, humidity effects, physiology instrumentation, temperature.

107. Wilkins, D. "Heat acclimatization in the Antarctic." *J Physiol Lond* 214, no. 1 (April 1971): 15P-16P. Topics discussed include: body temperature, cold climate effects, seasonality, perspiration, weather.

108. Wyndham, C. et al. "The heat reactions of Bantu males in various states of acclimatization. I. The sweat rate rectal temperature relationship." *Int Z Angew Physiol* 23, no. 1 (12 September 1966): 63-78. Topics discussed include: South Africa, thermoregulation, Negroid race, perspiration, tropical climates.

109. Wyss, V. "On the effects of radiant heat on different regions of the human body. 3. Behavior of cardiocirculatory and respiratory activity, oxygen consumption, oral and skin temperature during infrared radiation of different areas in acclimatized and non acclimatized subjects." *Med Lav* 57, no. 4 (April 1966): 262-85. Topics discussed include: infrared rays, blood circulation, blood pressure, effects on heart beat, oxygen consumption, respiration.

110. Yousef, M. "Effects of climatic stresses on thermoregulatory processes in man." *Experientia* 43, no. 1 (15 January 1987): 14-9. Topics discussed include: acclimation, cold climate effects, environment, heat, homeostasis, thermoregulation, climate.

111. Zalevskii, V. "The status of the temperature of children of various ages as a function of meteorological factors." *Gig Sanit* 31, no. 8 (August 1966): 37-41. Topics discussed include: acclimation, environmental exposure, Georgia Republic, Russia, Siberia, thermoregulation, climate.

112. Zunnunov, Z. "Physical work capacity as a criterion of human resistance under the hot climatic conditions of an arid zone." *Fiziol Zh SSSR* 78, no. 6 (June 1992): 18-24. Topics discussed include: acclimatization physiology, ethnic groups, exercise test, Uzbekistan, desert climate, exercise physiology, physical endurance physiology, tropical climates.

ALLERGIES

113. Aalberse, R. et al. "'Horoscope effect' not only for seasonal but also for non seasonal allergens." *Clin Exp Allergy* 22, no. 11 (November 1992): 1003-6. Topics discussed include: mites, pollen, pregnancy, immunology, labor.

114. Aberg, N. "Birth season variation in asthma and allergic rhinitis." *Clin Exp Allergy* 19, no. 6 (November 1989): 643-8. Topics discussed include: epidemiologic methods, Sweden, asthma, hay fever, rhinitis, perennial epidemiology.

115. Abidov, M. et al. "Various immunological indicators in patients with allergic dermatitis and systemic lupus erythematosus under the climatic conditions of Uzbekistan." *Vestn Dermatol Venerol* no. 2 (February 1982): 33-7. Topics discussed include: leukocyte count, tropical climate, dermatitis, atopic immunology, lup erythematos, systemic immunology, lymphopenia etiology.

116. Alemany-Vall, R. "Bioclimatological and Biometeorological Aspects of Allergic Diseases." *Intern. J. Bioclimatol. Biometeorol.* 2 (1958): IV C 4a. Topics discussed include: climate, meteorology, allergy, health.

117. Ando, M. "Type III and type IV allergy: summer type hypersensitivity pneumonitis." *Jpn J Med* 28, no. 4 (July 1989): 548-9. Topics discussed include: immunology, seasonality.

118. Arblaster, P. et al. "Obstructive airways disease: immediate and late results of treatment in an alpine climate." *Acta Allergol* 21, no. 1 (1966): 25-44. Topics discussed include: United Kingdom, respiratory functions, Switzerland, asthma, treatment, bronchitis.

119. Arnolt, R. et al. "Correlations between asthmatic crisis and meteorological conditions in Rosario, Argentina." *Allergol Immunopathol Madr* 12, no. 5 (September 1984): 387-95. Topics discussed include: air movements, atmospheric effects, seasonality, timing, asthma etiology.

120. Baldini, G. et al. "Geo climatic environment and allergic sensitization. Epidemiological study in 2 schoolchild populations in Tuscany." *Minerva Pediatr* 36, no. 15-16 (31 August 1984): 755-9. Topics discussed include: asthma, epidemiology, dermatitis, atopic epidemiology, hay fever, Italy, skin tests, hypersensitivity.

121. Bar, O. O. et al. "Effects of dry and humid climates on exercise induced asthma in children and preadolescents." *J Allergy Clin Immunol* 60, no. 3 (September 1977): 163-8. Topics discussed include: asthma, etiology, treatment, effects on heart beat, swimming, physiopathology, physical exertion, humidity.

122. Bauer, C. "Spring and summer with pollen allergies." *Krankenpfl J* 26, no. 5 (1 May 1988): 228-30. Topics discussed include: asthma etiology, seasonality, respiratory hypersensitivity etiology.

123. Beaumont, F. et al. "A volumetric aerobiologic study of seasonal fungus prevalence inside and outside dwellings of asthmatic patients living in northeast Netherlands." *Ann Allergy* 53, no. 6 (December 1984): 486-92. Topics discussed include: asthma, lung diseases, immunology, Netherlands.

124. Berard, P. et al. "Seasonal and chronic allergic conjunctivitis in adults with children. Multicenter evaluation of the collyrium N acetylaspartylglutamic acid (NAAGA) in ophthalmologic and allergy practices." *Bull Soc Ophtalmol Fr* 87, no. 4 (April 1987): 475-6, 479-83. Topics discussed include: chronic disease, dipeptides, magnesium, treatment using drugs.

125. Berg, T. et al. "In vitro diagnosis of atopic allergy. IV. Seasonal variations of IgE antibodies in children allergic to pollens." *Int Arch Allergy Appl Immunol* 41, no. 2 (1971): 452-62. Topics discussed include: absorption, antibody specificity, asthma, chromones, hay fever, hypersensitivity.

126. Bjerrum, P. et al. "Treatment of seasonal allergic rhinitis with budesonide and disodium cromoglycate. A double blind clinical comparison between budesonide and disodium cromoglycate." *Allergy* 40, no. 1 (January 1985): 65-9. Topics discussed include: airway resistance, clinical trials, hay fever physiopathology, hay fever treatment using drugs.

127. Blamoutier, J. et al. "Seasonal allergic respiratory manifestations." *Presse Therm Clim* 109, no. 2 (1972): 105-7. Topics discussed include: allergens isolation, purification, desensitization, immediate diagnosis, skin tests, respiratory hypersensitivity.

128. Borelli, S. et al. "Changes in the environment dependent allergic reactivity in patients with neurodermatitis constitutionalis following high altitude climate therapy." *Hautarzt* 18, no. 10 (October 1967): 456-8. Topics discussed include: acetylcholine diagnostic use, bradykinin, histamine, serotonin, skin tests, balneology, neurodermatitis treatment.

129. Borum, P. et al. "Seasonal allergic rhinitis and depot injection of a corticosteroid. Evaluation of the efficacy of medication early and late in the season based on detailed symptom recording." *Allergy* 42, no. 1 (January 1987): 26-32. Topics discussed include: clinical trials, delayed action preparations, injections, intramuscular, hay fever treatment using drugs, methylprednisolone, dosage.

130. Boulet, L. et al. "Asthma and increases in nonallergic bronchial responsiveness from seasonal pollen exposure." *J Allergy Clin Immunol* 71, no. 4 (April 1983): 399-406. Topics discussed include: antibodies, bronchial provocation, methacholine compounds, asthma, immunology.

131. Bourguet, J. et al. "Symptomatic treatment of allergies and seasonal infections of the field of otorhinolaryngology." *Ann Otolaryngol Chir Cervicofac* 91, no. 10-11 (October 1974): 633-8. Topics discussed include: drug tolerance, acetaminophen, ethylamines, hypersensitivity, treatment using drugs, infection, phenacetin, phenylpropanolamine.

132. Brandon, M. et al. "Clinical studies of terfenadine in seasonal allergic rhinitis." *Arzneimittelforschung* 32, no. 9a (1982): 1204-5. Topics discussed include: benzhydryl compounds, histamine h receptor blockaders, hay fever treatment using drugs.

133. Brostoff, J. "Cellular and humoral effects of hyposensitization in patients with summer hay fever." *Int Arch Allergy Appl Immunol* 45, no. 1 (1973): 162-9. Topics discussed include: antibodies, anti-idiotypic, autoradiography, hemagglutination tests, pollen, seasonality, skin tests, antibody formation, desensitization, immunology.

134. Brostoff, J. et al. "Co-seasonal prick desensitization in summer hay fever." *Acta Allergol* 23, no. 1 (February 1968): 35-8. Topics discussed include: histamine, placebos, plant extracts, pollen, timing, treatment using drugs.

135. Brown, H. et al. "Allergy to seasonal mould spores." *Allergol Immunopathol Madr.* 9 (1981): 137-42. Topics discussed include: air microbiology, asthma diagnosis, etiology, bronchial provocation tests, England, hay fever diagnosis, nasal provocation tests, spores.

136. Bruckner, A. "Maritime climate therapy. Supply of drugs." *Allerg Asthmaforsch* 6 (1969): 49-53. Topics discussed include: drug utilization, asthma treatment using drugs, eczema treatment using drugs, thalassotherapy.

137. Buisseret, P. "Seasonal allergic symptoms due to fungal spores." *Br Med J* 2, no. 6034 (28 August 1976): 507-8. Topics discussed include: asthma etiology, hay fever etiology, spores.

138. Butenko, G. et al. "Differentiated sanatorium and climate treatment of chronic bronchitis patients with clinical manifestations of allergy." *Vopr Kurortol Fizioter Lech Fiz Kult* no. 2 (March 1983): 16-9. Topics discussed include: autoimmune diseases, chronic disease, hypersensitivity, Russia, bronchitis, health resorts.

139. Carosso, A. et al. "The effect of birth season on pollenosis." *Ann Allergy* 56, no. 4 (April 1986): 300-3. Topics discussed include: dust, hypersensitivity etiology, Italy, hay fever epidemiology.

140. Carrasco, E. et al. "Geographic climatic characteristics and allergogenic flora of Chile." *Rev Med Chil* 101, no. 2 (February 1973): 155-62. Topics discussed include: Chile, pollen, seasonality, causes of allergenic reactions, geography, plants.

141. Castillo, J. et al. "Seasonal influence on serum and receptor bound IgE in pollinosis." *Ann Allergy* 64, no. 6 (June 1990): 542-6. Topics discussed include: antigens, hay fever immunology.

142. ---. "IgE receptors and serum IgE in pollinosis. The influence of season and action of ketotifen." *Allergol Immunopathol Madr* 15, no. 4 (July 1987): 179-83. Topics discussed include: asthma, treatment using drugs, hay fever, immunoglobulins, surface analysis, immunology, ketotifen.

143. Charpin, D. et al. "Seasonal allergic symptoms and their relation to pollen exposure in south east France." *Clin Exp Allergy* 23, no. 5 (May 1993): 435-9. Topics discussed include: asthma, immunology, environmental exposure statistics, France, grasses, hay fever, risk, trees, epidemiology.

144. Coffman, D. "A controlled trial of disodium cromoglycate in seasonal allergic rhinitis." *Br J Clin Pract* 25, no. 9 (September 1971): 403-6. Topics discussed include: asthma complications, grasses, pollen, skin tests, chromones, hay fever treatment using drugs.

145. Cookson, J. et al. "Seasonal asthma and the house dust mite in tropical Africa." *Clin Allergy* 5, no. 4 (December 1975): 375-80. Topics discussed include: humidity, skin tests, immunology.

146. Cooper, B. "Observations on the treatment of seasonal rhinitis with chemically modified pollen allergen." *Br J Clin Pract* 30, no. 11-12 (November 1976): 217-20. Topics discussed include: tyrosine, desensitization, hay fever treatment.

147. Crimi, E. et al. "Late asthmatic reaction to perennial and seasonal allergens." *J Allergy Clin Immunol* 85, no. 5 (May 1990): 885-90. Topics discussed include: antigenic determinants, mites, pollen, timing, immunology, asthma.

148. ---. "Effect of seasonal exposure to pollen on specific bronchial sensitivity in allergic patients." *J Allergy Clin Immunol* 85, no. 6 (June 1990): 1014-9. Topics discussed include: antibodies, bronchial provocation tests, skin tests, asthma.

149. Czubalski, K. et al. "Importance of the mental factor in hay fever and non-seasonal allergic rhinitis in the light of psychosomatic studies." *Pol Tyg Lek* 30, no. 19 (12 May 1975): 809-12. Topics discussed include: emotions, psychophysiologic disorders, stress, hay fever etiology.

150. Dahl, S. "Meteorologic studies in a clinical department for allergic bronchial asthma." *Allerg Asthma Leipz* 11, no. 3 (1965): 85-94. Topics discussed include: barometric pressure, humidity effects, Sweden, temperature, asthma epidemiology, seasonality, weather.

151. Dart, J. et al. "Perennial allergic conjunctivitis: definition, clinical characteristics and prevalence. A comparison with seasonal allergic conjunctivitis." *Trans Ophthalmol Soc U K* 105, no. Pt 5 (1986): 513-20. Topics discussed include: allergic immunology, pathology, dust, hay fever complications, rhinitis, perennial complications, skin tests, tears immunology, allergic diagnosis.

152. Delaitre, F. et al. "Respective value of thermal and climatic therapies and desensitization in treatment of respiratory allergy in children." *Presse Therm Clim* 105, no. 4 (1968): 240-6. Topics discussed include: balneology, desensitization, respiratory hypersensitivity treatment.

153. Diaz-Rubio, M. "Climate; Influence on Allergy." *Acta Allergologica* 5 (1952): 320-324. Topics discussed include: disease, health, allergy.

154. Dzhuraev, M. "Features of the course of allergic diseases in children in mountain climates." *Pediatriia no.* 4 (April 1984): 53-4. Topics discussed include: Tajikistan, altitude, hypersensitivity epidemiology.

155. Eichenberger, H. et al. "Co-seasonal desensitization of pollinosis with the scarification method of Blamoutier." *Acta Allergol* 21, no. 3 (1966): 261-7. Topics discussed include: asthma treatment, desensitization, hay fever treatment, pollen.

156. Ellis, R. "Allergic rhinitis seasonal and nonseasonal." *Geriatrics* 25, no. 9 (September 1970): 113-6. Topics discussed include: hay fever, diagnosis, treatment.

157. Enaud, B. et al. "Climatotherapy of allergic diseases in France." *Presse Therm Clim* 108, no. 1 (1971): 35-8. Topics discussed include: asthma treatment, France, hypersensitivity treatment.

158. Fein, B. et al. "The climatic factors in the etiology of allergic diseases." *J Asthma Res* 3, no. 1 (September 1965): 17-23. Topics discussed include: humidity, pollen, temperature, Texas, weather, hypersensitivity etiology.

159. Fennerty, A. et al. "Seasonal variation in suppressor T cell subsets and non specific suppressor cell function in hay fever sufferers." *Allergy* 44, no. 2 (February 1989): 103-7. Topics discussed include: antibodies, monoclonal immunology, concanavalin a pharmacology, drug effects, lymphocytes, suppressor effector physiology, blood, classification.

160. Frankland, A. "A clinical evaluation of repository therapy in Great Britain: seasonal allergic rhinitis." *Ann Allergy* 26, no. 11 (November 1968): 584-6. Topics discussed include: emulsions, grasses, United Kingdom, pollen, desensitization, hay fever treatment using drugs, pollen, dosage.

161. ---. "Seasonal allergic rhinitis." *Proc R Soc Med* 64, no. 4 (April 1971): 447-50. Topics discussed include: dust, grasses, mites, pollen, spores, hay fever.

162. Freidhoff, L. et al. "A study of the human immune response to Lolium perenne (rye) pollen and its components, Lol p I and Lol p II (Rye I and Rye II). II. Longitudinal variation of antibody levels in relation to symptomatology and pollen exposure and correction of seasonally elevated antibody levels to basal values." *J Allergy Clin Immunol* 80, no. 5 (November 1987): 646-55. Topics discussed include: hay fever immunology, plant extracts immunology, radioimmunoassay, skin tests, grasses immunology, lolium immunology, immunology.

163. Galan, L. I. et al. "Origin and distribution of seasonal allergies. A primary care view." *Aten Primaria* 10, no. 9 (December 1992): 1019-23. Topics discussed include: environment, grasses, primary health care, Spain, epidemiology, causes of allergenic reactions, hay fever, pollen.

164. Galvagno, V. J. et al. "Influence of Meteorologic Variations on Allergic Syndromes." *Allergia* 3 (November 1949): 97-101. Topics discussed include: weather, disease, health, allergy.

165. Gerblich, A. et al. "Seasonal variation of airway function in allergic rhinitis." *J Allergy Clin Immunol* 77, no. 5 (May 1986): 676-81. Topics discussed include: bronchi drug effects, bronchial provocation tests, hay fever, methacholine compounds, diagnostic use, pharmacology, physiopathology.

166. Gering, G. I. et al. "Genesis of the nasal discharge in seasonal allergic rhinopathy." *Vestn Otorinolaringol* 35, no. 6 (November 1973): 15-8. Topics discussed include: epithelium cytology, hay fever, pathology, physiopathology, nose secretion.

167. Giulev, A. et al. "Initial results of specific desensitization of seasonal allergic rhinitis in the Otolaryngologic Clinic of the I.P. Pavlov Medical Institute in Plovdiv, Bulgaria." *Zh Ushn Nos Gorl Bolezn* 29, no. 1 (January 1969): 70-3. Topics discussed include: allergens, Bulgaria, hospitals, desensitization, hay fever treatment.

168. Giusan, A. "Endonasal allergoid therapy of seasonal allergic rhinosinusitis." *Vestn Otorinolaringol* no. 6 (November 1985): 81-2. Topics discussed include: administration, intranasal, desensitization, immunologic methods, hay fever treatment, pollen, sinusitis treatment.

169. Giusan, A. et al. "Complex histaglobin and allergoid therapy of seasonal allergic diseases of the upper respiratory tract." *Vestn Otorinolaringol* no. 2 (March 1983): 68-9. Topics discussed include: drug combinations, causes of allergenic reactions, asthma, desensitization, gamma globulins, hay fever, histamine, dosage.

170. ---. "Comparative characteristics of the effectiveness of specific treatment methods in seasonal allergic rhinosinusitis." *Vestn Otorinolaringol* no. 3 (May 1980): 50-2. Topics discussed include: clinical trials, pollen, desensitization, immunologic methods, hay fever treatment, plant extracts, sinusitis.

171. Grammer, L. et al. "Absence of nasal priming as measured by rhinitis symptom scores of ragweed allergic patients during seasonal exposure to ragweed pollen." *Allergy Proc* 11, no. 5 (September 1990): 243-6. Topics discussed include: timing, hay fever, physiopathology, nasal mucosa, nasal provocation tests.

172. Graul, B. "Maritime climate therapy. Subjective well being during climate therapy in asthmatics dependent upon climates." *Allerg Asthmaforsch* 6 (1969): 171-9. Topics discussed include: asthma, infants, neurocirculatory asthenia etiology, prognosis, treatment, neurocirculatory asthenia treatment, thalassotherapy.

173. Haahtela, T. et al. "Influence of the pollen season on immediate skin test reactivity to common allergens." *Allergy* 35, no. 1 (January 1980): 15-21. Topics discussed include: dust, skin tests.

174. Harnack, K. "Maritime climate therapy. Maritime climate therapy in endogenous eczema. The importance of maritime climate therapy in endogenous eczema in comparison with other climates." *Allerg Asthmaforsch* 6 (1969): 54-62. Topics discussed include: eczema treatment, thalassotherapy.

175. Hasegawa, M. et al. "Seasonal variations of total histamine in patients with seasonal allergic rhinitis." *Clin Allergy* 13, no. 3 (May 1983): 277-86. Topics discussed include: chromatography, high pressure liquid, fluorescence, histidine analysis, spermidine analysis, hay fever, blood.

176. Henocq, E. et al. "Can the seasonal allergens favor the recurrence of chronic inflammatory attacks?" *Acta Allergol* 32, no. 3 (June 1977): 219-20. Topics discussed include: chronic disease, recurrence, inflammation.

177. Henriksen, J. "Exercise induced bronchoconstriction. Seasonal variation in children with asthma and in those with rhinitis." *Allergy* 41, no. 7 (September 1986): 499-506. Topics discussed include: asthma, blood, basophils, bronchial spasm, immunology, eosinophils, leukocyte count, pollen immunology, respiratory function tests, hay fever.

178. Hillas, J. et al. "A survey of airborne pollen and spores in Auckland: its use in the diagnosis of seasonal allergies." *N Z Med J* 89, no. 628 (24 January 1979): 37-40. Topics discussed include: cell count, New Zealand, weather, hay fever diagnosis, meteorological effects, spores.

179. Irander, K. et al. "Hyposensitization in allergic seasonal rhinitis as evaluated by rhinomanometry." *Acta Otolaryngol Suppl Stockh* 360 (1979): 24-6. Topics discussed include: airway resistance, causes of allergenic reactions, allergens diagnostic use, manometry, nasal cavity physiopathology, desensitization, hay fever treatment.

180. Jimenez-Diaz, C. et al. "Climatic Asthmas." *J. Allergy* 3 (1932): 396-403. Topics discussed include: climate, disease, health, asthma.

181. Jung, B. et al. "Comparison of the beneficial health resort effects in two groups of asthmatics with and without climate therapy in salt pits." *Allerg Asthma Leipz* 14, no. 1 (April 1968): 58-66. Topics discussed include: acetylcholine diagnostic use, Germany, isoproterenol diagnostic use, sodium chloride, spirometry, asthma treatment, balneology, health resorts.

182. Kalra, S. et al. "Absence of seasonal variation in concentrations of the house dust mite allergen Der p1 in south Manchester homes." *Thorax* 47, no. 11 (November 1992): 928-31. Topics discussed include: asthma etiology, environmental exposure, household articles, humidity effects, radioallergosorbent test, analysis, dust, mites immunology.

183. Karachevtseva, T. "Climatic treatment of children with respiratory allergy." *Vopr Kurortol Fizioter Lech Fiz Kult* no. 4 (July 1984): 58-64. Topics discussed include: acclimation, altitude, asthma treatment, balneology, health resorts, microclimate, physical treatment methods, USSR, weather.

184. Karjalainen, J. et al. "Seasonal variability of exercise induced asthma especially outdoors. Effect of birch pollen allergy." *Clin Exp Allergy* 19, no. 3 (May 1989): 273-8. Topics discussed include: cold, exercise test, effects on heart beat, histamine, temperature, vital capacity, asthma.

185. Khan, A. "The role of air pollution and weather changes in childhood asthma." *Ann Allergy* 39, no. 6 (December 1977): 397-400. Topics discussed include: carbon monoxide, ozone, pollen, regression analysis, sulfur dioxide, temperature, asthma etiology.

186. Kleinsorge, H. et al. "Results of the 1965 research study on high sea climatic cure in patients with bronchial asthma." *Allerg Asthma Leipz* 13, no. 5 (1967): 236-44. Topics discussed include: asthma treatment using drugs, Germany, lung physiopathology, psychological tests, recreation, asthma treatment, balneology.

187. Koivikko, A. "Aerobiology, climate and allergy." *Duodecim* 97, no. 9 (1981): 610-8. Topics discussed include: air microbiology, air pollution effects, causes of allergenic reactions, hypersensitivity etiology.

188. Kolesar, J. et al. "Changes of the immunoglobulin levels (IgG, IgA and IgM) during a high mountain climatic cure in asthmatic patients." *Allerg Immunol Leipz* 17, no. 3 (1971): 255-61. Topics discussed include: asthma, blood, altitude, immunology, immunoglobulins.

189. ---. "Influence of high mountain climatic cure on the histamine fixation in pollen allergy." *Allerg Asthma Leipz* 13, no. 5 (1967): 208-11. Topics discussed include: altitude, hay fever, blood, balneology, hay fever treatment, histamine blood.

190. Korner, E. et al. "Seasonal burning palate in allergy against grass pollen." *Z Hautkr* 54, no. 18 (15 September 1979): 817-20. Topics discussed include: grasses, hypersensitivity, immediate immunology, burning mouth syndrome immunology, mouth diseases immunology.

191. Kornienko, A. et al. "Functional morphological changes of the nasal mucosa in allergic seasonal pollen rhinitis." *Vestn Otorinolaringol no.* 3 (May 1974): 75-80. Topics discussed include: hay fever, pathology, physiopathology, nasal mucosa.

192. Kraszko, P. et al. "Grotto climate therapy in obstructive bronchitic syndrome." *Allerg Immunol Leipz* 20-21, no. 1 (1974): 47-54. Topics discussed include: asthma treatment, climate.

193. LaForce, C. F. et al. "Fluticasone propionate: an effective alternative treatment for seasonal allergic rhinitis in adults and adolescents." *J Fam Pract* 38, no. 2 (February 1994): 145-52. Topics discussed include: hay fever physiopathology.

194. Lau, S. et al. "Seasonal variations in house dust mite allergens significance for children with allergy." *Monatsschr Kinderheilkd* 138, no. 2 (February 1990): 58-61. Topics discussed include: Germany, mites analysis, respiratory hypersensitivity, epidemiology, rhinitis, perennial etiology.

195. Lazarus, P. "Maritime climate therapy. Pulmonary function in patients with bronchial asthma." *Allerg Asthmaforsch* 6 (1969): 141-55. Topics discussed include: acetylcholine, asthma treatment, isoproterenol, neurotic disorders, spirometry, ventilation, vital capacity.

196. Leino, M. et al. "The effect of sodium cromoglycate eyedrops compared to the effect of terfenadine on acute symptoms of seasonal allergic conjunctivitis." *Acta Ophthalmol Copenh* 70, no. 3 (June 1992): 341-5. Topics discussed include: acute disease.

197. Lelong, M. et al. "Is it necessary to test children having summer respiratory problems with cereal pollens?" *Allerg Immunol Paris* 21, no. 10 (December 1989): 394-5. Topics discussed include: allergens, cereals immunology, grasses immunology, hay fever diagnosis, seasonality, skin tests, hypersensitivity diagnosis, pollen.

198. Leppik, I. et al. "Seasonal incidence of phenytoin allergy unrelated to plasma levels." *Arch Neurol* 42, no. 2 (February 1985): 120-2. Topics discussed include: drugs, hypersensitivity, diagnosis.

199. Lindsay, M. A. "Group comparative trial of 2% sodium cromoglycate (Opticrom) with placebo in the treatment of seasonal allergic conjunctivitis." *Clin Allergy* 9, no. 3 (May 1979): 271-5. Topics discussed include: cromolyn sodium, placebos, pollen, treatment using drugs, hypersensitivity.

200. Lopez, M. R. et al. "Seasonal variation of allergy causing fungi in the southern part of Mexico City." *Allergol Immunopathol Madr* 14, no. 1 (January 1986): 43-8. Topics discussed include: humidity, hypersensitivity etiology, Mexico, urban health, air microbiology, purification.

201. Lowhagen, O. et al. "Modification of bronchial hyperreactivity after treatment with sodium cromoglycate during pollen season." *J Allergy Clin Immunol* 75, no. 4 (April 1985): 460-7. Topics discussed include: asthma, treatment using drugs, physiopathology, bronchi, bronchial provocation testshistamine pharmacology.

202. Machiels, J. et al. "Complexes of grass pollen allergens and specific antibodies reduce allergic symptoms and inhibit the seasonal increase of IgE antibody." *Clin Exp Allergy* 20, no. 6 (November 1990): 653-60. Topics discussed include: antibody specificity, hay fever, immunology, antigen antibody complex, desensitization, treatment.

203. ---. "Allergen antibody complexes can efficiently prevent seasonal rhinitis and asthma in grass pollen hypersensitive patients. Allergen antibody complex immunotherapy." *Allergy* 46, no. 5 (July 1991): 335-48. Topics discussed include: asthma treatment, desensitization, hay fever, immunology, hypersensitivity, prevention.

204. Maderazo, E. et al. "Granulocyte migration abnormality in patients suffering from seasonal allergic rhinitis: failure of treatment with cimetidine." *Int Arch Allergy Appl Immunol* 74, no. 3 (1984): 211-5. Topics discussed include: hay fever, treatment using drugs, physiology.

205. Madonini, E. et al. "Seasonal increase of bronchial reactivity in allergic rhinitis." *J Allergy Clin Immunol* 79, no. 2 (February 1987): 358-63. Topics discussed include: bronchial provocation tests, pollen, skin tests, hay fever physiopathology.

206. May, K. "Meteorological and climatic factors in bronchial asthma. A clinical dilemma." *Allergol Immunopathol Madr* 11, no. 6 (November 1983): 465-72. Topics discussed include: asthma physiopathology, atmospheric effects, cold, humidity effects, wind, asthma etiology.

207. Michailov, P. et al. "Changes in allergic reactivity of skin patients following altitude climatotherapy." *Allerg Asthma Leipz* 12, no. 5 (1966): 268-71. Topics discussed include: allergens, desensitization, histamine pharmacology, hypersensitivity, serotonin pharmacology, skin diseases diagnosis, skin tests.

208. ---. "Effect of high altitude climate therapy on skin reactivity to prostaglandins E1, E2 and F1 alpha in allergic dermatoses." *Dermatol Monatsschr* 169, no. 5 (May 1983): 305-10. Topics discussed include: dermatitis, atopic immunology, intradermal tests, atopic treatment.

209. ---. "Dynamics of lysosomal enzymes in patients with allergic dermatoses before and after high altitude climate therapy." *Hautarzt* 32, no. 8 (August 1981): 417-9. Topics discussed include: dermatitis, atopic treatment, urticaria treatment, atopic enzymology.

210. ---. "Enzymatic changes in mountain climate therapy of allergic dermatoses." *Allerg Asthma Leipz* 11, no. 6 (1965): 251-4. Topics discussed include: dermatitis, contact treatment, altitude, metabolism, contact metabolism.

211. Mikhalov, P. et al. "Therapeutic effectiveness of high mountain climate in treatment of patients with allergic dermatosis." *Vestn Dermatol Venerol no.* 7 (July 1977): 37-41. Topics discussed include: altitude, climate.

212. Mori, K. et al. "Seasonal fluctuations of substance P and vasoactive intestinal peptide concentrations in nasal secretions of patients with nasal allergy to Japanese cedar pollen." *Arerugi* 41, no. 12 (December 1992): 1651-7. Topics discussed include: allergens, nasal mucosa secretion, respiratory hypersensitivity etiology, substance analysis, vasoactive intestinal peptide analysis.

213. Moshkevich, V. et al. "Clinical picture and specific treatment of seasonal allergic rhinosinusopathy." *Vestn Otorinolaringol* 33, no. 1 (January 1971): 67-73. Topics discussed include: desensitization, hay fever diagnosis, treatment.

214. ---. "Seasonal allergic rhinosinusopathy in Alma Ata." *Zh Ushn Nos Gorl Bolezn* 30, no. 3 (May 1970): 82-7. Topics discussed include: Kazakhstan, hay fever epidemiology.

215. Murray, A. et al. "The seasonal variation of allergic respiratory symptoms induced by house dust mites." *Ann Allergy* 45, no. 6 (December 1980): 347-50. Topics discussed include: mites, skin tests, respiratory hypersensitivity diagnosis.

216. ---. "Sensitization to house dust mites in different climatic areas." *J Allergy Clin Immunol* 76, no. 1 (July 1985): 108-12. Topics discussed include: humidity effects, immunization, radioallergosorbent test, skin tests, dust, hypersensitivity immunology, mites immunology.

217. ---. "The seasonal variation in a population of house dust mites in a North American city." *J Allergy Clin Immunol* 64, no. 4 (October 1979): 266-9. Topics discussed include: asthma etiology, Canada, dust, humidity effects, mites immunology, temperature, timing.

218. Naito, K. et al. "Seasonal variations of nasal resistance in allergic rhinitis and environmental pollen counts." *Auris Nasus Larynx* 20, no. 1 (1993): 19-29. Topics discussed include: allergens isolation, purification, Canada, hay fever physiopathology, Japan, nose diseases diagnosis, airway resistance, nasal cavity.

219. Nordvall, S. et al. "Timothy specific IgG antibody levels vary with the pollen seasons." *Allergy* 41, no. 8 (November 1986): 575-80. Topics discussed include: antibody specificity, metabolism, hay fever immunology.

220. Novotny, Z. et al. "Cold allergy manifested in tropical climate." *An Otorrinolaringol Ibero Am* 2, no. 1 (1975): 28-34. Topics discussed include: Cuba, hypersensitivity, tropical climate.

221. Ordman, D. "The Climate Factor in Perennial Respiratory Allergy and Its Relation to House-Dust Sensitivity." *Intern. Arch. Appl. Immunol.* 9, no. 3, 4 (1956): 129-145. Topics discussed include: disease, health, allergy.

222. ---. "Respiratory allergy and the regional climate in Israel." *Ann Allergy* 25, no. 2 (February 1967): 106-14. Topics discussed include: desert, humidity effects, Israel, South Africa, Spain, temperature, respiratory hypersensitivity epidemiology.

223. ---. "Seasonal respiratory allergy in Windhoek: the pollen and fungus factors." *S Afr Med J* 44, no. 9 (28 February 1970): 250-3. Topics discussed include: Africa, air analysis, geography, causes of allergenic reactions, respiratory hypersensitivity epidemiology.

224. ---. "Seasonal respiratory allergy in Bulawayo, Rhodesia: the pollen and fungus factors." *S Afr Med J* 46, no. 48 (2 December 1972): 1898-901. Topics discussed include: air, grasses, spores, Zimbabwe, causes of allergenic reactions, respiratory hypersensitivity etiology.

225. Osterballe, O. et al. "Immunotherapy in hay fever with two major allergens 19, 25 and partially purified extract of timothy grass pollen. A controlled double blind study. In vitro variables, season i." *Allergy* 36, no. 3 (April 1981): 183-99. Topics discussed include: clinical trials, grasses, hay fever immunology, desensitization, immunologic methods, hay fever treatment, immunology.

226. Ostergaard, P. "Frequency of perennial allergic respiratory symptoms in children referred primarily for seasonal symptoms. Judged on the basis of routine skin tests." *Ugeskr Laeger* 143, no. 52 (21 December 1981): 3539-41. Topics discussed include: respiratory hypersensitivity, skin tests, asthma, rhinitis, perennial epidemiology.

227. Paggiaro, P. et al. "Pattern of airway response to allergen extract of Phleum pratensis in asthmatic patients during and outside the pollen season." *Respiration* 57, no. 1 (1990): 51-6. Topics discussed include: grasses, asthma physiopathology, bronchi physiopathology.

228. Palma, C. A. "Co seasonal treatment of pollinoses in Portugal with the method of cutaneous grids." *Rev Fr Allergol* 7, no. 2 (April 1967): 92-5. Topics discussed include: Portugal, skin tests, desensitization, hay fever treatment, pollen.

229. Parker, W. J. et al. "The relationships between late cutaneous responses and specific antibody responses with outcome of immunotherapy for seasonal allergic rhinitis." *J Allergy Clin Immunol* 84, no. 5 Pt 1 (November 1989): 667-77. Topics discussed include: hay fever immunology, trees, desensitization, skin tests.

230. Paskova, Z. "Climatotherapy and otorhinolaryngology in allergic patients." *Cas Lek Cesk* 107, no. 36 (1968): 1105-8. Topics discussed include: climate, hay fever treatment, balneology, hypersensitivity treatment, otorhinolaryngologic diseases treatment.

231. Paulus, H. et al. "Association of allergic bronchial asthma with certain air pollutants and weather parameters." *Int J Biometeorol* 11, no. 1 (March 1967): 119-27. Topics discussed include: cereals, Minnesota, air pollution effects, asthma epidemiology, hypersensitivity complications.

232. Pavlik, I. et al. "Neural factors in the weather sensitivity of asthmatics." *Allerg Immunol Leipz* 18, no. 2 (1972): 85-96. Topics discussed include: arousal, autonomic nervous system physiopathology, circadian rhythm, cough, humidity effects, asthma physiopathology, atmospheric effects.

233. Peat, J. et al. "Sensitivity to common allergens: relation to respiratory symptoms and bronchial hyper responsiveness in children from three different climatic areas of Australia." *Clin Exp Allergy* 21, no. 5 (September 1991): 573-81. Topics discussed include: Australia epidemiology, bronchial hyperreactivity epidemiology, hypersensitivity, hypersensitivity, skin tests, asthma, climate.

234. Pedersen, P. et al. "Seasonal variation of asthma and allergic rhinitis. Consultation pattern in general practice related to pollen and spore counts and to five indicators of air pollution." *Allergy* 39, no. 3 (April 1984): 165-70. Topics discussed include: grasses, spores immunology, trees, asthma physiopathology, hay fever physiopathology.

235. Petrovska, I. et al. "Changes in the clinico functional and immunological indices of patients with chronic obstructive lung disease, the allergic form, following combined climatotherapy at Sandanski health resort." *Vutr Boles* 30, no. 1 (1991): 41-4. Topics discussed include: balneology, Bulgaria, combined modality treatment, immunology, respiratory hypersensitivity, rehabilitation, health resorts.

236. Piccolo, M. et al. "Outbreaks of asthma attacks and meteorologic parameters in Bahia Blanca, Argentina." *Ann Allergy* 60, no. 2 (February 1988): 107-10. Topics discussed include: Argentina, climatic effects, asthma epidemiology, disease outbreaks.

237. Platts, M. T. et al. "Seasonal variation in dust mite and grass pollen allergens in dust from the houses of patients with asthma." *J Allergy Clin Immunol* 79, no. 5 (May 1987): 781-91. Topics discussed include: dust, asthma immunology, mites, immunology.

238. Primault, B. "Meteorologic conditions leading to crisis situations in allergic patients." *Schweiz Med Wochenschr* 110, no. 12 (22 March 1980): 459-63. Topics discussed include: allergens, dust, pollen, hypersensitivity etiology.

239. Rasmuson, M. et al. "Allergic reaction patterns in relation to age, sex, season, eosinophilia and some genetic marker systems." *Acta Allergol* 28, no. 5 (December 1973): 365-400. Topics discussed

include: age effects, agglutination tests, blood cell count, hypersensitivity blood, gender effects, skin tests, causes of allergenic reactions, immunology.

240. Reddoch, A. et al. "Seasonal allergic pollenosis." *Arctic Med Res.* 47 Supp, no. 1 (1988): 105-9. Topics discussed include: Alaska, cold climates, hay fever etiology, pollen analysis.

241. Reid, M. et al. "Seasonal asthma in northern California: allergic causes and efficacy of immunotherapy." *J Allergy Clin Immunol* 78, no. 4 Pt 1 (October 1986): 590-600. Topics discussed include: asthma immunology, asthma treatment, California, dose response relationship, rhinitis epidemiology, asthma etiology, desensitization, grasses immunology.

242. Reynolds, G. "Summer allergy." *Queens Nurs J* 18, no. 4 (July 1975): 104-6. Topics discussed include: antigen antibody reactions, hay fever, histamine liberation, urticaria, causes of allergenic reactions, hypersensitivity immunology, seasonality.

243. Ribon, A. et al. "Bronchial asthma in children and its occurrence in relation to weather and air pollution." *Ann Allergy* 30, no. 5 (May 1972): 276-81. Topics discussed include: barometric pressure, emergency service, hospital, hospitalization, humidity effects, New York City, seasonality, temperature, asthma epidemiology.

244. Rogala, B. et al. "Effect of specific immunotherapy on IgE immune complexes in seasonal allergic rhinitis." *Pneumonol Alergol Pol.* 60 Supp, no. 1 (1992): 57-62. Topics discussed include: drug combinations, hay fever, skin tests methods, desensitization, immunologic methods, immunology, plant extracts, dosage.

245. Sale, C. "Humidification during the cold weather to assist perennial allergic rhinitis patients." *Ann Allergy* 29, no. 7 (July 1971): 356-7. Topics discussed include: environment, heating, methods, seasonality, hay fever prevention, humidity.

246. Salkie, M. et al. "The influence of season and of sex on the serum level of total IgE and on the distribution of allergen specific IgE." *Clin Biochem* 17, no. 6 (December 1984): 362-6. Topics discussed include: antibody specificity, gender effects, immunology, hypersensitivity, immediate immunology.

247. Salvaggio, J. et al. "New Orleans asthma. II. Relationship of climatologic and seasonal factors to outbreaks." *J Allergy* 45, no. 5 (May 1970): 257-65. Topics discussed include: air pollution, atmospheric effects, bronchitis, hospitalization, humidity effects, Louisiana, respiratory hypersensitivity, temperature, asthma epidemiology.

248. Schaaf, L. et al. "Suppression of seasonal allergic rhinitis symptoms with daily hydroxyzine." *J Allergy Clin Immunol* 63, no. 2 (February 1979): 129-33. Topics discussed include: immunosuppression, hay fever treatment using drugs, hydroxyzine therapeutic use.

249. Schachter, E. et al. "The protective effect of a cold weather mask on exercised induced asthma." *Ann Allergy* 46, no. 1 (January 1981): 12-6. Topics discussed include: respiratory functions, asthma, exercise.

250. Scherr, M. et al. "Three year evaluation of biofeedback techniques in the treatment of children with chronic asthma in a summer camp environment." *Ann Allergy* 41, no. 5 (November 1978): 288-92. Topics discussed include: chronic disease, electromyography, peak expiratory flow rate, asthma treatment, biofeedback psychology, camping, seasonality.

251. Schmidt, W. I. et al. "Effect of a stay in the North Sea climate on lymphocyte subpopulations in the peripheral blood of patients with exogenous allergic bronchial asthma and chronic bronchitis." *Pneumologie.* 44 Supp, no. 1 (February 1990): 241-2. Topics discussed include: bronchitis, smoking, asthma, health resorts.

252. Schmutzler, W. et al. "Seasonal variation of mast cell function in human adenoid tissues." *Int Arch Allergy Appl Immunol* 82, no. 3-4 (1987): 277-8. Topics discussed include: concanavalin a pharmacology, histamine liberation drug effects, leukocyte count, mast cells pathology, adenoids pathology.

253. Shapiro, R. et al. "The importance of field studies and meteorologic factors in pollen and mold surveys. A three year report of Southern California." *Ann Allergy* 23, no. 10 (October 1965): 484-91. Topics discussed include: California, respiratory hypersensitivity epidemiology, spores, air microbiology, air pollution effects, seasonality, weather.

254. Shibasaki, M. et al. "Premenstrual asthma with seasonal variation." *Ann Allergy* 68, no. 4 (April 1992): 315-8. Topics discussed include: inhalation, causes of allergenic reactions, antibodies analysis, treatment using drugs, bronchial hyperreactivity, mites.

255. Simon, H. et al. "Disease specific and non specific seasonal parameter changes in hay fever patients." *Allerg Immunol Leipz* 36, no. 1 (1990): 3-10. Topics discussed include: blood, hay fever physiopathology.

256. Sotomayor, H. et al. "Seasonal increase of carbachol airway responsiveness in patients allergic to grass pollen. Reversal by corticosteroids." *Am Rev Respir Dis* 130, no. 1 (July 1984): 56-8. Topics discussed include: grasses, respiratory hypersensitivity treatment using drugs, airway resistance drug effects, pharmacology, immunology.

257. Stafford, W. et al. "Cromolyn modifies histamine bronchial reactivity in symptomatic seasonal asthma." *Ann Allergy* 52, no. 6 (June 1984): 401-5. Topics discussed include: asthma, treatment using drugs, immunology, peak expiratory flow rate, physiopathology, bronchial provocation tests.

258. Storm van Leeuwen, W. "Allergic Diseases in Relation to Climate (Allergenproof Chamber)." *Inrtern. Clin.* 2, no. Jue (June 1927): 1-24. Topics discussed include: health, allergy.

259. Subiza, J. et al. "Prediction of annual variations in atmospheric concentrations of grass pollen. A method based on meteorological factors and grain crop estimates." *Clin Exp Allergy* 22, no. 5 (May 1992): 540-6. Topics discussed include: agriculture, allergens analysis, asthma etiology, grasses immunology, hay fever etiology, humidity effects, rain, seasonality, Spain, immunology.

260. Sutkovoi, D. et al. "State of lipid peroxidation and thymus dependent immunity system in patients with allergic diseases of the respiratory organs during rehabilitation in a mountain climate." *Fiziol Zh* 31, no. 3 (May 1985): 287-91. Topics discussed include: adaptation, altitude, immunity, rehabilitation.

261. Tanay, A. et al. "Astemizole in perennial allergic rhinitis with seasonal exacerbations: a placebo controlled double blind study." *Ann Allergy* 63, no. 6 Pt 1 (December 1989): 493-4. Topics discussed include: rhinitis, perennial treatment using drugs.

262. Thomas, J. et al. "Double blind, controlled study of clemastine fumarate, chlorpheniramine and placebo in the symptomatic treatment of seasonal allergic rhinitis in desensitized and nondesensitized patients." *Ann Allergy* 38, no. 3 (March 1977): 175-81. Topics discussed include: pharmacology, hay fever, physiopathology, placebos, timing, desensitization, treatment using drugs.

263. Topilsky, M. et al. "Disodium cromoglycate in the treatment of seasonal and perennial rhinitis." *Ann Allergy* 36, no. 4 (April 1976): 246-8. Topics discussed include: aerosols, hay fever etiology, pollen, timing.

264. Tromp, S. "Influence of weather and climate on asthma and bronchitis." *Rev Allergy* 22, no. 11 (November 1968): 1027-44. Topics discussed include: allergens, asthma, etiology, treatment, atmospheric effects, autonomic nervous system, thermoregulation, bronchitis, Netherlands, epidemiology.

265. Tseng, R. et al. "Seasonal asthma in Hong Kong and its management implications." *Ann Allergy* 63, no. 3 (September 1989): 247-50. Topics discussed include: age effects, asthma, etiology, diagnosis, Hong Kong, hospitalization statistics, meteorology.

266. Tsetsarskii, B. et al. "Allergoid therapy effectiveness in seasonal allergic diseases of the upper respiratory tracts." *Vestn Otorinolaringol no.* 1 (January 1982): 37-40. Topics discussed include: immunologic methods, hay fever treatment, plant extracts, treatment.

267. van der Heide, S. et al. "Seasonal variation in airway hyperresponsiveness and natural exposure to house dust mite allergens in patients with asthma." *J Allergy Clin Immunol* 93, no. 2 (February 1994): 470-5. Topics discussed include: asthma, physiopathology, bronchial hyperreactivity, mites, immunology, seasonality, etiology, dust.

268. Verdiani, P. et al. "Different prevalence and degree of nonspecific bronchial hyperreactivity between seasonal and perennial rhinitis." *J Allergy Clin Immunol* 86, no. 4 Pt 1 (October 1990): 576-82. Topics discussed include: age effects, chronic disease, Italy, risk, skin tests, bronchi physiopathology, hay fever, epidemiology.

269. Viander, M. et al. "The seasonal symptoms of hyposensitized and untreated hay fever patients in relation to birch pollen counts: correlations with nasal sensitivity, prick tests and RAST." *Clin Allergy* 8, no. 4 (July 1978): 387-96. Topics discussed include: skin tests, desensitization, hay fever diagnosis, immunology, trees.

270. Videbaek, J. "Respiratory allergy and climatic conditions. I. Review." *Acta Allergol* 23, no. 6 (December 1968): 441-54. Topics discussed include: barometric pressure, electricity, humidity effects, respiratory hypersensitivity etiology.

271. Viegas, M. et al. "Changes in nasal mast cell numbers in and out of the pollen season." *Int Arch Allergy Appl Immunol* 82, no. 3-4 (1987): 275-6. Topics discussed include: biopsy, grasses, leukocyte count, hay fever pathology, mast cells pathology, nasal mucosa pathology.

272. Vuurman, E. et al. "Seasonal allergic rhinitis and antihistamine effects on children's learning." *Ann Allergy* 71, no. 2 (August 1993): 121-6. Topics discussed include: cognition, hay fever psychology, immediate treatment using drugs, hypersensitivity, immediate psychology, learning drug effects.

273. Wagner, C. et al. "Relation between positive small air ions, weather fronts and pulmonary function in patients with bronchial asthma." *Ann Allergy* 51, no. 4 (October 1983): 430-5. Topics discussed include: aerosols, air pollution effects, peak expiratory flow rate, air pollution effects, asthma physiopathology, lung physiopathology.

274. Walshaw, M. et al. "The effect of seasonal and domestic factors on the distribution of Euroglyphus maynei in the homes of Dermatophagoides pteronyssinus allergic patients." *Clin Allergy* 17, no. 1 (January 1987): 7-14. Topics discussed include: beds, dust, England, humidity effects, mites isolation, purification, socioeconomic factors, sodium analysis, asthma etiology, mites immunology.

275. Wedler, E. "Balneological and climate therapy and allergy in childhood a critical view." *Kinderarztl Prax* 38, no. 2 (February 1970): 73-8. Topics discussed include: asthma treatment, health resorts, altitude, baths, hypersensitivity treatment.

276. Weyer, A. et al. "Grass pollen hyposensitization versus placebo therapy. I. Clinical effectiveness and methodological aspects of a pre seasonal course of desensitization with a four grass pollen extract." *Allergy* 36, no. 5 (July 1981): 309-17. Topics discussed include: clinical trials, hay fever treatment using drugs, plant extracts, skin tests, timing, desensitization, placebos.

277. Wortmann, F. "Co-seasonal treatment of pollinosis and other respiratory allergies by quadrillage." *Schweiz Med Wochenschr* 95, no. 34 (21 August 1965): 1141-3. Topics discussed include: desensitization, pollen, respiratory hypersensitivity treatment.

278. Yoshida, K. et al. "Environmental mycological studies on the causative agent of summer type hypersensitivity pneumonitis." *J Allergy Clin Immunol* 81, no. 2 (February 1988): 475-83. Topics discussed include: antibodies, microbiology, mycoses, immunology, alveolitis, housing, seasonality, isolation, purification.

DIET & NUTRITION

279. Aldashev, A. et al. "Indices of the nutritional status of workers in the oil and gas production industry adapting to the extreme conditions of an arid zone." *Vopr Pitan no.* 3 (May 1986): 25-8. Topics discussed include: caloric intake, diet, blood, vitamins, adaptation, desert climates.

280. Behrman, J. R. "Nutrition, Health, Birth Order and Seasonality: Intrahousehold Allocation among Children in Rural India." *Journal of Development Economics* 28, no. 1 (February 1988): 43-62. Topics discussed include: subsidy programs, demography.

281. Bojilen, K. et al. "The influence of climate and nutrition on age at menarche: a historical review and a modern hypothesis." *Hum Biol* 40, no. 1 (February 1968): 69-85. Topics discussed include: proteins, tropical climates, age effects, menstruation.

282. Brown, K. et al. "Effects of season and illness on the dietary intake of weanlings during longitudinal studies in rural Bangladesh." *Am J Clin Nutr* 41, no. 2 (February 1985): 343-55. Topics discussed include: age effects, breast feeding, caloric intake, fever physiopathology, gender effects, diet, health, infant nutrition.

283. ---. "Seasonal changes in nutritional status and the prevalence of malnutrition in a longitudinal study of young children in rural Bangladesh." *Am J Clin Nutr* 36, no. 2 (August 1982): 303-13. Topics discussed include: body height, body weight, infant nutrition, protein, energy, malnutrition.

284. Bryson, R. et al. *Climates of Hunger*. Madison, WI: University of Wisconsin Press, 1977. Topics discussed include: climate, economy, biology, agriculture, nutrition.

285. Bryson, R. A. et al. *Climates of Hunger: Mankind and the World's Changing Weather*. Madison, WI: University of Wisconsin Press, 1979. Topics discussed include: climate, economy, biology, agriculture, nutrition.

286. De Castro, J. *Geography of Hunger*. London: Gollancz, 1952. Topics discussed include: geography, biology, economy, nutrition.

287. Devgun, M. et al. "Vitamin D nutrition in relation to season and occupation." *Am J Clin Nutr* 34, no. 8 (August 1981): 1501-4. Topics discussed include: hospitalization, ultraviolet rays, urban health, blood, occupations.

288. Fehilly, A. et al. "A weighed dietary survey of men in Caerphilly, South Wales." *Hum Nutr Appl Nutr* 38, no. 4 (August 1984): 270-6. Topics discussed include: carbohydrates, fats, proteins, England, fatty acids, Scotland, seasonality, Wales, diet, nutrition.

289. Ferro-Luzzi, A. "Seasonal energy stress in marginally nourished rural women: interpretation and integrated conclusions of a multicentre study in three developing countries." *European Journal of Clinical Nutrition* 44 suppl (1990): 41-6. Topics discussed include: public health, developing countries, biology, malnutrition, women physiology, bioenergetics.

290. Jelliffe, D. "Infant nutrition in the subtropics and tropics." *World Health Organ Monogr Ser* 29 (1968): 1-335. Topics discussed include: Africa, Asia, diseases, proteins, health education, rickets, socioeconomics, South America.

291. Juuti, M. et al. "Seasonal variation in urinary excretion of calcium, oxalate, magnesium and phosphate on free and standard mineral diet in men with urolithiasis." *Scand J Urol Nephrol* 15, no. 2 (1981): 137-41. Topics discussed include: urinary therapy, calcium, magnesium, oxalates, phosphates.

292. Karpe, H. J. et al. "Climate and Development: Climatic Change and Variability and the Resulting Social, Economic and Technological Implications: Overview: Report of the Hamburg Congress." *Climate and development: Climatic change and variability and the resulting social, economic and technological implications*, 3-14. eds H. J. Karpe, D. Otten, and S. C. Trinidade. New York; Berlin; London and

Tokyo: Springer in cooperation with the United Nations, 1990. Topics discussed include: conservation, pollution, energy, technological change, innovation.

293. Kight, M. et al. "Serum fatty acids of women living in a semi desert climate." *J Am Diet Assoc* 61, no. 2 (August 1972): 152-4. Topics discussed include: fasting, linoleic acids, lipids blood, nutrition, blood.

294. Krasnopevtsev, V. et al. "Characteristics of the nutrition of adolescents at state occupational technical schools for construction workers in different climatic zones of the Soviet Union." *Gig Sanit* 37, no. 10 (October 1972): 39-42. Topics discussed include: Moscow, seasonality, Siberia, health services, education.

295. Lawson, D. et al. "Relative contributions of diet and sunlight to vitamin D state in the elderly." *Br Med J* 2, no. 6185 (4 August 1979): 303-5. Topics discussed include: blood, seasonality, diet, vitamins, metabolism.

296. Leonard, W. "Nutritional determinants of high altitude growth in Nunoa, Peru." *Am J Phys Anthropol* 80, no. 3 (November 1989): 341-52. Topics discussed include: body height, body weight, caloric intake, diet, eating, Peru, seasonality, skinfold thickness, socioeconomics, growth, nutrition.

297. ---. "Protection of children from seasonal nutritional stress in an Andean agricultural community." *Eur J Clin Nutr* 43, no. 9 (September 1989): 597-602. Topics discussed include: caloric intake, diet, Peru, child nutrition, disorders.

298. Loutan, L. et al. "Seasonal variations in nutrition among a group of nomadic pastoralists in Niger." *Lancet* 1, no. 8383 (28 April 1984): 945-7. Topics discussed include: body weight, Niger, rain, transients, migrants, water supply, diet.

299. Mavlanov, K. et al. "Seasonal dynamics of the indices of humoral immunity in children aged 3 months to 1 year in relation to nutrition in an arid zone." *Vopr Pitan 1992* no. 4 (July 1992): 26-9. Topics discussed include: Caucasoid race, Mongoloid race, Turkmenistan ethnology, desert climates.

300. Mitchell, H. H. et al. *Nutrition and climatic stress, with particular reference to man.* 234 pages. Springfield, Ill.: Thomas, 1951. Topics discussed include: climate.

301. Mitkova, N. et al. "Optimal nutrition and drinking diet of the crews in tropical conditions of sailing." *Biul Inst Med Morsk Gdansk* 17, no. 3 (1966): 411-8. Topics discussed include: Bulgaria, ships, diet, naval medicine, tropical climate.

302. Nelson, M. et al. "Seasonal variations in dietary iodine intake and thyrotoxicosis." *Hum Nutr Appl Nutr* 39, no. 3 (June 1985): 213-6. Topics discussed include: United Kingdom, epidemiology, iodine analysis, milk.

303. Nethery, V. M. "The effect of heat stress and work on total plasma protein concentration, albumin concentration and hematocrit in endurance trained and sedentary adult females." M.S. diss., University of Oregon, College of Human Development and Performance, 1982. Topics discussed include: exercise, women, physiology, blood.

304. Nutter, J. "Seasonal changes in female athletes' diets." *Int J Sport Nutr* 1, no. 4 (December 1991): 395-407. Topics discussed include: caloric intake, carbohydrates, fats, proteins, golf, hockey, minerals, running, tennis, vitamins, diet, sports.

305. Poskitt, E. et al. "Diet, sunlight, and 25 hydroxy vitamin D in healthy children and adults." *Br Med J* 1, no. 6158 (27 January 1979): 221-3. Topics discussed include: blood, United Kingdom, seasonality, ultraviolet rays, vitamins, diet.

306. Pudel, V. "The four season cure. Concept, development, execution." *Bibl Nutr Dieta* no. 46 (1990): 124-7. Topics discussed include: diet, Germany, nutrition.

307. Schelp, F. et al. "Seasonal variation of wasting and stunting in preschool children during a three year community based nutritional intervention study in northeast Thailand published erratum appears in Trop Med Parasitol 1990 Dec;41(4):450." *Trop Med Parasitol* 41, no. 3 (September 1990): 279-85. Topics discussed include: age effects, body height, body weight, nutrition, education.

308. Schofield, S. "Seasonal Factors Affecting Nutrition in Different Age Groups and Especially Preschool Children." *Journal of Development Studies* 11, no. 1 (October 1974): 22-40. Topics discussed include: age groups, supplies, labor demands, storage, villages, social services.

309. Shetty, P. S. et al. "Role of the sympathetic nervous system in adaptation to seasonal energy deficiency." *European Journal of Clinical Nutrition* 44 suppl (1990): 47-53. Topics discussed include: nervous system, adaptation, biology.

310. Specker, B. et al. "Seasonal differences in serum vitamin D binding protein in exclusively breast fed infants: negative relationship to sunshine exposure and 25 hydroxyvitamin D." *J Pediatr Gastroenterol Nutr* 5, no. 2 (March 1986): 290-4. Topics discussed include: blood, breast feeding, sunlight, vitamin, protein.

311. Srivastava, K. et al. "Human nutrition in cold and high terrestrial altitudes." *Int J Biometeorol* 36, no. 1 (March 1992): 10-3. Topics discussed include: caloric intake, cold climate, exertion, stress, diet, physiopathology, nutritional requirements.

312. Strachan, D. et al. "Ventilatory function and winter fresh fruit consumption in a random sample of British adults." *Thorax* 46, no. 9 (September 1991): 624-9. Topics discussed include: age effects, England, Scotland, seasonality, gender effects, smoking physiopathology, Wales, diet, physiology, fruit.

313. Sundararaj, R. et al. "Seasonal variation in the diets of pre school children in a village (North Arcot District). II. Intake of vitamins and minerals." *Indian J Med Res* 57, no. 2 (February 1969): 375-83. Topics discussed include: India, minerals, vitamins, child nutrition, diet.

314. Sundarraj, R. et al. "Seasonal variation in the diets of pre school children in a village (North Arcot District). I. Intake of calories, protein and fat." *Indian J Med Res* 57, no. 2 (February 1969): 249-59. Topics discussed include: carbohydrates, fats, proteins, India, child nutrition, diet.

315. Tranchant, D. et al. "Tropical ataxia. Clinical and nutritional study." *Med Trop Mars* 45, no. 2 (April 1985): 175-7. Topics discussed include: ataxia diagnosis, diet, fats, tropical climate, ataxia etiology, protein, energy, malnutrition.

316. Valverde, V. et al. "Seasonality and nutritional status. A review of findings from developed and developing countries." *Arch Latinoam Nutr* 32, no. 3 (September 1982): 521-40. Topics discussed include: body weight, caloric intake, developing countries, Scotland, Sweden, growth, nutrition.

317. Wassermann, M. et al. "Problems of toxicology facing occupational health in developing countries. II." *IMS Ind Med Surg* 38, no. 7 (July 1969): 220-3. Topics discussed include: genotype, government, infection, nutrition, parasitic diseases, medicine, tropical climate.

318. Whitemore RW et al. "Non-occupational exposures to pesticides for residents of two U.S. cities." *Arch Environ Contam Toxicol* 26, no. 1 (January 1994): 47-59. Topics discussed include: environmental exposure, pesticides, diet, dust, Florida, Massachusetts, risk.

319. Wijn, J. D. et al. "Study of the daily activities and diet of 10 male and 10 female students during the winter months." *Ned Tijdschr Geneeskd* 109, no. 38 (18 September 1965): 1754-9. Topics discussed include: diet, exertion physiology, seasonality.

320. Wilmsen, E. N. "Seasonal Effects of Dietary Intake on Kalahari." *San Fed Proc* 37 (1978): 65-72. Topics discussed include: seasonality.

321. Zheng, S. et al. "Nutritional status in Linxian, China: effects of season and supplementation." *Int J Vitam Nutr Res* 59, no. 2 (1989): 190-9. Topics discussed include: anthropometry, China, diet, epidemiology, etiology, physiology, minerals, blood, vitamins.

DISEASE & ILLNESS

322. Adler, J. et al. "Perforated peptic ulcer a seasonal disease?" *Aust N Z J Surg* 54, no. 1 (February 1984): 59-61. Topics discussed include: Atralia, duodenal ulcer epidemiology, patient admission trends, stomach ulcer epidemiology.

323. Allan, T. M. et al. "Seasonal variation in health and diseases with section on effects of weather and temperature : a bibliography." England: London, New York: Mansell Pub. Ltd., 1994. Topics discussed include: medical climatology.

324. Amelung, W. "Climate and Cardial Diseases." *Exptl Med Surg* 9 (1951): 238-247. Topics discussed include: health issues.

325. Anders, H. S. "The Relation of Sunshine to the Prevalence of Influenza." *Baln Climat J London* (1902):.Topics discussed include: weather, temperature, disease, health issues.

326. Anders, J. M. "Temperature and Atmospheric Humidity and Their Relation to Health and Disease." *Trans Am Climatological Assoc* 5 (1887): 127-135. Topics discussed include: weather, health issues.

327. Andreenko, G. et al. "Seasonal changes in the blood coagulating and anticoagulating system indices in men at the preclinical stage of ischemic heart disease." *Kardiologiia* 20, no. 3 (March 1980): 61-4. Topics discussed include: fatigue, mental physiopathology, risk, stress, psychological physiopathology.

328. Anonymous. "Discussion at the Eastborne Public Health Congress: Soil and Climate in Relation to Rheumatic Diseases." *Brit Med J* (1933): 1060-1062. Topics discussed include: environment, health issues, rheumatism.

329. Ansari, A. et al. "Climatic factors in ischemic heart disease." *South Med J* 62, no. 5 (May 1969): 579-82. Topics discussed include: air conditioning, angina pectoris, coronary treatment, electrocardiography, environmental exposure, heat, humidity, epidemiology.

330. Antunez, D. M. et al. "Bacterial myositis, a tropical disease. Report of a case in temperate climate." *Rev Clin Esp* 171, no. 2 (31 October 1983): 151-2. Topics discussed include: myositis microbiology, Spain, abscess diagnosis, myositis diagnosis.

331. Apperly, F. L. "The Possible Influence of Climate on the Incidence of Peptic Ulcer in Australia." *Med J Australia* 1 (1930): 779-781. Topics discussed include: disease, health issues, ulcer.

332. Armstrong, Ch. "Seasonal Distribution of Poliomyelitis." *Am J Publ Health* 40, no. 10 (1950): 1296. Topics discussed include: disease, health issues.

333. Aycock, W. L. "A Study of the Significance of Geographical and Seasonal Variations in the Incidence of Poliomyelitis." *J Preventive Med* 3 (1929): 245. Topics discussed include: geography, disease, health issues.

334. Aznaur'Ian, M. et al. "Experience in work organization for the prevention of the meteotropic reaction on at the K.E. Voroshilov Sanatorium in Sochi." *Vopr Kurortol Fizioter Lech Fiz Kult no.* 4 (July 1975): 332-5. Topics discussed include: climatic effects, health issues resorts, hypertension, neurasthenia, rheumatic heart disease, treatment.

335. Babiker, A. et al. "Focality and seasonality of Schistosoma mansoni transmission in the Gezira Irrigated Area, Sudan." *J Trop Med Hyg* 88, no. 2 (April 1985): 57-63. Topics discussed include: agriculture, disease reservoirs, ecology, health issues, education, mollcacides, epidemiology, water supply standards, biomphalaria parasitology.

336. Becker, F. "Can weather cause disease?" *Med Monatsschr Pharm* 8, no. 1 (January 1985): 8-14. Topics discussed include: Europe, meteorology, physiopathology.

337. Beliakov, V. "Epidemiological surveillance the basis of the modern organization of epidemic control work." *Zh Mikrobiol Epidemiol Immunobiol no.* 5 (May 1985): 53-8. Topics discussed include: communicable diseases epidemiology, communicable diseases transmission, administration, seasonality, disease outbreaks prevention, control.

338. Bhopal, R. et al. "Seasonal variation of Legionnaires' disease in Scotland." *J Infect* 22, no. 2 (March 1991): 153-60. Topics discussed include: diagnosis, epidemiology, timing, travel.

339. Boles, R. S. et al. "Seasonal Incidence and Precipitating Causes of Hemorrhage from Peptic Ulcer." *J Am Med Assoc* 156, no. 15 (1954): 1379-1383. Topics discussed include: disease, health issues.

340. Brandt, C. et al. "Infections in 18,000 infants and children in a controlled study of respiratory tract disease. II. Variation in adenovirus infections by year and season." *Am J Epidemiol* 95, no. 3 (March 1972): 218-27. Topics discussed include: immunology, microbiology, District of Columbia, epidemiologic methods, New York City, socioeconomics, respiratory tract.

341. Braverman, D. et al. "Is duodenal ulcer a seasonal disease? A retrospective endoscopic study of 3105 patients." *Am J Gastroenterol* 87, no. 11 (November 1992): 1591-3. Topics discussed include: diagnosis, duodenoscopy, Israel, epidemiology.

342. Buckley, Ch. W. et al. "Climate Influences in Health and Rheumatic Diseases." *Proc Roy Soc Med, London, Sect Balneol Climatol* 23 (1930): 31-39. Topics discussed include: health issues, rheumatism.

343. Cave, D. R. et al. "Seasonal Variation in the Clinical Presentation of Crohn's Disease and Ulcerative Colitis." *Int J Epidemiol* 4 (1975): 317-320. Topics discussed include: health issues, illness.

344. Celestino, D. et al. "Seasonal recurrence of acute episodes of Meniere's disease." *Acta Otorhinolaryngol Belg* 41, no. 6 (1987): 951-7. Topics discussed include: periodicity.

345. Clarke, J. F. "The Geographical and Climatic Distribution of Rheumatic Fever." *Symp Rheumatism, Med J Record* (1928): 4. Topics discussed include: geography, disease, health issues, rheumatism.

346. Cline, I. M. "The Climatic Causation of Disease; with Chart Showing Pathological Distribution of Climate in the United States." *Trans Texas State Med Assoc* (1895). Topics discussed include: health issues.

347. Coenen, C. et al. "Is gastroduodenal ulcer a seasonal disease?" *Med Klin* 81, no. 17 (22 August 1986): 555-9. Topics discussed include: cross-sectional analysis, cycles, peptic ulcer epidemiology.

348. Cromer, T. "Seasonal control for an endemic disease with seasonal fluctuations." *Theor Popul Biol* 33, no. 2 (April 1988): 115-25. Topics discussed include: cross-sectional analysis, gonorrhea epidemiology, mass screening, propagation prevention.

349. Cucu, F. et al. "Seasonal variations of serum cholesterol detected in the Bucharest Multifactorial Prevention Trial of Coronary Heart Disease. Ten years follow up (1971-1982)." *Med Interne* 29, no. 1-2 (January 1991): 15-21. Topics discussed include: blood, epidemiology, Romania, cholesterol.

350. Cuisinier, R. J. et al. "Tropical diabetes mellitus, a new nutritional indicator?" *Med Trop Mars* 45, no. 2 (April 1985): 179-84. Topics discussed include: Africa, chronic disease, diet treatment, epidemiology, India, Jamaica, tropical climates, developing countries, etiology.

351. Danilovic, V. et al. "L'Influence du Climat Marin de la Mer Adriatique sur l'Asthme Infantile." *Congr Intern d'Hydroclimatisme et de Thalassothérapie* 2 (1954): 55-60. Topics discussed include: climate, disease, health issues, asthma.

352. De Rudder, B. "Wet and Dry Climates and Weather in the Causation of Disease." Summaries Principal Repts Intern Congr Rheumatism and Hydrology, London and Oxford, (March 25-31. 1938). Topics discussed include: climate, health issues.

353. Dean, A. et al. "Seasonal gastroenteritis and malabsorption at an American military base in the Philippines. I. Clinical and epidemiologic investigations of the acute illness." *Am J Epidemiol* 95, no. 2 (February 1972): 111-27. Topics discussed include: acute disease, diarrhea, causes, feces microbiology, food supply, nausea, temperature, timing, epidemiology, military medicine.

354. Deitz, K. "The Incidence of Infectious Diseases under the Influence of Seasonal Fluctuations." in *Mathematical Models in Medicine*, 1-15. editor J. Berger. Berlin: Springer-Verlag, 1976. Topics discussed include: health issues.

355. Dolezalova, V. "Seasonal rhythm of infectious eye disease in North Africa." *Ophthalmologica* 177, no. 4 (1978): 181-7. Topics discussed include: temperature, Tunisia, water, epidemiology.

356. Dorado, P. S. et al. "The influence of climate factors and air pollution on exacerbations of obstructive chronic pulmonary disease." *Med Clin Barc* 78, no. 6 (16 March 1982): 226-30. Topics discussed include: humidity, Spain, sulfur dioxide, temperature, air pollutants, etiology, meteorology.

357. Doroshenko, I. et al. "Temporary loss of work capacity in connection with rheumatic diseases in workers of the agricultural industrial complex." *Revmatologiia Mosk* no. 3 (July 1991): 8-11. Topics discussed include: age effects, Russia, seasonality, absenteeism, epidemiology.

358. Douglas, A. "Seasonality of hip fracture and haemorrhagic disease of the newborn." *Scott Med J* 38, no. 2 (April 1993): 37-40. Topics discussed include: age effects, infants, gender effects, neck fractures.

359. Dukes-Dobos, F. N. et al. "Proceedings of a NIOSH Workshop on Recommended Heat Stress Standards: sponsored by National Institute for Occupational Safety and Health, Cincinnati, Ohio, September 1979." NIOSH Workshop on Recommended Heat Stress Standards (2nd: 1979: Cincinnati, OH), and National Institute for Occupational Safety and Health. U.S. Dept. of Health and Human Services, Public Health Service, Centers for Disease Control, National Institute for Occupational Safety and Health; For sale by the Supt. of Docs., U.S. G.P.O., Cincinnati, OH; Washington, DC, 1980. Topics discussed include: heat, physiological effects, industrial safety.

360. Dysken, M. et al. "Seasonal distribution of births in Alzheimer's disease." *Int Psychogeriatr* 3, no. 1 (March 1991): 53-8. Topics discussed include: diagnosis, causes, brain pathology, Minnesota, risk, epidemiology.

361. Ebringer, R. et al. "Seasonal variation of acute anterior uveitis: differences between HLA B27 positive and HLA B27 negative disease." *Br J Ophthalmol* 69, no. 3 (March 1985): 202-4. Topics discussed include: Australia, immunology, antigens, epidemiology.

362. Edwards, J. H. "Seasonal Incidence of Congenital Disease in Birmingham." *Ann Hum Genet* 25 (1961): 89-93.

363. Einhorn, M. "Seasonal Incidence and Study of Factors Influencing the Production of One Thousand Recurrences of Gastroduodenal Ulcers in Eight Hundred Patients." *Am J Med Sci* 179 (1930): 259-264. Topics discussed include: disease, health issues, ulcer.

364. Eldredge, E. A. "Drought, Famine and Disease in Nineteenth-Century Lesotho." *African Economic History* no. 16 (October 1987): 61-93. Topics discussed include: history, Africa, product prices, markets.

365. Erpf, S. F. "Dental Caries and Paradental Disturbances: II The Seasonal Incidence of Dental Caries." *J Am Dent Assoc* 25, no. 5 (1938): 681-693. Topics discussed include: illness, dental problems.

366. Farrag, N. et al. "Pyomyositis an under reported disease in temperate climates." *J Infect* 17, no. 3 (November 1988): 259-63. Topics discussed include: climate, England, muscles pathology, myositis pathology, streptococcus isolation, microbiology.

367. Frouchtman, R. "Asthma and Climate." *Medicina (Méx)* 32, no. 25 (May 1952): 221-231. Topics discussed include: disease, health issues, asthma.

368. Gafafer, W. "Upper Respiratory Disease (Common Cold) and the Weather, Baltimore 1928-1930." *Am J Hyg* 13 (1931): 771. Topics discussed include: health issues.

369. Gibinski, K. "A review of seasonal periodicity in peptic ulcer disease." *Chronobiol Int* 4, no. 1 (1987): 91-9. Topics discussed include: seasons, peptic ulcers, cycles.

370. Gibinski, K. et al. "Seasonal occurrence of abdominal pain and endoscopic findings in patients with gastric and duodenal ulcer disease." *Scand J Gastroenterol* 17, no. 4 (June 1982): 481-5. Topics discussed include: duodenal ulcer, pathology, endoscopy, stomach ulcer, diagnosis, pain.

371. Gill, J. et al. "Wind chill and the seasonal variation of cerebrovascular disease." *J Clin Epidemiol* 41, no. 3 (1988): 225-30. Topics discussed include: England, patient admission, wind, epidemiology, cold climates.

372. Goldberger, J. et al. "Pellagra Incidence in Relation to Sex, Age, Season, Occupation and 'Disabling Sickness' in Seven Cotton-Mill Villages of South Carolina During 1916." *Publ Health Rept* 35 (1920): 1650-1664. Topics discussed include: disease, health issues.

373. Greenwood, B. et al. "Meningococcal disease and season in sub-Saharan Africa." *Lancet* 1, no. 8390 (16 June 1984): 1339-42. Topics discussed include: transmission, Nigeria, rain, temperature, wind, epidemiology.

374. Grinfel'D, A. et al. "Study of the sources and ways of spread of epidemic hepatitis (Botkin's disease). I. Seasonal prevalence and periodicity of epidemic hepatitis for 11 years (1953-1963) in various age groups of the Odessa population." *Zh Mikrobiol Epidemiol Immunobiol* 42, no. 12 (December 1965): 29-34. Topics discussed include: aging, Ukraine, hepatitis epidemiology.

375. Harmon, G. E. "Seasonal Incidence of Whooping Cough in the United States." *Am J Publ Health* 22 (1932): 831. Topics discussed include: disease, health issues.

376. Hartwich, G. et al. "Seasonal variation in the incidence of initial manifestations in Hodgkin's disease." *Dtsch Med Wochenschr* 95, no. 26 (26 June 1970): 1387-9. Topics discussed include: Germany, causes, epidemiology.

377. Heath, D. et al. *High-altitude medicine and pathology*. 352 pages. London, Boston: Butterworths, 1989.

378. Heitz, J. "Climatic Indications in Cardio-Vascular Disease." *J Med Chir Prat* 98 (1927): 355-360. Topics discussed include: climate, health issues.

379. Hellier, F. "Seasonal variation in the incidence of skin disease." *Br J Dermatol* 81, no. 11 (November 1969): 867. Topics discussed include: epidemiology, psoriasis epidemiology, skin epidemiology.

380. Henderson, A. et al. "Season of birth for Alzheimer's disease in the Southern Hemisphere." *Psychol Med* 21, no. 2 (May 1991): 371-4. Topics discussed include: diagnosis, causes, genetics, Australia, epidemiology, risk.

381. Henschen, F. *The History and Geography of Diseases*. New York: Delacorte Press, 1966. Topics discussed include: health issues, geography.

382. Hill, L. "Climate and Rheumatic Troubles." *Brit J Rheumatol* I (1938): 11-14. Topics discussed include: disease, health issues, rheumatism.

383. ---. "Rheumatism and Climate." *Brit Med J* 2 (1939): 276-278. Topics discussed include: disease, health issues, rheumatism.

384. Hindmann, S. M. et al. "Seasonal Distribution of Measles, Scarlet Fever and Diphteria for Periods of High and Low Incidence." *Am J Hyg* 20 (1934): 555. Topics discussed include: disease, health issues.

385. Howe, G. M. *A World Geography of Human Diseases.* New York: Academic Press, 1977. Topics discussed include: geography, health issues.

386. Hunter, J. "The summer disease. Some field evidence on seasonality in childhood lead poisoning." *Soc Sci Med* 12, no. 2D (June 1978): 85-94. Topics discussed include: preschool children, ethnic groups, blood, mass screening, maternal fetal exchange, Michigan, pregnancy, risk, skin pigmentation, epidemiology.

387. Ivanchenko, G. et al. "Experience in the treatment of ulcer disease at a climatic sanatorium." *Vrach Delo* no. 1 (January 1984): 35-6. Topics discussed include: Ukraine, duodenal ulcer treatment, health issues, resorts, stomach ulcer treatment.

388. Jaremin, B. "Acute non inflammatory chemical diarrhoeas among employees of Polish building enterprise in a tropical country." *Bull Inst Marit Trop Med Gdynia* 37, no. 3-4 (1986): 229-37. Topics discussed include: acute disease, epidemiology, diarrhea treatment, Iraq, ethnology, seasonality, water supply.

389. ---. "Limited, intragroup epidemics of acute infectious diseases of the upper respiratory tracts and throat among workers of a Polish construction site in a tropical country." *Bull Inst Marit Trop Med Gdynia* 38, no. 1-2 (1987): 42-9. Topics discussed include: Iraq, work-related drug treatment, ethnology, respiratory tract infections, seasonality, tropical climates, propagation.

390. Johansson, M. et al. "Influence of treatment and change of climate in women with rheumatoid arthritis. A controlled prospective study of psychological, medical and social effects." *Scand J Rheumatol Suppl* no. 9 (1975): 1-193. Topics discussed include: cost, disability evaluation, emotions, psychology, psychotreatment, socioeconomics, Spain, Sweden, travel, climate.

391. Kasuya, S. et al. "Studies on tsutsugamushi disease in Gifu Prefecture. 3. Seasonal fluctuation in positive rates of Rickettsia tsutsugamushi in wild rodents and number of patients." *Kansenshogaku Zasshi* 60, no. 9 (September 1986): 1022-6. Topics discussed include: Japan, muridae, scrub typhus veterinary, rodent epidemiology, scrub typhus epidemiology.

392. Kerpel-Fronius, E. et al. "Cause and Significance of Seasonal Variation in the Haemorragic Tendency in the Newborn." *Arch Disease Childhood* 23 (1948): 87-92. Topics discussed include: health issues, birth.

393. Khoda, J. et al. "Is enterolithiasis a disease occurring more frequently in desert climate? Apropos of 8 cases." *J Chir Paris* 129, no. 2 (February 1992): 111-4. Topics discussed include: etiology, surgery, intestinal complications, Israel, tuberculosis.

394. Khomik, S. et al. "Dynamics of typhoid bacteria discharge in chronic carriers according to the season of the year and its significance in the epidemiology of the disease." *Zh Mikrobiol Epidemiol Immunobiol* 47, no. 1 (January 1970): 31-5. Topics discussed include: microbiology, typhoid fever.

395. Khomin, L. et al. "Effect of seasonal antirheumatic measures on the course of rheumatic heart disease." *Vrach Delo* no. 3 (March 1980): 57. Topics discussed include: drug combinations, timing, drug treatment, penicillin.

396. Kinkade, J. H. "Climatologic Factors in Ear, Nose and Throat-Disease." *Laryngoscope* 63 (1953): 311-320. Topics discussed include: climate.

397. Klumpp, R. et al. "Focal, seasonal and behavioural patterns of infection and transmission of Schistosoma haematobium in a farming village at the Volta Lake, Ghana." *J Trop Med Hyg* 90, no. 5 (October 1987): 265-81. Topics discussed include: age effects, epidemiologic methods, infants, rural populations, gender effects, snails.

398. Kochar, M. et al. "Effect of seasonal temperature change on blood pressure in a treated hypertensive population." *J Clin Hypertens* 1, no. 1 (March 1985): 49-52. Topics discussed include: body weight, hypertension physiopathology.

399. Komarov, F. et al. "Status of gastric mucosa in duodenal ulcer at various stages of the disease in different seasons." *Sov Med* no. 8 (1990): 12-6. Topics discussed include: pathology.

400. Kondo, K. et al. "Is risk to motor neuron disease influenced by the season of birth?" *Jinrui Idengaku Zasshi* 34, no. 3 (September 1989): 243-6. Topics discussed include: Japan, nervous system pathology, risk, motor neurons pathology.

401. Kozhevnikova, D. et al. "Cardiovascular disorders in Addison's disease under the mountain climate conditions of Kirghizia." *Zdravookhr Kirg* no. 4 (July 1976): 21-4. Topics discussed include: physiopathology, heart rate effects, Kyrgyzstan, altitude, causes, climate.

402. Latarjet, M. et al. "Medical treatment of bronchiectasis disease in childhood and adolescence. (Experience at a specialized climatic center)." *Therapeutique* 46, no. 4 (April 1970): 393-7. Topics discussed include: long term care, bronchiectasis treatment.

403. Lawther, P. J. "Climate, Air Pollution and Chronic Bronchitis." *Symp on Weather and Disease, Proc Royal Soc of Med* 51, no. 4 (April 1958): 262-264. Topics discussed include: environment, health issues.

404. Lefebvre, P. "A seasonal neurologic disease: paralysis following tick bites." *Sem Hop Paris* 57, no. 43-44 (1981 November 18-25): 1809-14. Topics discussed include: tick paralysis, diagnosis, immunology, physiopathology.

405. Levashov, N. et al. "Basic surgical diseases and the work experience in tropical Africa (Uganda)." *Sov Med* no. 11 (November 1978): 124-8. Topics discussed include: anesthesia, Uganda, USSR, surgery, tropical climates.

406. Lindtjorn, B. "Kala azar in south west Ethiopia: seasonal variation in disease occurrence." *Trans R Soc Trop Med Hyg* 78, no. 6 (1984): 790-1. Topics discussed include: Ethiopia, visceral transmission, rain, epidemiology.

407. Loevinsohn, M. E. "Climatic warming and increased malaria incidence in Rwanda." *Lancet* 343, no. 8899 (19 March 1994): 714-718. Topics discussed include: medicine.

408. Macpherson, R. K. *Physiological responses to hot environments; an account of work done in Singapore, 1948-1953, at the Royal Naval Tropical Research Unit,* with an appendix on preliminary work done at the National Hospital for Nervous Diseases, London, 323 pages. Medical Research Council (Great Britain) Royal Naval Personnel Research Committee. Medical Research Council (Great Britain) Special report series, no. 298, London: H.M. Stationery Off., 1960. Topics discussed include: heat physiology.

409. Mann, I. C. *Culture, race, climate, and eye disease; an introduction to the study of geographical ophthalmology.* 580 pages. Springfield, Ill.: C. C. Thomas, 1966. Topics discussed include: geography, ethnology, culture.

410. Marian, I. et al. "Climate and Rheumatism." *Romania Med* 17 (1939): 169-172. Topics discussed include: disease, health issues, rheumatism.

411. Marshall, W. A. et al. "Seasonal Varriations in Growth Rates of Normal and Blind Children." *Hum Biol* 43 (1971): 502-516. Topics discussed include: health issues, biology, disease.

412. McKeown, Th. et al. "Seasonal Incidence of Congenital Malformations of the Central Nervous System." *Lancet* 260, no. 6648 (1951): 192-196. Topics discussed include: disease, health issues, birth.

413. McLeod, R. "Heat illness in early season football practice." *J Ky Med Assoc* 70, no. 8 (August 1972): 613-4. Topics discussed include: seasonality, exhaustion, sports medicine.

414. Melichar, F. "Civilization diseases in tropical climate." *Med Welt* 11 (13 March 1971): 431-3. Topics discussed include: civilization, Vietnam, arteriosclerosis epidemiology, diabetes mellitus epidemiology, hypertension epidemiology, tropical medicine.

415. Mhalu, F. et al. "Seasonality of cholera in Tanzania: possible role of rainfall in disease transmission." *East Afr Med J* 64, no. 6 (June 1987): 378-87. Topics discussed include: cross-sectional analysis, Tanzania, cholera transmission, rain, weather.

416. Michaud, L. "Alpine Climate and Diseases of Heart." *J State Med* 38 (1930): 322-332. Topics discussed include: health issues, heart.

417. Mills, C. A. "Climate Environment in Health and Disease." *Am Inst MedClimate* 3 (1957): 979-995. Topics discussed include: environment, health issues.

418. ---. "Infection Frequency and Host Resistance as Influenced by Climate and the Weather." Proc 6th Pacific Science Congr, 473-492. 1939. Stanford and San Francisco: Berkeley, 1939. Topics discussed include: disease, health issues.

419. ---. *Medical Climatology: Climatic and Weather Influences in Health and Disease*. Springfield: IL: Charles C. Thomas, 1939. Topics discussed include: health issues.

420. ---. "Seasonal and Regional Factors in Acute Rheumatic Fever and Rheumatic Heart Disease." *J Lab Clin Med* 24, no. 1 (1938): 53-60. Topics discussed include: geography, health issues, rheumatism, heart.

421. Milstein, V. et al. "Manic depressive illness: onset, diurnal temperature and season of birth." *Diseases of the Nervous System* 37, no. 7 (1976): 373-375. Topics discussed include: epidemiology, bipolar manic depression.

422. Mimica, M. et al. "The prevalence of chronic nonspecific pulmonary disease in relation to air pollution, climate and smoking." *Lijec Vjesn* 100, no. 4 (April 1978): 213-7. Topics discussed include: Yugoslavia, respiratory tract epidemiology, smoking.

423. Mitchell, A. "The constancy of the relations of the weather to disease." *Trans Sanitary Inst Great Britain* 15 (1882): 190-203. Topics discussed include: health issues.

424. Mitiuk, A. et al. "Temporary loss of working capacity in agricultural workers suffering from cardiovascular diseases depending on climatic and metereological factors." *Vrach Delo* no. 5 (May 1977): 95-7. Topics discussed include: USSR, epidemiology, disability, weather, work.

425. Mitropol'Skii, A. et al. "Characteristics of the course of Botkin's disease in different climatic conditions." *Voen Med Zh* 8 (August 1972): 48-50. Topics discussed include: cold climates, hepatitis, tropical climates.

426. Momiyama, M. "The Chronological Changes in the Calendar of Seasonal Diseases." *Kose no Sihyo* 14 (1963): 1-11. Topics discussed include: climate, health issues.

427. ---. "A Geographical Study of Seasonal Disease Calendar Model by Period and Country." *Pap Met Geophys* 14 (1963): 1-11. Topics discussed include: geography, climatic effects, health issues.

428. Moodley, D. et al. "Cryptosporidium infections in children in Durban. Seasonal variation, age distribution and disease status." *S Afr Med J* 79, no. 6 (16 March 1991): 295-7. Topics discussed include: age effects, feces, infants, South Africa, epidemiology.

429. Mori, S. "Evidence of Seasonal Variations of Umbilical Hernia." *Orient J Diseases Infants* 17 (1935): 25-28. Topics discussed include: health issues.

430. Muller, W. et al. "Geographic and seasonal distribution of disease pictures with possible arthropod etiology. Investigations on the archives of the Wurzburg Neurological University Clinic from the time period 1945 1964 in regard to the question of disease pictures in Franconia caused by arthropods." *Dtsch Z Nervenheilkd* 189, no. 3 (14 September 1966): 240-58. Topics discussed include: etiology, geography, Germany, ticks, arbovirus infections.

431. Newell, G. "Seasonal onset of Hodgkin's disease." *Lancet* 1, no. 758 (6 May 1972): 1024-5. Topics discussed include: health issues.

432. Newell, G. et al. "Seasonal diagnosis of Hodgkin's disease among young adults." *J Natl Cancer Inst* 74, no. 1 (January 1985): 53-6. Topics discussed include: gender effects, epidemiology.

433. Nichols, E. "Climate and Its Relation to Acute Respiratory Conditions." *U S Dep Agric Weather Bureau, Monthly Weather Review* 48 (September 1920): 495-509. Topics discussed include: health issues, disease, biology.

434. Nikberg, I. et al. "Effect of weather conditions on the condition of patients with ischemic heart disease." *Vrach Delo* no. 7 (July 1976): 85-9. Topics discussed include: myocardial infarction.

435. Niv, Y. et al. "Seasonal variation in duodenal ulcer disease in the Upper Galilee." *Harefuah* 119, no. 7-8 (October 1990): 193-5. Topics discussed include: diagnosis, Israel, epidemiology.

436. North, C. E. "Seasonal Diseases and Seasonal Temperatures." *American Journal of Public Health* 3 (1913): 228-232. Topics discussed include: health issues, climate.

437. Oganov, R. et al. "Epidemiological characteristics of dyslipoproteinaemia and certain other risk factors of atherosclerosis and ischaemic heart disease in 11 and 14 year children in different climatogeographic zones. Results of a cooperative study." *Cor Vasa* 30, no. 4 (1988): 248-56. Topics discussed include: climatic effects, exercise, hypertension, obesity, risk, smoking, USSR.

438. Oranskii, I. et al. "Seasonal rhythm of cardiac loading tolerance and of immune reactivity in ischemic heart disease." *Sov Med* no. 7 (1981): 11-4. Topics discussed include: antibodies, coronary immunology, exercise, biological clocks, physiopathology.

439. Ostensen, H. et al. "Seasonal periodicity of peptic ulcer disease. A prospective radiologic study." *Scand J Gastroenterol* 20, no. 10 (December 1985): 1281-4. Topics discussed include: Norway, duodenal ulcers, dyspepsia, epidemiology, cycles, stomach ulcers.

440. Owens, I. B. et al. "Influence of Season on the Severity of Diabetes and Its Sclerotic Complications." *Am J Med Sci* 199 (1940): 705-708. Topics discussed include: disease, health issues.

441. Paul, J. R. et al. "Climate and Rheumatic Heart Disease; Survey among American Indians School Children in Northern and Southern Localities." *J Am Med Assoc* 108 (1937): 2096-2100. Topics discussed include: health issues, rheumatism, heart.

442. Peters, O. H. "Season and Disease." *Proc Roy Soc Med* Sect.1. (1909). Topics discussed include: health issues.

443. Petersen, W. F. *The Patient and Weather.* Ann Arbor: 4 vols. MI: Edwards Brothers, 1938. Topics discussed include: health issues, climatic effects, disease.

444. Petersen, W. F. et al. "Scarlet Fever and the Meteorological Environment." *Arch Pediat* 55 (1938): 682. Topics discussed include: weather, disease, health issues.

445. Pham, Q. et al. "Follow up study of construction workers with obstructive lung disease." *Scand J Respir Dis* 58, no. 4 (August 1977): 215-26. Topics discussed include: age effects, bronchitis, exertion, France, oxygen consumption, seasonality, smoking, work-related medicine.

446. Philpot, M. et al. "Season of birth in Alzheimer's disease." *Br J Psychiatry* 155, .(November 1989): 662-6. Topics discussed include: age effects, causes, epidemiology, risk.

447. Piesman, J. et al. "Seasonal variation of transmission risk of Lyme disease and human babesiosis." *Am J Epidemiol* 126, no. 6 (December 1987): 1187-9. Topics discussed include: Massachusetts, risk, ticks, microbiology, parasitology.

448. Pin, N. et al. "Dehydration from outdoor work and urinary stones in a tropical environment." *Occup Med* 42, no. 1 (February 1992): 30-2. Topics discussed include: disease, dehydration, risk, Singapore, epidemiology, work-related diseases.

449. Pinfold, J. et al. "Seasonal effects on the reported incidence of acute diarrhoeal disease in northeast Thailand." *Int J Epidemiol* 20, no. 3 (September 1991): 777-86. Topics discussed include: diarrhea, causes, epidemiologic methods, Thailand, water supply, weather, diarrhea epidemiology.

450. Port, T. "The effect of the weather on general and stomatologic disease pictures." *DDZ* 20, no. 10 (18 May 1966): 304-8. Topics discussed include: the mouth, teeth.

451. Porter, M. "Seasonal change and its effect on the prevalence of infectious skin disease in a Gambian village." *Trans R Soc Trop Med Hyg* 74, no. 2 (1980): 162-8. Topics discussed include: Gambia, epidemiology, scabies, tinea versicolor, skin, infectious epidemiology.

452. Poskitt, E. M. E. "Seasonal Variation in Infection and Malnutrition at a Rural Paediatric Clinic in Uganda." *Trans Roy Soc Trop Med Hyg* 66 (1972): 931-936. Topics discussed include: disease, health issues, nutrition, agriculture.

453. Post, J. D. "Food shortage, climatic variability, and epidemic disease in preindustrial Europe." *The Economic History Review* s2 39 (November 1986): 662-3.

454. Postema, C. et al. "Meningococcal disease in the winter season 1988 1989." *Ned Tijdschr Geneeskd* 134, no. 12 (24 March 1990): 609-13. Topics discussed include: Netherlands, epidemiology, meningitis, meningococcal epidemiology.

455. Poulton, E. P. "Wet and Dry Climates and Weather in the Causation of Disease." Summaries Principal Reports Intern Congr Rheumatism and Hydrology, London and Oxford, 1938 March 25-31. 1938. Topics discussed include: climate, health issues.

456. Prasad, S. "Observations on seasonal, socio economic, and heredo familial factors of rheumatic heart disease in children." *Indian Pediatr* 4, no. 7 (July 1967): 290-7. Topics discussed include: rheumatic heart genetics, socioeconomics.

457. Ravindranathan, T. et al. "Seasonal variation in microfilaraemia (bancrofti) density and filaria disease." *J Commun Dis* 12, no. 2 (June 1980): 73-7. Topics discussed include: blood, India, population density.

458. Redwood, A. et al. "Climate and painful crisis of sickle cell disease in Jamaica." *Br Med J* 1, no. 6001 (10 January 1976): 66-8. Topics discussed include: hospitalization, Jamaica, pain, rain, temperature, anemia, sickle cell epidemiology, climate.

459. Riemerschmid, G. "Some Effects of Solar Radiation and Other Factors of Climatic Environment on Human Beings." *S Afr Med J* 18 (1944): 365-368. Topics discussed include: climate, biology, health issues, disease.

460. Rosati, G. et al. "The geographical distribution of multiple sclerosis, rheumatoid arthritis, rheumatic heart disease and poststreptococcal nephritis in Sardinia: climatic and socioeconomic factors." *J Neurol* 219, no. 1 (12 September 1978): 27-35. Topics discussed include: geography, Italy, socioeconomics, arthritis, multiple sclerosis, nephritis epidemiology.

461. Rusznyak. "Diseases of the Seasons." *Wien Arch inn Med* 3 (1922): 379-396. Topics discussed include: health issues.

462. Sargent, F. "Weekly and Seasonal Trends of Upper Respiratory Infections in a Group of 2000 Students." *Am J Publ Hyg* 30 (1940): 533. Topics discussed include: disease, health issues, weather.

463. Sargent, F. et al. "Medical Progress. Season, Nutrition and Pellagra." *New England J Med* 242 (1950): 447-453, 507-514. Topics discussed include: health issues, disease.

464. Schnitzler, L. "Besnier Boeck Schaumann disease. Actinic seasonal sarcoidosis: 18 year course." *Ann Dermatol Venereol* 112, no. 10 (1985): 831-4. Topics discussed include: lung diagnosis, timing, sarcoidosis pathology, skin pathology.

465. Scott, J. C. et al. "Climatic Effects on Cardiac Output and the Circulation in Man." *Am J Physiol* 129 (1940): 102-122. Topics discussed include: climate, health issues, disease, illness, physionomy, biology.

466. Sedgwick, W. T. et al. "Statistical Studies on the Seasonal Prevalence of Typhoid in Various Countries and itd Relation to Seasonal Temperature." *Memoires of the American Academy of Arts and Sciences* (August 1902): 12. Topics discussed include: statistics, health issues, disease.

467. Sekerova, N. "Temporary disability to work caused by diseases of the upper respiratory tract in the tobacco industry of Plovdiv." *Folia Med Plovdiv* 11, no. 1 (1969): 46-54. Topics discussed include: age effects, Bulgaria, disability evaluation, work-related medicine, seasonality, gender effects, timing, work-related epidemiology, tobacco.

468. Shakhnazarov, A. "Effect of sanatorium and climatic treatment on autonomic tonus in hypertension and ischemic heart disease." *Vrach Delo* no. 7 (July 1980): 18-20. Topics discussed include: coronary physiopathology, hypertension physiopathology, Russia, autonomic nervous system physiopathology, treatment, health issues resorts.

469. Sharp, W. B. et al. "Climate and Streptococcus Rheumatism Relationship." *J Infectious Diseases* 60 (1937): 15-24. Topics discussed include: health issues, rheumatism.

470. Shavakhabov, S. et al. "Experience in organizing the work of oral rehydration centers in a hot climate." *Ter Arkh* 63, no. 11 (1991): 12-3. Topics discussed include: acute disease, Asia, diarrhea, infant treatment, community health centers, fluid treatment, tropical climates.

471. Shiffman, M. et al. "Seasonality in water related intestinal disease in Guatemala." *Int J Biometeorol* 20, no. 3 (October 1976): 223-9. Topics discussed include: Guatemala, intestinal epidemiology, water supply.

472. Shoenfeld, Y. et al. "Cholinergic urticaria. A seasonal disease." *Arch Intern Med* 141, no. 8 (July 1981): 1029-30. Topics discussed include: climate, urticaria etiology.

473. Skorobogatyi, A. "The seasonality of disease exacerbation studied for the planning of chronoprophylaxis." *Vrach Delo* no. 12 (December 1989): 76-8. Topics discussed include: timing, health issues, care, cycles, preventive medicine.

474. Slobodeniuk, A. et al. "Effectiveness of the vaccine prevention of influenza in industrial enterprises in the epidemic seasons of 1980-1983." *Vrach Delo* no. 7 (July 1984): 122-3. Topics discussed include: Siberia, disease propagation, influenza vaccines, work-related diseases.

475. Smallman, R. M. et al. "Seasonality in tropical AIDS: a geographical analysis." *Int J Epidemiol* 21, no. 3 (June 1992): 547-56. Topics discussed include: cluster analysis, nutrition, tropical climates, Uganda, epidemiology, agriculture, disease propagation.

476. Stanek, J. et al. "Epidemiological analysis of acute respiratory disease (ARD) and characteristics of the influenza epidemic in Bohemia during the season 1986, 1987." *J Hyg Epidemiol Microbiol Immunol* 34, no. 2 (1990): 155-63. Topics discussed include: Czechoslovakia, epidemiology, propagation statistics, influenza epidemiology, respiratory tract epidemiology.

477. Stanishevskaia, I. et al. "Effect of a shift in climatic and geographic conditions on the results of the sanatorium and health resort treatment of residents of Siberia with ischemic heart disease." *Vopr Kurortol Fizioter Lech Fiz Kult no.* 4 (July 1987): 17-20. Topics discussed include: adaptation, balneology, coronary physiopathology, geography, Siberia, timing, coronary treatment, health issues resorts.

478. Stintzing, G. et al. "Aetiology of acute diarrhoeal disease in infancy and childhood, during the peak season Addis Ababa 1977: a preliminary report." *Ethiop Med J* 15, no. 4 (October 1977): 141-6. Topics discussed include: diarrhea, epidemiology, Ethiopia, diarrhea, children.

479. Sung, S. "The relationship between seasonal prevalence of Keshan disease and hair selenium level of local inhabitants." *Chung Hua Yu Fang I Hsueh Tsa Chih* 14, no. 1 (February 1980): 17-9. Topics discussed include: China, hair analysis, causes, selenium analysis.

480. Sutton, R. N. P. "Acute Respiratory Disease and Climate." Proceedings of Third International Biometeorological Congress, editors S. W. Tromp, and W. H. Weihe, 41. Oxford, 1963. Oxford: Pergamon Press, 1963. Topics discussed include: health issues, weather.

481. Sygit, M. et al. "Temporary work incapacity (absenteeism from work) because of disease in naval industry workers. II. An analysis of the occupational influence in a group of workers with various naval industry jobs on the proportion and nature of the work incapacity (absenteeism from work because of disease)." *Sante Publique Bucur* 30, no. 2 (April 1987): 65-75. Topics discussed include: etiology, Poland, seasonality, ships, absenteeism, naval medicine, epidemiology, work-related medicine.

482. Thompson, H. E. "Climate and Rheumatic Disease." *Arizona Med* 8 (1951): 31-34. Topics discussed include: health issues, rheumatism.

483. Tisdall, F. F. et al. "Seasonal Variation in the Antirachitic Effect of Sunshine." *Am J Diseases Children* 42 (1931): 1144. Topics discussed include: temperature, weather, health issues.

484. Toker, B. et al. "Seasonal Variations in Incidence of Peptic Ulcer." *Türk Tip Cemiyeti Mecmuasi* 13 (1947): 137-145. Topics discussed include: disease, health issues, ulcer.

485. Topcjan, Z. et al. "Mechanism of action of climatic balneologic factors at the Dzhermuk Spa in ischemic heart disease." *Fysiatr Revmatol Vestn* 59, no. 4 (August 1981): 203-7. Topics discussed include: altitude, balneology, coronary treatment.

486. Touitou, Y. et al. "Age related changes in both circadian and seasonal rhythms of rectal temperature with special reference to senile dementia of Alzheimer type." *Gerontology* 32, no. 2 (1986): 110-8. Topics discussed include: circadian rhythm, hydrocortisone blood, gender effects, alzheimer's disease physiopathology, cycles.

487. Tromp, S. W. "The Relationship of Weather and Climate to Health and Disease." in *Environmental Medicine*, 72-79. editors G. M. Howe, and J. A. Loraine. London: W. Heinemann, 1973. Topics discussed include: health issues.

488. Trowbridge, F. et al. "Seasonal changes in malnutrition and diarrheal disease among preschool children in El Salvador." *Am J Trop Med Hyg* 28, no. 1 (January 1979): 136-41. Topics discussed include: body height, body weight, diarrhea, children, epidemiology, El Salvador, nutrition, protein, malnutrition.

489. Vallee, F. "Stresses of change and mental health among the Canadian Eskimos." *Arch Environ Health* 17, no. 4 (October 1968): 565-70. Topics discussed include: acculturation, anxiety, bipolar disorder, Canada, cold climates, disease propagation, hysteria, adaptation, Eskimos, mental disorders, stress.

490. Volosevich, A. et al. "Characteristics of the course of Reiter's disease in a hot, humid climate." *Voen Med Zh 1988* no. 9 (September 1988): 30-2. Topics discussed include: adaptation, timing, USSR, humidity, tropical climate.

491. Waddy, B. B. "Climate and Respiratory Infection." *Lancet* 2 (1952): 674-677. Topics discussed include: health issues, disease, weather, biology.

492. Wannenwetsch, E. "Altitude climate therapy in skin diseases. Its significance for social income insurance." *Ther Ggw* 108, no. 4 (April 1969): 575 passim. Topics discussed include: Germany, income, insurance, social medicine, skin treatment.

493. Warner, E. C. "The Relation of Soil and Climate to Rheumatic Disease." *J State Med* 41 (1933): 621-642. Topics discussed include: environment, health issues, rheumatism.

494. Webster, J. "The Influence of Weather on Disease and on the Human Frame." *Lancet* 1 (1859): 588-589. Topics discussed include: health issues, physiology, biology.

495. Weihe, W. "Climate, Health and Disease." in Proceedings of the World Climate Conference, 313-368. Geneva: World Meteorological Organization, 1979. Topics discussed include: health issues.

496. Weil, M. P. "Rhumatismes et Climats." *Rev Méd* 51 (1934): 252-262. Topics discussed include: climate, disease, health issues, rhumatism.

497. Wemmer, U. "Croup syndrome and harmful agents in respiratory air. A statistical analysis of seasonal fluctuations of emission values in relation to the incidence of disease." *Fortschr Med* 102, no. 34 (13 September 1984): 835-7. Topics discussed include: dust, Germany, ozone, air pollution, croup epidemiology, laryngitis.

498. Wickstrom, G. "Effect of work on degenerative back disease. A review." *Scand J Work Environ Health* 4 Supp, no. 1 (1978): 1-12. Topics discussed include: back pain, genetics, climatic effects, environmental exposure, posture, stress, work-related epidemiology.

499. Zunnunov, Z. "The central hemodynamic reaction of patients with ischemic heart disease to the climatic and weather conditions of an arid zone." *Vopr Kurortol Fizioter Lech Fiz Kult no.* 3 (May 1991): 19-22. Topics discussed include: angina pectoris, physiopathology, treatment, Asia, coronary treatment, exertion, desert climates, hemodynamics physiology.

HORMONES

500. Behall, K. et al. "Seasonal variation in plasma glucose and hormone levels in adult men and women." *Am J Clin Nutr* 40, no. 6 Suppl (December 1984): 1352-6. Topics discussed include: age effects, caloric intake, carbohydrates, glucagon, blood, gender effects, glucose.

501. Campbell, I. et al. "The plasma insulin and growth hormone response to oral glucose: diurnal and seasonal observations in the Antarctic." *Diabetologia* 11, no. 2 (April 1975): 147-50. Topics discussed include: Antarctic regions, blood, glucose, light, insulin.

502. Casper, R. et al. "In vitro fertilization: diurnal and seasonal variation in luteinizing hormone surge onset and pregnancy rates." *Fertil Steril* 49, no. 4 (April 1988): 644-8. Topics discussed include: blood, urine, menotropins, ovulation, pregnancy, circadian rhythm, metabolism.

503. Elebute, E. et al. "The metabolic response to surgical trauma in a tropical environment a preliminary report." *West Afr Med J Niger Pract* 19, no. 1 (February 1970): 18-21. Topics discussed include: adrenal cortex hormones, Nigeria, nutrition, urine, sodium, thyroid diseases, surgery.

504. Fletcher, H. et al. "Survey of asthma deaths in the Northern region, 1970 85." *Arch Dis Child* 65, no. 2 (February 1990): 163-7. Topics discussed include: drug treatment, emergencies, epidemiology, health services, seasonality.

505. Francesconi, R. P. et al. *Hypohydration and Heat Acclimation: Plasma Renin and Aldosterone during Exercise*, Performer: Army Research Inst. of Environmental Medicine, Natick, MA., 1983. Topics discussed include: exercise, acclimation, stress, hydration, response biology, body fluids, hormones, medicine.

506. Gertner, A. et al. "Thyroid hormones in chronic heat exposed men." *Int J Biometeorol* 27, no. 1 (May 1983): 75-82. Topics discussed include: desert climates, seasonality, thyroxine blood, triiodothyronine blood.

507. Hackney, A. et al. "Thyroid hormone changes during military field operations: effects of cold exposure in the Arctic." *Aviat Space Environ Med* 63, no. 7 (July 1992): 606-11. Topics discussed include: Arctic regions, darkness, Norway, cold climates, military personnel, blood.

508. Hakimowa, S. "On the question of locally conditioned peculiarities in the metabolism and excretion of steroid hormones in women under the climatic geographic conditions of the Tadzhikistan Republic." *Endokrinologie* 53, no. 5 (1968): 355-64. Topics discussed include: androsterone, urine, anoxia, etiocholanolone, menstruation, ovulation, prasterone, altitude.

509. Ivanova, G. et al. "Effect of climatic conditions of the arid zone (Turkmenia) on corticosteroid metabolism." *Probl Endokrinol Mosk* 18, no. 2 (March 1972): 10-4. Topics discussed include: adrenal glands secretion, adrenocorticotropic hormone, timing, Turkmenistan, urine.

510. Kauppila, A. et al. "Inverse seasonal relationship between melatonin and ovarian activity in humans in a region with a strong seasonal contrast in luminosity." *J Clin Endocrinol Metab* 65, no. 5 (November 1987): 823-8. Topics discussed include: melatonin, blood, urine, ovary metabolism, pituitary hormones, anterior blood, sex hormones, testosterone, light, metabolism.

511. ---. "The season, melatonin and reproduction." *Duodecim* 104, no. 21 (1988): 1687-93. Topics discussed include: sex hormones, melatonin, physiology, cycles, reproduction.

512. ---. "The effect of season on the circulating concentrations of anterior pituitary, ovarian and adrenal cortex hormones and hormone binding proteins in the subarctic area; evidence of increased activity of the pituitary ovarian axis in spring." *Gynecol Endocrinol* 1, no. 2 (June 1987): 137-50. Topics discussed include: blood, Finland, menstrual cycle, progesterone, testosterone, adrenal cortex, proteins, metabolism, light, sex hormones.

513. Krall, E. et al. "Effect of vitamin D intake on seasonal variations in parathyroid hormone secretion in postmenopausal women." *N Engl J Med* 321, no. 26 (28 December 1989): 1777-83. Topics discussed include: calcium, blood, dose response, Massachusetts, menopause.

514. Lichtenstein, P. et al. "Calcium regulating hormones and minerals from birth to 18 months of age: a cross sectional study. I. Effects of sex, race, age, season, and diet on vitamin D status." *Pediatrics* 77, no. 6 (June 1986): 883-90. Topics discussed include: analysis of variance, bottle feeding, breast feeding, blood, calcium, infant food, Negroid race, vitamins, protein.

515. Lips, P. et al. "Seasonal variation in serum concentrations of parathyroid hormone in elderly people." *J Clin Endocrinol Metab* 57, no. 1 (July 1983): 204-6. Topics discussed include: calcium, metabolism, neck fractures, vitamins, blood.

516. Meikle, A. et al. "Hereditary and environmental influences on the variation of thyroid hormones in normal male twins." *J Clin Endocrinol Metab* 66, no. 3 (March 1988): 588-92. Topics discussed include: obesity, blood, twins.

517. Meller, Y. et al. "Seasonal variation in serum levels of vitamin D metabolites and parathormone in geriatric patients with fractures in Southern Israel." *Isr J Med Sci* 22, no. 1 (January 1986): 8-11. Topics discussed include: blood, Israel, sunlight, timing, fractures, metabolism, hormones, vitamins, metabolism.

518. Meulmeester, J. et al. "Vitamin D status, parathyroid hormone and sunlight in Turkish, Moroccan and Caucasian children in The Netherlands." *Eur J Clin Nutr* 44, no. 6 (June 1990): 461-70. Topics discussed include: emigration, immigration, nutrition, risk, vitamins, seasonality.

519. Oddie, T. et al. "Variation in values for iodothyronine hormones, thyrotropin, and thyroxine binding globulin in normal umbilical cord serum with season and duration of storage." *Clin Chem* 25, no. 7 (July 1979): 1251-3. Topics discussed include: blood, drug stability, pregnancy, timing.

520. Oliveri, M. et al. "Seasonal variations of 25 hydroxyvitamin D and parathyroid hormone in Ushuaia (Argentina), the southernmost city of the world." *Bone Miner* 20, no. 1 (January 1993): 99-108. Topics discussed include: blood, Argentina, epidemiology, bone density, calcium, rickets, vitamins, cycles.

521. Pasquali, R. et al. "Seasonal variations of total and free thyroid hormones in healthy men: a chronobiological study." *Acta Endocrinol Copenh* 107, no. 1 (September 1984): 42-8. Topics discussed

include: acclimation, body weight, caloric intake, physical labor, gender effects, temperature, blood, metabolism.

522. Rapoport, S. et al. "The content of gastrointestinal hormones and pepsinogen I in the blood of patients with duodenal peptic ulcer at different seasons of the year." *Klin Med Mosk* 68, no. 10 (October 1990): 71-2. Topics discussed include: blood.

523. Reinberg, A. et al. "Circadian and seasonal changes in ACTH induced effects in healthy young men." *Eur J Clin Pharmacol* 25, no. 5 (1983): 657-65. Topics discussed include: body temperature, fatigue, urine, hormone pharmacology.

524. Rubio, P. P. et al. "Seasonal variations of the plasma levels of thyroid hormones." *Rev Clin Esp* 156, no. 4 (29 February 1980): 245-7. Topics discussed include: seasonality, blood.

525. Rudnicki, M. et al. "Effects of age, sex, season and diet on serum ionized calcium, parathyroid hormone and vitamin D in a random population." *J Intern Med* 234, no. 2 (August 1993): 195-200. Topics discussed include: blood, diet, phosphates, gender differences, calcium, vitamins.

526. Saljukov, V. et al. "The role of hereditary factors in phenotypic variability of hormone levels in the population genetically adapted to circumpolar environment." *Arctic Med Res* 51, no. 3 (July 1992): 143-9. Topics discussed include: acclimation, blood, seasonality, Siberia, cold climates.

527. Shephard, R. "Some consequences of polar stress: data from a transpolar ski trek." *Arctic Med Res* 50, no. 1 (January 1991): 25-9. Topics discussed include: Arctic regions, diet, frostbite, hormones, hypothermia, stress, ultraviolet rays, cold climates, metabolism.

528. Shvareva, N. "Hormonal support for the processes of long term adaptation of the female body to northern latitudes." *Probl Endokrinol Mosk* 30, no. 4 (July 1984): 25-31. Topics discussed include: hormones, menstruation, Siberia, blood, timing, urban populations, adaptation, cold climates.

529. Skvarko, S. "Seasonal changes in secretion of adrenal cortex hormones in children of early school age under normal school routine." *Pediatr Akush Ginekol* 5 (September 1969): 18-9. Topics discussed include: health services, Ukraine, urine.

530. Specker, B. et al. "Calcium regulating hormones and minerals from birth to 18 months of age: a cross sectional study. II. Effects of sex, race, age, season, and diet on serum minerals, parathyroid hormone, and calcitonin." *Pediatrics* 77, no. 6 (June 1986): 891-6. Topics discussed include: age effects, bottle feeding, breast feeding, calcium, infant food, magnesium, Negroid race, gender effects, diet, minerals, blood.

531. Stumpf, W. et al. "The steroid hormone of sunlight soltriol (vitamin D) as a seasonal regulator of biological activities and photoperiodic rhythms." *J Steroid Biochem Mol Biol* 39, no. 2 (August 1991): 283-9. Topics discussed include: brain physiology, radiation effects, retina physiology, skin, periodicity.

532. Testart, J. et al. "Seasonal influence of diurnal rhythms in the onset of the plasma luteinizing hormone surge in women." *J Clin Endocrinol Metab* 55, no. 2 (August 1982): 374-7. Topics discussed include: blood, circadian rhythm.

533. Touitou, Y. et al. "Adrenocortical hormones, ageing and mental condition: seasonal and circadian rhythms of plasma 18 hydroxy 11 deoxycorticosterone, total and free cortisol and urinary corticosteroids." *J Endocrinol* 96, no. 1 (January 1983): 53-64. Topics discussed include: gender effects, urine, circadian rhythm, blood.

534. Van, D. W. L. et al. "Unusual sex hormone patterns among desert dwelling hunter gatherers." *J Clin Endocrinol Metab* 46, no. 4 (April 1978): 658-63. Topics discussed include: age effects, Botswana, menopause, Negroid race, puberty, gender effects, desert climates, blood, testosterone.

535. Vanderschueren, D. et al. "Seasonal variation in bone metabolism in young healthy subjects." *Calcif Tissue Int* 49, no. 2 (August 1991): 84-9. Topics discussed include: hormones, pharmacology, vitamins, protein, blood, bones, calcium, vitamins.

536. Webb, A. et al. "An evaluation of the relative contributions of exposure to sunlight and of diet to the circulating concentrations of 25 hydroxyvitamin D in an elderly nursing home population in Boston." *Am J Clin Nutr* 51, no. 6 (June 1990): 1075-81. Topics discussed include: daily activities, Boston, hormones, blood, seasonality, aging, diet, nursing homes.

537. Weitzman, E. et al. "Seasonal patterns of sleep stages and secretion of cortisol and growth hormone during 24 hour periods in northern Norway." *Acta Endocrinol Copenh* 78, no. 1 (January 1975): 65-76. Topics discussed include: Arctic regions, Norway, protein, timing, circadian rhythm, sleep stages.

538. Wetterberg, L. "Lighting: Nonvisual effects." *Scandinavian Journal of Work, Environment and Health* 16, no. Suppl 1 (1990): 26-28. Topics discussed include: rhythmic variations, hormones, light treatment, environment, biological rhythms, depression.

METABOLISM

539. Abe, T. et al. "Seasonal variations in skin temperature, skin pH, evaporative water loss and skin surface lipid values on human skin." *Chem Pharm Bull Tokyo* 28, no. 2 (February 1980): 387-92. Topics discussed include: body regulation, surface properties, skin metabolism, water metabolism.

540. Abrarov, A. et al. "Protein and lipid metabolic characteristics of persons in a hot climate." *Vopr Pitan no.* 3 (May 1979): 35-9. Topics discussed include: blood, diet, cycles, reference values, seasonality, Uzbekistan.

541. Adolph, E. F. *Physiology of man in the desert.* 357 pages. Monographs in the physiological sciences, New York: Hafner Pub. Co., 1969. Topics discussed include: body temperature, heat, water metabolism.

542. Aitken, J. et al. "The physiological regulation of the seasonal variations in phosphorus and calcium metabolism." *J Physiol Lond* 226, no. 2 (October 1972): 93P-94P. Topics discussed include: calcium, blood, vitamins.

543. Akanji, A. et al. "The effect of ambient temperature on glucose tolerance and its implications for the tropics." *Trop Geogr Med* 43, no. 3 (July 1991): 283-7. Topics discussed include: blood glucose, body weight, diabetes, etiology, hospitals, metabolism.

544. Akhmedov, R. et al. "Seasonal dynamics of general metabolism and some autonomic functions in man under the climatic conditions of Uzbekistan." *Hum Physiol* 6, no. 5 (September 1980): 341-8. Topics discussed include: body temperature, carbon dioxide, oxygen consumption, acclimation, physiology.

545. Albrecht, E. et al. "Metabolism during O2 deficiency in a mountain climate. Therapeutic influence of acclimatization." *Pflugers Arch Gesamte Physiol Menschen Tiere* 293, no. 1 (1967): 1-18. Topics discussed include: blood, body temperature, colorimetry, mountaineering, respiration.

546. Angelov, I. "Effect of climatic factors on carbohydrate metabolism in patients with diabetes mellitus and obesity." *Vopr Kurortol Fizioter Lech Fiz Kult* 33, no. 2 (March 1968): 135-8. Topics discussed include: health resorts, carbohydrates, diabetes, obesity.

547. Anghelova, O. et al. "Studies in serotonine metabolism under the conditions of sea and mountain climates as well as of experimental ultraviolet irradiation for the purpose of deriving inferences for how to make the most of their prophylactic effect." *Nauchni Tr Vissh Med Inst Sofiia* 47, no. 1 (1968): 61-9. Topics discussed include: altitude, humidity, temperature, radiation effects.

548. Arayne, M. et al. "Effect of environments on human functions: studies on water metabolism." *JPMA J Pak Med Assoc* 27, no. 11 (November 1977): 430-4. Topics discussed include: adaptation, cardiovascular, diseases etiology, cold climates, desert climates, tropical climates, water deprivation, water supply standards.

549. Arendt, J. "Assay of melatonin and its metabolites: results in normal and unusual environments." *J Neural Transm Suppl* 21 (1986): 11-33. Topics discussed include: arctic regions, chromatography, circadian rhythm, melatonin analogs, derivatives, blood, pineal body physiology.

550. Audran, M. et al. "Seasonal variations of vitamin D metabolites in man." *Presse Med* 13, no. 11 (17 March 1984): 661-3. Topics discussed include: ultraviolet rays, blood, radiation effects, metabolism.

551. Babadzhanov, S. "Status of mineral metabolism in hypertensive disease in hot climate." *Ter Arkh* 40, no. 8 (August 1968): 100-2. Topics discussed include: calcium, blood, chlorides, hypertension, phosphor, potassium, sodium.

552. Baker, S. et al. "Prevalence, pathogenesis and prophylaxis of iron deficiency in the tropics." *In: Kief H, etal, ed Iron metabolism and its disorders Amsterdam, Excerpta Medica, W3 EX89* no. 366 (1975): 145-158. Topics discussed include: anemia, epidemiology, treatment, blood, pregnancy, protein binding.

553. Balaz, V. et al. "Seasonal periodicity of lipid metabolism in healthy subjects." *Vnitr Lek* 16, no. 9 (September 1970): 841-6. Topics discussed include: heparin pharmacology, lipids blood, cycles.

554. Bates, C. et al. "Seasonal variations in ascorbic acid status and breast milk ascorbic acid levels in rural Gambian women in relation to dietary intake." *Trans R Soc Trop Med Hyg* 76, no. 3 (1982): 341-7. Topics discussed include: fruit, Gambia, lactation, rural population, metabolism, diet, pregnancy.

555. Baumgartner, A. "Effect of climate and weather on sleep." *Internist Berl* 25, no. 9 (September 1984): 531-5. Topics discussed include: thermoregulation, metabolism, homeostasis, sleep disorders etiology, sleep stages.

556. Belakovskii, M. et al. "Nutrition in high altitude ascents." *Kosm Biol Aviakosm Med* 17, no. 2 (March 1983): 14-22. Topics discussed include: anoxia, metabolism, body weight, caloric intake, climatic effects, diarrhea, diet, digestion, taste physiology, mountaineering.

557. Berdyshev, V. "Body supply of vitamin C in sailors in tropical climates." *Voen Med Zh no.* 2 (February 1985): 51-3. Topics discussed include: ascorbic acid, etiology, metabolism, naval medicine.

558. Beutler, E. "Environmental modification of red cell metabolism." *Blood* 36, no. 2 (August 1970): 263-5. Topics discussed include: adaptation, Caucasoid race, rickets, tropical climates, vitamins, blood, environment.

559. Brabant, C. "Heat exposure standards and women's work: equitable or debatable?" *Women Health* 18, no. 3 (1992): 119-30. Topics discussed include: energy metabolism, heart, laundering, microclimates, seasonality, work-related exposure, stress physiopathology, women.

560. Brown, W. "Heat and cold in farm workers." *Occup Med* 6, no. 3 (July 1991): 371-89. Topics discussed include: acclimation, physiology, diseases, etiology, metabolism, body temperature, heart rate, health issues.

561. Buskirk, E. et al. "Nutrition, environment and work performance with special reference to altitude." *Fed Proc* 26, no. 6 (November 1967): 1760-7. Topics discussed include: adaptation, thermoregulation physiology, body weight, calorimetry, climatic effects, military medicine, temperature, basal metabolism, physical work.

562. Campbell, I. "Nutrition in adverse environments, 2. Energy balance under polar conditions." *Hum Nutr Appl Nutr* 36, no. 3 (June 1982): 165-78. Topics discussed include: acclimation, adaptation, psychological physiology, antarctic regions, seasonality, caloric intake, cold climates, diet, metabolism.

563. Claremont, A. et al. "Comparison of metabolic, temperature, heart rate and ventilatory responses to exercise at extreme ambient temperatures (0 degrees and 35 degrees C.)." *Med Sci Sports* 7, no. 2 (June 1975): 150-4. Topics discussed include: body, cold climates, environment, controlled, lactates blood, perspiration, tropical climates, physical work, oxygen consumption, respiration.

564. Collins, K. J. et al. *Capacity for work in the tropics.* 297 pages. Society for the Study of Human Biology symposium series; 26, Cambridge England ; New York: Cambridge University Press, 1988. Topics discussed include: physiological issues, tropical medicine, metabolism, physical work, nutrition, malnutrition.

565. Collins, K. et al. "Nutritional and environmental studies on an ocean going oil tanker. 2. Heat acclimatization and nutrient balances." *Br J Ind Med* 28, no. 3 (July 1971): 246-58. Topics discussed include: Arabia, thermoregulation, diet, blood, metabolism, perspiration, water, urine, naval medicine, nutrition, tropical climates.

566. Damshkaln, L. "Indices of lipid metabolism in sailors during voyages in tropical latitudes." *Vopr Pitan* 30, no. 5 (September 1971): 35-40. Topics discussed include: cholesterol, blood, fatty acids, naval medicine, tropical climates.

567. Elebute, E. et al. "Proceedings: the metabolic response to abdominal surgery in a tropical environment." *Br J Surg* 60, no. 11 (November 1973): 902. Topics discussed include: amino acids, metabolism, temperature, the abdomen.

568. Goromaru, T. et al. "Seasonal variation in urinary excretion of aminopyrine and its metabolities in man." *Chem Pharm Bull Tokyo* 27, no. 2 (February 1979): 563-7. Topics discussed include: urine.

569. Gustafson, F. L. et al. "The Seasonal Variation in Basal Metabolism." *Am J Physiol* 86 (1928): 93. Topics discussed include: biology.

570. Haas, J. "Studies on working capacity and total metabolic rate of man performing industrial and traditional tasks i a tropicl climate. II. Discussion and results." *Z Tropenmed Parasitol* 21, no. 4 (December 1970): 446-58. Topics discussed include: adaptation, disability evaluation, Europe, Liberia, work-related medicine, physical education, training, sports medicine, tropical medicine.

571. Hale, H. et al. "Seasonal changes among endocrine metabolic indices of men residing in a subtropical climate. SAM TR 66 114." *Tech Rep SAM TR* (November 1966): 1-14. Topics discussed include: Texas, urine analysis, adrenal glands physiology, adrenal medulla physiology, heat, metabolism.

572. Hassi, J. et al. "Cold weather work." *Duodecim* 105, no. 5 (1989): 481-92. Topics discussed include: thermoregulation, metabolism, work-related diseases, physiopathology, etiology, cold climate.

573. Heller, H. C. *Central Nervous Mechanisms Regulating Body Temperature*, Performer: Stanford Univ., CA. Dept. of Biological Sciences., October 1981. Topics discussed include: microwaves, radiation effects, central nervous system, skin anatomy, body weight, metabolism, medicine.

574. Hori, S. et al. "Study on seasonal variations of heat tolerance in young Japanese males and effects of physical training thereon." *Jpn J Physiol* 24, no. 5 (October 1974): 463-74. Topics discussed include: body temperature, body water metabolism, body weight, Japan, mathematics, perspiration, timing, physical education, training.

575. Hsyoung, C. "Studies on human adaptability to climatic conditions. Report 4. Seasonal variations in basal metabolic rate on civilian Japanese and Caucasians in Japan." *Nippon Eiseigaku Zasshi* 31, no. 3 (August 1976): 404-16. Topics discussed include: Japan, acclimation, basal metabolism, Caucasoid race, Mongoloid race.

576. Huang, S. Y. et al. *Increased Metabolism and Dead Space as Components of Ventilation at High Altitude*, Performer: Colorado Univ. Health Sciences Center, Denver., 12 July 1983. Topics discussed include: acclimation, exercise physiology, hyperventilation, carbon dioxide, oxygen, respiration, medicine, stress.

577. Jones, T. et al. "Seasonal gastroenteritis and malabsorption at an American military base in the Philippines. II. Malabsorption following the acute illness." *Am J Epidemiol* 95, no. 2 (February 1972): 128-39. Topics discussed include: acute disease, diarrhea, etiology, metabolism, epidemiology, tropical climates, etiology, sucrase metabolism, timing, military medicine.

578. Juttmann, J. et al. "Seasonal fluctuations in serum concentrations of vitamin D metabolites in normal subjects." *Br Med J Clin Res Ed* 282, no. 6273 (25 April 1981): 1349-52. Topics discussed include: circadian rhythm, blood, metabolism, cycles.

579. Kashiwazaki, H. "Seasonal fluctuation of BMR in populations not exposed to limitations in food availability: reality or illusion?" *Eur J Clin Nutr* 44 Supp, no. 1 (1990): 85-93. Topics discussed include: temperature, basal metabolism, physiology, body composition, caloric intake, food supply.

580. Kashiwazaki, H. et al. "Influence of upper and lower thermoneutral room temperatures (20 degrees C and 25 degrees C) on fasting and post prandial resting metabolism under different outdoor temperatures." *Eur J Clin Nutr* 44, no. 5 (May 1990): 405-13. Topics discussed include: body composition, physiology, intestinal absorption, seasonality, eating, fasting, temperature.

581. Kitzing, J. et al. "Metabolism and body temperature during light exertion in warm and cold environment." *Int Z Angew Physiol* 23, no. 2 (17 November 1966): 159-65. Topics discussed include: cold effects, environmental exposure, heat, muscles physiology, oxygen consumption, climatic effects, physical work.

582. Klausen, K. et al. "Metabolic reactions to work in the desert." *J Appl Physiol* 22, no. 2 (February 1967): 292-6. Topics discussed include: California, heat, Indiana, lactates blood, Nevada, oxygen consumption, desert climates, exertion physiology, metabolism, respiration.

583. Kovtun, V. "Protein metabolic characteristics of open bone fractures in temperate and hot dry climates." *Voen Med Zh no.* 10 (October 1986): 48-9. Topics discussed include: desert climates, timing, USSR, blood, proteins, metabolism, fractures.

584. Krylov, A. et al. "Effect of health resort factors and clinical treatment in various seasons of the year on indices of lipid metabolism in persons with predisposition for atherosclerosis and in patients with ischemic heart disease." *Ter Arkh* 49, no. 1 (1977): 34-7. Topics discussed include: age effects, atherosclerosis treatment, cholesterol, coronary treatment, USSR, arteriosclerosis, atherosclerosis, health resorts.

585. ---. "Seasonal fluctuations in lipid metabolism indices." *Voen Med Zh no.* 1 (January 1975): 38-41. Topics discussed include: cholesterol, blood, coronary disease, military medicine, myocardial infarction, cycles, USSR.

586. Krylov, V. et al. "Metabolic indicators in children depending on their national ethnic background and climatic geographical zones of residence." *Lab Delo* no. 1 (1978): 25-8. Topics discussed include: arctic regions, Siberia, amino acids, blood, glucose analysis, cold climates, electrolytes, lipids.

587. Kudrin, I. et al. "Endocrine metabolic changes during adaptation to high temperature." *Voen Med Zh no.* 8 (August 1987): 40-3. Topics discussed include: energy metabolism, timing, tropical climates, USSR, adaptation, endocrine glands metabolism, heat.

588. Kulin, L. et al. "Effect of climatic conditions on energy metabolism in infants with different constitutions." *Acta Paediatr Acad Sci Hung* 6, no. 3 (1965): 395-410. Topics discussed include: body height, thermoregulation, body weight, oxygen consumption.

589. Kuno, Y. et al. "Metabolic adaptations to temperature and altitude. Concluding remarks." *Fed Proc* 25, no. 4 (July 1966): 1427-33. Topics discussed include: adipose tissue metabolism, body regulation, catecholamines, enzymes, fatty acids, nonesterified, glycolysis, oxygen consumption, potassium metabolism.

590. Kuwano, K. "Studies on human adaptability to climatic conditions reprt 7. Basal metabolic rate and its seasonal variations of residents in Okinawa." *Nippon Eiseigaku Zasshi* 33, no. 4 (October 1978): 638-46. Topics discussed include: Japan, acclimation, basal metabolism, climate.

591. Lane, H. et al. "Effect of physical activity on human potassium metabolism in a hot and humid environment." *Am J Clin Nutr* 31, no. 5 (May 1978): 838-43. Topics discussed include: Florida, heat, humidity, perspiration, physical work, potassium, sports medicine, weather.

592. Lawrence, M. et al. "Fat gain during pregnancy in rural African women: the effect of season and dietary status." *Am J Clin Nutr* 45, no. 6 (June 1987): 1442-50. Topics discussed include: body weight, food supply, Gambia, pregnancy metabolism, rural population, adipose tissue anatomy, histology, nutrition.

593. Leonard, W. "Household level strategies for protecting children from seasonal food scarcity." *Soc Sci Med* 33, no. 10 (1991): 1127-33. Topics discussed include: age effects, agriculture, anthropometry, caloric intake, diet, metabolism, physical work, family composition, nutrition, Peru.

594. Lewis, R. "Proceedings: Sudden change from tropical to temperate climate and electrolyte metabolism." *West Afr J Pharmacol Drug Res* 2, no. 1 (December 1974): 57P. Topics discussed include: cold effects, electrolytes, tropical climates.

595. Linden, M. et al. "Seasonal variation in the phagocytic activity and arachidonic acid metabolism of human blood monocytes in healthy non smokers, smokers and chronic bronchitics." *Agents Actions Suppl* 30 (1990):121-32. Topics discussed include: chronic diseases, blood, bronchitis, smoking.

596. Lizarenko, P. et al. "Energy expenditures in the physical work of winter swimming enthusiasts." *Fiziol Cheloveka* 12, no. 6 (November 1986): 1036-8. Topics discussed include: adaptation, cold effects, Moscow, metabolism, physical work, seasonality, swimming.

597. Lloyd, R. et al. "Does the elimination rate of cesium in humans change with the seasons?" *Health Phys* 18, no. 6 (June 1970): 623-9. Topics discussed include: urine, feces, metabolism, environment, temperature.

598. Lovely, R. H. et al. *Subtle Consequences of Exposure to Weak Microwave Fields: Are There Non Thermal Effects*, Performer: University Hospital, Seattle, WA. Bioelectromagnetics Research Lab., 1983. Topics discussed include: body temperature, food consumption, body weight, mass, metabolism, thermoregulation, medicine, biology, radiobiology, pollution.

599. Malhotra, M. et al. "Dietary intake and energy requirement of Indian submariners in tropical waters." *Ergonomics* 19, no. 2 (March 1976): 141-8. Topics discussed include: India, nutritional requirements, diet, metabolism, submarine medicine, tropical climates.

600. Masawe, A. "Nutritional anaemias. Part 1: Tropical Africa." *Clin Haematol* 10, no. 3 (October 1981): 815-42. Topics discussed include: epidemiology, hypochromic, immunology, metabolism, treatment, iron, pregnancy, protein deficiency, tropical climates, deficiency diseases.

601. Mason, E. et al. "Variations in basal metabolic rate responses to changes between tropical and temperate climates." *Hum Biol* 44, no. 1 (February 1972): 141-72. Topics discussed include: age effects, calorimetry, ethnic groups, India, oxygen consumption, residential mobility, timing, acclimation.

602. Mathew, L. et al. "Body temperature and basal metabolic changes during acclimatization to altitude (3,500 m) in man." *Indian J Physiol Pharmacol* 20, no. 4 (October 1976): 197-202. Topics discussed include: oxygen consumption, skin, basal metabolism, body regulation.

603. Menger, W. et al. "Effect of the weather on metabolic reactions in acetonemic vomiting under different climatic conditions." *Monatsschr Kinderheilkd* 116, no. 7 (July 1968): 427-33. Topics discussed include: Germany, humidity, blood, urine, cycles, seasonality, vomiting, etiology.

604. Morimoto, T. et al. "Seasonal difference in responses of body fluids to heat stress." *Jpn J Physiol* 24, no. 3 (June 1974): 249-62. Topics discussed include: blood, proteins, metabolism, body weight, chlorides, dehydration, osmolar concentration, plasma, perspiration.

605. Nag, A. et al. "Physical activities of women in farming and non farming seasons." *Journal of Human Ergology* 14, no. 2 (December 1985): 65-70. Topics discussed include: food intake, energy, India, agricultural worker, exertion, metabolism, India, nutrition.

606. Nakamura, M. et al. "Local specificity of the seasonal variation in the basal metabolic rate of Japanese." *Tohoku J Exp Med* 99, no. 2 (October 1969): 171-8. Topics discussed include: Japan, Mongoloid race, occupations, temperature, basal metabolism.

607. Neer, R. "Environmental light: effects on vitamin D synthesis and calcium metabolism in humans." *Ann N Y Acad Sci* 453 (1985): 14-20. Topics discussed include: blood, seasonality, skin, ultraviolet rays, calcium, sunlight, vitamins, biosynthesis.

608. Nikberg, I. et al. "Frequency and seasonal biorhythm of metabolic indices and vascular lesions in diabetics." *Vrach Delo* no. 11 (November 1984): 57-60. Topics discussed include: coronary disease, diabetes, hypertension epidemiology, biological cycles, metabolism.

609. Novozhilov, G. "Basal metabolism in navigation in the temperate and lower latitudes." *Voen Med Zh* 4 (April 1968): 72-5. Topics discussed include: USSR, climatic effects, naval medicine, tropical climates.

610. Palca, J. W. "Effects of Cold Ambient Temperature on Human Sleep, Metabolism, and Thermoregulation." Ph.D. diss., University of California, Santa Cruz, 1982. Topics discussed include: thermoregulation, psychology, biology; available from UMI.

611. Pandolf, K. "Aging and heat tolerance at rest or during work." *Exp Aging Res* 17, no. 3 (1991): 189-204. Topics discussed include: acute disease, body water metabolism, physical endurance, rest, stress etiology, acclimation, aging, work-related medicine.

612. Panin, L. et al. "Role of lipids in the energy supply of human adaptive reactions at high latitudes." *Vestn Akad Med Nauk SSSR* no. 6 (1979): 49-55. Topics discussed include: arctic regions, arteries, Siberia, timing, veins, adaptation, cold climates, metabolism, blood.

613. Park, H. et al. "Seasonal variations in the basal metabolic rate of the Korean." *Yonsei Med J* 10, no. 2 (1969): 139-45. Topics discussed include: Korea, thyroid function tests, basal metabolism.

614. Petrasek, R. "Influence of climatic conditions on energy and nutrient requirements." *Prog Food Nutr Sci* 2, no. 11-12 (1978): 505-14. Topics discussed include: adaptation, cold effects, temperature, metabolism, nutrition.

615. Rastogi, G. et al. "Thyroid function in changing weather in a subtropical region." *Metabolism* 25, no. 8 (August 1976): 903-8. Topics discussed include: adaptation, India, seasonality, temperature, thyroid function tests, physiology, blood, tropical climates.

616. Reiter, E. et al. "Vitamin D metabolites in adolescents and young adults with cystic fibrosis: effects of sun and season." *J Pediatr* 106, no. 1 (January 1985): 21-6. Topics discussed include: Arizona, bones, metabolism, calcium, Massachusetts, minerals, blood, sunlight.

617. Roberts, D. F. "Basal Metabolism, Race and Climate." *J Roy Anthropol Inst* 82 (1952): 169-183. Topics discussed include: biology, physionomy, anthropology.

618. Salamone L. M. et al. "Contributions of vitamin D intake and seasonal sunlight exposure to plasma-25 hydroxyvitamin-D concentration in elderly women." *American Journal of Clinical Nutrition* 59, no. 1 (January 1994): 80-86. Topics discussed include: dietetics, menopause, women, skin, bone loss, cutaneous production, biochemical, geriatric populations.

619. Samueloff, S. "Metabolic aspects of desert adaptation (man)." *Adv Metab Disord* 7, no. 0 (1974): 95-138. Topics discussed include: body temperature, calorimetry, environment, physical work, feeding behavior, heat, nutrition, physiology, respiration, perspiration, temperature, blood.

620. Sargent, F. "A Critical Homeostasis: Séason and Metabolism." *Arch Meteorol Geophys Bioklimatol* Serie B, no. 3 (1951): 289-296. Topics discussed include: biology, health.

621. ---. "Season and the Metabolism of Fat and Carbohydrate: A Study of Vestigial Physiology." *Meteorol Monographs* 2, no. 8 (1954): 68-80.

622. Sasaki, T. et al. "Proceedings: Seasonal variation in basal metabolism in the recent years." *Nippon Seirigaku Zasshi* 36, no. 8-9 (1 September 1974): 389. Topics discussed include: Japan.

623. Schmauz, R. et al. "Pulse rate, core temperature, water and electrolyte metabolism during a 10 month stay in a natural tropical climate." *Z Tropenmed Parasitol* 20, no. 3 (September 1969): 364-73. Topics discussed include: physical work, Nigeria, urine, perspiration, body regulation, tropical water, electrolytes.

624. Schrier, R. et al. "Renal, metabolic, and circulatory responses to heat and exercise. Studies in military recruits during summer training, with implications for acute renal failure." *Ann Intern Med* 73, no. 2 (August 1970): 213-23. Topics discussed include: acclimation, blood, body weight, heart rate, military medicine, proteins, respiration, seasonality.

625. Schultink, W. et al. "Body weight changes and basal metabolic rates of rural Beninese women during seasons with different energy intakes." *Eur J Clin Nutr* 44 Supp, no. 1 (1990): 31-40. Topics discussed include: adaptation, Benin, rural populations, basal metabolism, body weight, caloric intake.

626. Senay, L. et al. "Body fluid responses of heat tolerant and intolerant men to work in a hot wet environment." *J Appl Physiol* 40, no. 1 (January 1976): 55-9. Topics discussed include: acclimation, blood, protein, metabolism, heart rate, humidity, South Africa, perspiration, body physiology, thermoregulation, physical labor.

627. Shishliannikova, G. et al. "Functional state of the body and various specific functions in women engaged in construction work in hot climate." *Gig Tr Prof Zabol no.* 4 (April 1983): 22-6. Topics discussed include: energy metabolism, heart rate, work-related diseases, pregnancy, skin temperature, heat.

628. Shvartz, E. et al. "Thermal responses during training in a temperate climate." *J Appl Physiol* 36, no. 5 (May 1974): 572-6. Topics discussed include: body composition, body temperature, body weight, metabolism, heart rate, oxygen consumption, skin temperature, perspiration, physical work.

629. Sladki, E. et al. "Effect of weather and climate on the mucopolysaccharides metabolism in connective tissue, on the origin of some internal diseases and contemporary possibilities of treatment." *Przegl Lek* 24, no. 5 (31 May 1968): 461-4. Topics discussed include: etiology.

630. Slavkov, B. et al. "Studies on the effect of climatic factors on the metabolism and adaptation mechanism in summer residents vacationing at a sea shore resort, their dietary requirements and daily routine." *Nauchni Tr Vissh Med Inst Sofiia* 45, no. 3 (1966): 55-61. Topics discussed include: adaptation, Bulgaria, diet, health resorts.

631. Sollberger, A. "Meteorological effects on short term rhythmic biological phenomena in man." *Prog Biometeorol* 1Pt 1B (1974): 487-495, 699-702. Topics discussed include: age effects, astronomy, body temperature, calcium metabolism, diagnosis, cycles, timing, circadian rhythm, environment.

632. Stout, R. et al. "Seasonal variations in fibrinogen concentrations among elderly people." *Lancet* 338, no. 8758 (6 July 1991): 9-13. Topics discussed include: blood viscosity, body temperature, cardiovascular diseases, proteins, cholesterol, temperature, metabolism.

633. Sundstroem, E. S. *Studies on adaptation of man to high altitudes.* 8 volumes. University of California publications in physiology. v.5, no.5-12, Berkeley: University of California Press, 1919. Topics discussed include: man, adaptation biology.

634. Terashima, S. "Clinical observation on aseptic diarrhea in infants in winter, with special reference to the metabolism of water electrolytes." *Nippon Shonika Gakkai Zasshi* 73, no. 8 (1 August 1969): 1092-104. Topics discussed include: age effects, etiology, infant nutrition, potassium, seasonality, sodium, diarrhea, water electrolyte balance.

635. Thorp, J. "Effects of seasonal variation lipid metabolism in animals and man." *Biochem Soc Symp* 24 (1963): 163-169. Topics discussed include: adrenal glands physiology, cholesterol, fatty acids, blood, physiology, lipids.

636. Tikhomirov, I. "Characteristics of the food ration of winterers in Antarctic stations." *Vopr Pitan 1973* no. 6 (November 1973): 31-7. Topics discussed include: adaptation, Antarctic regions, dental caries, metabolism, mineral waters, seasonality, USSR, vitamins, acclimation, cold climates, diet, nutrition.

637. Tin, M. T. et al. "Energy intake and energy output of Burmese farmers at different seasons." *Hum Nutr Clin Nutr* 39, no. 1 (January 1985): 7-15. Topics discussed include: body constitution, caloric intake, food, Myanmar, skinfold thickness, agriculture, metabolism.

638. Troup, J. T. "The effect of acetylsalicylic acid administration on metabolic, cardiovascular, and thermoregulatory function : in young males during acute exercise in hot and neutral environments." Ph.D. diss., Temple University, 1983. Topics discussed include: aspirin, physiology, heat, exercise.

639. Umidova, Z. et al. "Changes in mineral metabolism in hypertension under the climatic conditions of Uzbekistan." *Ter Arkh* 39, no. 9 (September 1967): 66-9. Topics discussed include: Uzbekistan, hypertension, water electrolyte balance.

640. Vallery, M. J. et al. "Can a protracted stay in the tropics permanently lower basal metabolic rates of European expatriates?" *Ann Hum Biol* 7, no. 3 (May 1980): 267-71. Topics discussed include: Africa, Asia, Caucasoid race, France, ethnology, timing, adaptation, basal metabolism, tropical climate.

641. Vashchenkova, A. "Seasonal changes in electrolyte metabolism of patients with psoriasis." *Vestn Dermatol Venerol 1976* no. 10 (October 1976): 37-40. Topics discussed include: potassium, psoriasis, sodium.

642. Verity, C. et al. "Seasonal changes in perinatal vitamin D metabolism: maternal and cord blood biochemistry in normal pregnancies." *Arch Dis Child* 56, no. 12 (December 1981): 943-8. Topics discussed include: blood, infants, pregnancy.

643. Volovich, V. et al. "Some problems of water metabolism in man under tropical oceanic conditions." *Voen Med Zh* 6 (June 1967): 50-2. Topics discussed include: USSR, naval medicine, tropical climates, water supply.

644. Vorob'Eva, A. et al. "The relationship between changes in the character of iron balance in the bodies of children, seasonal nutrition and additional iron loading." *Pediatriia* 50, no. 5 (May 1971): 59-61. Topics discussed include: diet, iron metabolism.

645. Weiner, J. "Work capacity, thermal responses and lung function: United Kingdom studies in the L.B.P." *Philos Trans R Soc Lond Biol* 274, no. 934 (1 July 1976): 457-72. Topics discussed include: body water, metabolism, body weight, heart rate, heat, Israel, New Guinea, Nigeria, gender effects, Sweden, thermoregulation, climatic effects, physical labor.

646. Weiner, J. et al. "The effect of work level and dietary intake on sweat nitrogen losses in a hot climate." *Br J Nutr* 27, no. 3 (May 1972): 543-52. Topics discussed include: dietary proteins, nitrogen, urine, perspiration, Tanzania, diet, physical work, nitrogen, metabolism.

647. Wheeler, E. et al. "The effect of work level and dietary intake on water balance and the excretion of sodium, potassium and iron in a hot climate." *Br J Nutr* 30, no. 1 (July 1973): 127-37. Topics discussed include: acclimation, adaptation, metabolism, heat, intestinal absorption, urine, perspiration, timing, diet, work.

648. Wong, H. "Some aspects of tropical endocrine and metabolic pediatrics." *J Singapore Paediatr Soc* 30, no. 1-2 (1988): 31-45. Topics discussed include: infants, endocrine diseases, etiology, metabolic diseases, tropical climates.

649. Yamazaki, T. et al. "Changes in body temperature and metabolism in children and infants during ether and halothane anesthesia. II. Variation of the skin temperature." *Masui* 22, no. 11 (October 1973): 1277-84. Topics discussed include: anesthesia, inhalation, seasonality, pharmacology, skin.

650. Yoshida, Y. "Studies on human adaptability of climatic conditions. 1. Comparison of the seasonal variations in the basal metabolic rate of U.S. Military Personnel and Japan Self Defence Forces in Japan." *Nippon Eiseigaku Zasshi* 26, no. 6 (February 1972): 505-12. Topics discussed include: Caucasoid race, Japan, Mongoloid race, acclimation, basal metabolism, military medicine.

651. Yoshimura, H. et al. "Seasonal variations of basal metabolism and body fluid." *Prog Biometeorol* 1Pt 1B (1974): 549-556,709-711. Topics discussed include: blood volume, body water, climatic effects, diet, Japan, temperature, physiology, body fluids.

652. Yoshimura, M. et al. "Quality of diet and human adaptability to heat and cold." *Nippon Rinsho* 28, no. 7 (July 1970): 1988-93. Topics discussed include: basal metabolism, seasonality, physiology, acclimation.

653. ---. "Climatic adaptation of basal metabolism." *Fed Proc* 25, no. 4 (July 1966): 1169-76. Topics discussed include: adaptation, Canada, Caucasoid race, Japan, Mongoloid race, seasonality, physiology, climate.

THERMOREGULATION

654. Akerstedt, T. "Field studies of shiftwork: II. Temporal patterns in psychophysiological activation in workers alternating between night and day work." *Ergonomics* 20, no. 6 (November 1977): 621-631. Topics discussed include: catecholamines, biological rhythms, conscioness states, reaction time, physiological aroal, body temperature, biness, indtrial personnel.

655. Allen, J. K. "Urinary Biogenetic Amine Patterns of Response to Heat Stress on Human Exercise." Ph.D. diss., Arizona State University, 1988. Topics discussed include: chemistry, biology, temperature, thermoregulation; available from UMI.

656. Andersen, I. "The effects of moderate heat stress on patients with ischemic heart disease." *Scand J Work Environ Health* 2, no. 4 (December 1976): 256-68. Topics discussed include: thermoregulation, body weight, hemodynamics, task performance, mental processes physiology.

657. Azhaev, A. "Hygienic standards for the microclimate of manufacturing enterprises with an elevated heat release." *Gig Sanit no.* 6 (June 1985): 68-70. Topics discussed include: thermoregulation, exhation physiopathology, temperature, climatic effects, microclimate.

658. ---. "Changes in human heat exchange and nutritional status after moving to a region with a hot climate." *Fiziol Cheloveka* 16, no. 4 (July 1990): 120-4. Topics discussed include: body temperature, physiology, body weight, seasonality, skin temperature, urine chemistry.

659. ---. "Physical functional state of pilots in a hot climate." *Voen Med Zh no.* 10 (October 1980): 54-6. Topics discussed include: body temperature, skin temperature, temperature, aerospace medicine.

660. Babaev, A. "The nature of the work schedule of construction workers under hot climate conditions." *Gig Tr Prof Zabol* 11, no. 4 (April 1967): 45-7. Topics discussed include: thermoregulation, environmental exposure, environmental health, Tajikistan, heat exhation.

661. Baker, P. "Temperature regulation at high altitude: Quechua Indians and U. S. whites during total body cold exposure." *Hum Biol* 39, no. 2 (May 1967): 154-69. Topics discussed include: South America, thermoregulation, Caucasoid race, cold climates.

662. Bee, V. "Temperature rhythms in native Alaskan and Caucasian children." *Biol Psychol* 24, no. 2 (April 1987): 101-4. Topics discussed include: Alaska, Caucasoid race, darkness, light, seasonality, circadian rhythms, Mongoloid race.

663. Bergh, U. "Human power at subnormal body temperatures." Ph.D. diss., Karolinska Institutet, Stockholm, 1980. Topics discussed include: cold effects, hypothermia, thermoregulation, physical labor.

664. Better, O. "Impaired fluid and electrolyte balance in hot climates." *Kidney Int Suppl* 21 (August 1987): 97-101. Topics discussed include: thermoregulation, heat, kidney calculi etiology, water electrolyte imbalance etiology.

665. Blatteis, C. M. "Effect of Altitude Exposure on Thermoregulatory Response of Man to Cold." *J Appl Physiol* 41 (1976): 848-858. Topics discussed include: environment, thermoregulation.

666. Blavatskii, V. "Consideration of body heat regulation in conducting climatic procedures with children." *Vopr Kurortol Fizioter Lech Fiz Kult* 34, no. 2 (March 1969): 139-45. Topics discussed include: air, sunlight, swimming, thermoregulation, cold effects, health resorts.

667. Blazejczyk, K. "The influence of climatic conditions on the heat balance of the human body." *Int J Biometeorol* 35, no. 2 (September 1991): 103-6. Topics discussed include: thermoregulation physiology, climate.

668. Bochkareva, L. "Effect of living in multi story buildings on various physiological functions of the body in the IV climatic region." *Gig Sanit no.* 8 (August 1990): 79-81. Topics discussed include: circadian rhythms, seasonality, Uzbekistan, altitude, thermoregulation, physiology, population characteristics, tropical climates.

669. Bothorel, B. "Thermal and sweating responses in normal and atopic subjects under internal and moderate external heat stress." *Arch Dermatol Res* 284, no. 3 (1992): 135-40. Topics discussed include: heart rate, thermoregulation, physiology, hypersensitivity physiopathology, perspiration.

670. Buguet, A. "EEG patterns and body temperatures in man during sleep in arctic winter nights." *Int J Biometeorol* 20, no. 1 (March 1976): 61-9. Topics discussed include: circadian rhythms, heart rate, thermoregulation, cold effects, sleep physiology.

671. Burse, R. L. "Sex differences in human thermoregulatory response to heat and cold stress." *Human Factors* 21, no. 6 (December 1979): 687-699. Topics discussed include: a literature review, body weight, physiological stress, morphology.

672. Chekina, N. "Heat balance of streetcar passengers in winter." *Gig Sanit no.* 8 (August 1981): 73-4. Topics discussed include: microclimates, Russia, skin temperature, thermoregulation, railroads, seasonality.

673. Chubinskii, S. "Effect of weather elements on the human body." *Vopr Kurortol Fizioter Lech Fiz Kult no.* 6 (November 1980): 57-60. Topics discussed include: adaptation, atmospheric pressure, blood pressure, thermoregulation, humidity, seasonality, temperature.

674. Cisse, F. "Central temperature during chronic exposure to a warm climate." *Arch Int Physiol Biochim Biophys* 99, no. 4 (August 1991): 315-21. Topics discussed include: adaptation, Africa, thermoregulation, exercise physiology, Negroid race.

675. ---. "Circadian cycles of central temperature in hot climate in man." *Arch Int Physiol Biochim Biophys* 99, no. 2 (April 1991): 155-9. Topics discussed include: Senegal, adaptation, physiology, body regulation.

676. Clark, R. P. *Man and his thermal environment.* 253 pages. London: E. Arnold, 1985. Topics discussed include: temperature physiology, thermoregulation, cold physiology, heat.

677. Davy, J. "Observations on the temperature of man and other animals." 300-311. Edinburgh: Printed for A. Constable, 1825. Topics discussed include: animals, heat, vertebrates, anatomy.

678. Dedenko, I. "Interrelations between functional changes and health status and climatic factors of Extreme North (review)." *Gig Sanit no.* 7 (July 1990): 4-9. Topics discussed include: USSR, thermoregulation physiology, cardiovascular diseases, cold climates, nervous system diseases, respiratory tract diseases.

679. DeLue, N. "Climate and environmental concepts." *Clin Perinatol* 3, no. 2 (September 1976): 425-30. Topics discussed include: acoustic stimulation, thermoregulation, infants, lighting, nurseries, hospitals.

680. Drinkwater, B. "Thermoregulatory response of women to intermittent work in the heat." *J Appl Physiol* 41, no. 1 (July 1976): 57-61. Topics discussed include: blood pressure, environment, heart rate, oxygen consumption, respiration, skin temperature, timing, thermoregulation, physical labor.

681. Eisenman, P. "Hot weather, exercise, old age, and the kidneys." *Geriatrics* 41, no. 5 (May 1986): 108-10, 113-4. Topics discussed include: thermoregulation, prevention, physiopathology, heat exhaustion, hemodynamics, water electrolyte balance, physical labor, kidney physiology.

682. Ejezie, G. "Malaria parasite density and body temperature in children under 10 years of age in Calabar, Nigeria." *Trop Geogr Med* 44, no. 1-2 (January 1992): 97-101. Topics discussed include: emergency service, hospitals, fever diagnosis, causes, infants, epidemiology, seasonality.

683. Epstein, Y. "Fluid balance in hot climates: sweating, water intake, and prevention of dehydration." *Public Health Rev* 13, no. 1-2 (1985): 115-37. Topics discussed include: body fluids, physiology, thermoregulation, desert climates, drinking, heat, perspiration, water electrolyte balance.

684. Faria, I. E. "Circadian changes in resting heart rate and body temperature, maximal oxygen consumption and perceived exertion." *Ergonomics* 25, no. 5 (May 1982): 381-386. Topics discussed include: oxygen uptake, biological rhythms, respiration, energy.

685. Gaedeken, P. "Contribution Statistique à la Réaction de l'Organisme sous l'Influence Physio-Chimique des Agents Météorologiques." *Arch Anthrop Crim* 24 (1909): 173. Topics discussed include: statistics, weather, climatic effects, biology.

686. Gander, P. "Adaptation of sleep and circadian rhythms to the Antarctic summer: a question of zeitgeber strength." *Aviat Space Environ Med* 62, no. 11 (November 1991): 1019-25. Topics discussed include: physiology, body temperature, fatigue physiopathology, heart rate, New Zealand, sunlight, adaptation.

687. Girling, F. "Seasonal changes in the physiological response of man to an acute cold stress." *Can J Physiol Pharmacol* 45, no. 1 (January 1967): 13-27. Topics discussed include: environmental exposure, shivering, thermoregulation, respiration.

688. Givoni, B. "Estimation of the effect of climate on man; development of a new thermal index." Ph.D. diss., Hebrew University, Jerusalem, Technion-Israel Institute of Technology, 1963. Topics discussed include: heat, physiology.

689. Gopfert, H. "Functional relationship between skin surface and reflex muscle tonus in various body regions and under various climatic conditions." *Verh Dtsch Ges Rheumatol* 2 Suppl, no. 2 (1972): 164-71. Topics discussed include: adaptation, thermoregulation, extremities, mechanoreceptors, skin blood supply, thermoreceptors, body surface area, muscle tonus.

690. Gordon, C. "Reassessment of the neural control of body temperature: importance of oscillating neural and motor components." *Comp Biochem Physiol A* 74, no. 3 (1983): 479-89. Topics discussed include: action potentials, feedback, kinetics, motor neurons, muscles, seasonality, shivering, skin, perspiration, thermoregulation, nervous system.

691. Gregory, W. B. "Thermoregulatory boundaries to exercise in hot, humid environments." Ph.D. diss., Texas A & M University, 1979. Topics discussed include: heat, physiology, humidity, thermoregulation.

692. Gunby, S. "Hot weather blues: heat related illness." *Ga Nurse* 45, no. 4 (July 1985): 1, 3. Topics discussed include: thermoregulation, exhaustion, physiopathology.

693. Gupta, J. "Physiological responses during work in hot humid environments." *Indian J Physiol Pharmacol* 25, no. 4 (October 1981): 339-47. Topics discussed include: thermoregulation, heart rate, humidity, oxygen consumption, perspiration, physical labor, tropical climates.

694. Hampton, W. "Hypothermia in winter and high altitude sports." *Conn Med* 45, no. 10 (October 1981): 633-6. Topics discussed include: thermoregulation, hypothermia physiopathology, treatment, causes, seasonality.

695. Heerd, E. "Studies of the water vapor loss of small skin areas in human. 3. The seasonal effect on insensible perspiration." *Pflugers Arch Gesamte Physiol Menschen Tiere* 291, no. 2 (1966): 174-86. Topics discussed include: climate, thermoregulation, skin physiology, perspiration physiology.

696. Hensel, H. *Thermoreception and temperature regulation*. 321 pages. Monographs of the Physiological Society; no. 38, London ; New York: Academic Press, 1981. Topics discussed include: body regulation, thermoreceptors.

697. Honma, K. "Seasonal variation in the human circadian rhythm: dissociation between sleep and temperature rhythm." *Am J Physiol* 262, no. 5 Pt 2 (May 1992): R885-91. Topics discussed include: melatonin, blood, sleep, circadian rhythms, cycles, physiology.

698. Horne, J. "Seasonal changes in the circadian variation of oral temperature during wakefulness." *Experientia* 31, no. 11 (15 November 1975): 1296-8. Topics discussed include: circadian rhythm.

699. Hunt, E. H. "The Regulation of Body Temperature in Extremes of Dry Heat." *J Hyg* 12 (1912): 479-488. Topics discussed include: biology, environment.

700. Issaeva, L. "Seasonal changes of the chemical thermoregulation in early childhood." *Dokl Bolg Akad Nauk* 22, no. 6 (1969): 709-12. Topics discussed include: thermoregulation, oxygen consumption.

701. Ivanov, I. "Physiological and hygienic criteria for the medical selection of servicemen for work in a hot climate." *Voen Med Zh* no. 4 (April 1984): 42-4. Topics discussed include: adaptation, thermoregulation, physical examination, perspiration, military medicine, personnel management, employee selection.

702. Ivashkin, V. "The causes and mechanisms of the development of the body weight deficiency syndrome in servicemen in a hot climate and in a mountain desert locale." *Voen Med Zh no.* 7 (July 1991): 21-5. Topics discussed include: adaptation, body constitution, seasonality, temperature, timing, Turkmenistan, altitude, desert, military personnel, tropical weight loss.

703. Jauregui, E. "Wet bulb temperature and discomfort index distribution in Mexico." *Int J Biometeorol* 11, no. 1 (March 1967): 21-8. Topics discussed include: Mexico, body regulation, climate.

704. Jensen, R. C. "Relationships between several prominent heat stress indices." National Institute for Occupational Safety and Health. Division of Biomedical and Behavioral Science. U. S. Dept. of Health, Education, and Welfare, Public Health Service, Center for Disease Control, National Institute for Occupational Safety and Health Division of Biomedical and Behavioral Science, Cincinnati, 1976. Topics discussed include: heat physiology, stress physiology, body temperature, man.

705. Johnson, K. G. "Human thermoregulatory behavior during a conflict between cold discomfort and money." *Physiology and Behavior* 30, no. 1 (January 1983): 145-150. Topics discussed include: temperature, monetary incentives, thermoregulation, cold effects.

706. Juzwiak, I. "Health conditions, heart rate and body temperature in women working in hot microclimate." *Pol Tyg Lek* 33, no. 33 (14 August 1978): 1303-6. Topics discussed include: health, climatic effects, heart rate, heat, microclimates, work-related medicine.

707. Kamimura, K. "Diurnal rhythm of body temperature in night shift workers." *Nippon Eiseigaku Zasshi* 24, no. 5 (January 1970): 497-501. Topics discussed include: seasonality, timing, circadian rhythms, work.

708. Kamon, E. "Physiologic limits to work in the heat and evaporative coefficient for women." *J Appl Physiol* 41, no. 1 (July 1976): 71-6. Topics discussed include: body water, heart rate, oxygen consumption, skin temperature, adaptation, thermoregulation, physical labor, perspiration.

709. Kasparov, A. "Physiological hygiene basis for the optimum air temperature at different enclosure temperatures." *Gig Sanit no.* 10 (October 1982): 29-31. Topics discussed include: body regulation, microclimates, seasonality, air.

710. Kawasaki, T. "Circadian, circa monthly and circannual rhythm of blood pressure, oral temperature and pulse rate in a clinically healthy woman." *Nippon Naika Gakkai Zasshi* 69, no. 10 (10 October 1980): 1293-8. Topics discussed include: circadian rhythms, regression analysis, seasonality, blood pressure, cycles.

711. Khudaiberdiev, M. "Elevated skin temperature as a criterion of adaptation to the high temperature of an arid zone." *Fiziol Cheloveka* 15, no. 6 (November 1989): 78-82. Topics discussed include: physical labor, seasonality, adaptation, desert climates, skin physiology.

712. ---. "Body temperature, calcium, sodium and glucose concentrations of the blood and the blood osmotic pressure during human adaptation to high temperature." *Fiziol Cheloveka* 16, no. 4 (July 1990): 125-9. Topics discussed include: seasonality, Turkmenistan, adaptation, physiology, blood, calcium.

713. ---. "Characteristics of human body temperature regulation and the functional activity of the skin thermoreceptors during seasonal adaptation to the high temperature of an arid zone." *Fiziol Cheloveka* 15, no. 4 (July 1989): 87-91. Topics discussed include: skin physiology, Turkmenistan, adaptation, thermoregulation, desert climates.

714. ---. "Heat perception during the seasonal adaptation to high temperature." *Fiziol Cheloveka* 17, no. 2 (March 1991): 125-30. Topics discussed include: body physiology, psychophysics, skin physiology, adaptation, sense physiology.

715. ---. "The perception of a temperature rise in human seasonal heat adaptation." *Fiziol Zh SSSR* 77, no. 1 (January 1991): 116-21. Topics discussed include: humidity, psychophysics, adaptation, physiology, thermoregulation, sense physiology.

716. Kleinhanss, G. "Recording and processing of various climatic parameters." *Biomed Tech Berlin* 25, no. 1-2 (January 1980): 12-26. Topics discussed include: air movements, body temperature, heart rate, humidity, mathematics, perspiration, temperature, climate.

717. Klinker, L. "Seasonal influences on the effect of treatment in patients with functional and organic heart disease." *Arch Phys Ther Leipz* 22, no. 3 (May 1970): 135-42. Topics discussed include: body weight, cardiovascular rehabilitation, cholesterol blood, circadian rhythms, diet, sunlight, temperature, hypertension.

718. Klopov, V. "Comparative evaluation of the human thermal state under the conditions of the central Arctic and under a temporate climate." *Gig Sanit no.* 4 (April 1976): 12-5. Topics discussed include: Arctic regions, body temperature, Russia, seasonality, skin temperature, thermal conductivity, thermoregulation, cold climates.

719. Koba, F. "Nursing report. A study on the temperature and humidity in the space between the body and bedding." *Kango Tenbo* 4, no. 4 (April 1979): 29-36. Topics discussed include: humidity, beds, linens, climatic effects, microclimates.

720. ---. "A study on the temperature and humidity in the space between the body and the bedding. (2)." *Kango Tenbo* 3, no. 7 (July 1978): 31-9. Topics discussed include: beds, linens, climatic effects, humidity, microclimates.

721. Kolka, M. A. *Peripheral Changes in Regional Sweating Responses to Exercise in Hypobaric Environments*, Performer: Army Research Inst. of Environmental Medicine, Natick, MA., February 1985. Topics discussed include: perspiration, hypoxia, exercise, physiology, hyperbaric conditions, altitude.

722. Kolmodin, H. B. "Problems related to shift work. A field study of Swedish railroad workers with irregular work hours." *Scand J Work Environ Health* 1, no. 4 (December 1975): 254-62. Topics discussed include: body temperature, circadian rhythms, urine, fatigue etiology, seasonality, sleep, sodium, Sweden, timing, work-related medicine, transportation.

723. Korenevskaia, E. "The microclimate of classrooms and the thermal status of childrren in schools with various enclosure designs under conditions of the 4th climatic zone." *Gig Sanit* 32, no. 10 (October 1967): 22-8. Topics discussed include: thermoregulation, environmental exposure, USSR, health services.

724. ---. "Hygienic standardization of the microclimate in rooms for children living in various climatic zones." *Gig Sanit* 36, no. 6 (June 1971): 41-5. Topics discussed include: thermoregulation, USSR, health services.

725. Kovalenko, V. "The effect of accelerated adaptation on the work capacity of servicemen in a hot climate." *Voen Med Zh no.* 11 (November 1990): 48-51. Topics discussed include: Asia, body temperature, physical education, training methods, military personnel, tropical work.

726. Kozyreva, T. "Effect of adaptation to a hot climate on thermoreceptor function." *Fiziol Cheloveka* 10, no. 1 (January 1984): 117-9. Topics discussed include: thermoregulation, Siberia, skin temperature, Turkmenistan, urban population, adaptation, thermoreceptors physiology.

727. Krause, M. "Thermoregulation during work in high temperature conditions." *Acta Physiological Polonica* 24, no. 3 (1973): 91-100.

728. Kuindzhi, N. "Assessment of the adaptation of schoolchildren to a change in circadian stereotype in relation to the switch from summer to winter and from winter to summer time." *Gig Sanit no.* 5 (May 1983): 31-4. Topics discussed include: blood pressure, thermoregulation, heart rate, Moscow, nervous system physiology, adaptation, circadian rhythms, seasonality.

729. Kuksinskaia, T. "The significance of meteorological conditions in work of medium strain." *Gig Sanit* 31, no. 5 (May 1966): 19-24. Topics discussed include: thermoregulation, body weight, environmental exposure, work-related medicine, air, stress, temperature, weather.

730. Kulin, L. "Influence of climate in th physiologic vital functions in infantile atrophy. Milieu therapy, pathogenesis, pathophysiology, adaptation mechanism." *Arch Kinderheilkd* 82, no. 2 (January 1971): 137-46. Topics discussed include: adaptation, air conditioning, thermoregulation, heat, diseases treatment, milieu treatment, causes, muscular atrophy.

731. Kurian, G. "Diurnal variation in body temperature and performance on a spatial task." *Journal of Indian Psychology* 7, no. 1 (January 1988): 20-24. Topics discussed include: diurnal variation, task peformance, India, biological rhythms.

732. Lamke, L. "Water loss through the skin in different climatic states." *Nord Med* 84, no. 42 (15 October 1970): 1345. Topics discussed include: heat, humidity, secretory rate, sweat, water electrolyte balance, thermoregulation, skin physiology, perspiration.

733. Lee, P. "Independence of seasonal variation of growth from temperature change." *Growth* 44, no. 1 (March 1980): 54-7. Topics discussed include: child development physiology, body height, growth.

734. Lewis, D. A. "Heat, exercise and adrenergic blockade : cardiovascular and thermoregulatory responses of six young men." Ph.D. diss., Pensylvania State University, 1988. Topics discussed include: sympatholytic agents, physiology, thermoregulation, exercise.

735. Liakh, G. "Thermoregulatory characteristics in exposure to high temperature combined with low air humidity." *Gig Tr Prof Zabol no.* 8 (1992): 32-5. Topics discussed include: desert climates, Kazakhstan, microclimates, work-related exposure, perspiration, timing, thermoregulation, heat, humidity.

736. Lomax, P. *Environment, Drugs and Thermoregulation, International Symposium on the Pharmacology of Thermoregulation (5th), Saint Paul de Vence, 15 November 1982*, Performer: California Univ., Los Angeles. Dept. of Pharmacology, 1983. Topics discussed include: hyperthermia, stress, exposure physiology, low temperatures, drugs, chemotreatment, body temperature, pharmacology, thermoregulation.

737. Lukin, V. "Change in the functional state of the body of operators working in a hot, moist climate." *Voen Med Zh* 4 (April 1973): 62-3. Topics discussed include: body temperature, heat, humidity, work-related medicine.

738. Machablishvili, O. "Effect of the industrial microclimate on the thermal state of the body of women under high temperatures and humidity." *Vrach Delo* no. 6 (June 1979): 96-8. Topics discussed include: Georgia Republic, gender effects, thermoregulation, climatic effects, humidity, microclimates, textile industry.

739. Maczynski, B. "The efficiency of thermoregulation in adult patients with bronchial asthma and chronic bronchitis after climatic treatment at the Baltic seaside." *Bull Pol Med Sci Hist* 13, no. 4 (October 1970): 179-81. Topics discussed include: bathing beaches, chronic disease, treatment, Poland, skin temperature, asthma, thermoregulation, bronchitis, health resorts.

740. ---. "Thermoregulating efficiency in adult patients with bronchial asthma and chronic bronchitis after climatic treatment on the Baltic coast." *Przegl Lek* 26, no. 2 (1970): 313-6. Topics discussed include: asthma physiopathology, balneology, bronchitis physiopathology, chronic disease, temperature, asthma treatment, bronchitis treatment, climate.

741. Magbagbeola, J. "The effect of atropine premedication on body temperature of children in the tropics." *Br J Anaesth* 45, no. 11 (November 1973): 1139-42. Topics discussed include: humidity, infants, atropine pharmacology, body drug effects, preanesthetic medication, tropical climates.

742. Magomedov, M. "Seasonal changes in the thermoregulation system reactivity to temperature agents." *Vopr Kurortol Fizioter Lech Fiz Kult* 30, no. 6 (November 1965): 533-40. Topics discussed include: body regulation physiology.

743. Maruta, N. "Seasonal changes of circadian pattern in human rectal temperature rhythm under semi natural conditions." *Experientia* 43, no. 3 (15 March 1987): 294-6. Topics discussed include: circadian rhythm.

744. Matiukhin, V. "Physical development of adolescents coming to Siberia from other climatic zones." *Gig Sanit no.* 7 (July 1982): 85-7. Topics discussed include: adaptation, anthropometry, body temperature, oxygen consumption, Siberia, growth.

745. ---. "Some indices of the seasonal changes in the physical development and physiological state of the body of adolescents after moving to western Siberia." *Gig Sanit no.* 9 (September 1975): 57-60. Topics discussed include: adaptation, arctic regions, body temperature, body weight, respiration, Russia, Siberia, growth, physiology.

746. Meciani, L. "Comments on cold. How and why seasonal diseases appear that are caused by so called 'colds' (excluding infections caused by influenza viruses)." *Minerva Med* 61, no. 24 (24 March 1970): 1209-22. Topics discussed include: thermoregulation, central nervous system physiopathology, vasomotor system, causes, respiratory tract infections.

747. Metz, B. "Quantitative prediction of physiological and psychological effects of thermal environment in man : actes du colloque international, Strasbourg, 9-13 juillet 1973." Editions du Centre national de la recherche scientifique, Paris, 1975. Topics discussed include: heat, physiology, psychology, thermoregulation.

748. Millard-Stafford, M. "Female athletes may have advantage in hot weather." *American Family Physician* 46 (September 1992): 878. Topics discussed include: body temperature, women, gender differences.

749. Moe, K. "Healthy elderly women and men have different entrained circadian temperature rhythms." *J Am Geriatr Soc* 39, no. 4 (April 1991): 383-7. Topics discussed include: epidemiology, menstrual cycles, seasonality, circadian rhythms, gender issues, sleep.

750. Molnar, G. "Hypoglycemia and body temperature." *JAMA* 227, no. 8 (25 February 1974): 916-21. Topics discussed include: humidity, hypothermia, insulin pharmacology, seasonality, skin, sweat glands, perspiration, timing, toes, hypoglycemia.

751. Murakami, H. "Seasonal changes of a man's response to prolonged walks; to Rokko Mountains Traverse." *J Sports Med Phys Fitness* 22, no. 2 (June 1982): 237-44. Topics discussed include: body temperature, eating, timing, locomotion, mountaineering.

752. Nadel, E. "Peripheral involvement in thermoregulatory response to an imposed heat debt in man." *J Appl Physiol* 27, no. 4 (October 1969): 484-8. Topics discussed include: body temperature, cold effects, forearm blood supply, ice cream, thermoregulation, blood flow, skin physiology.

753. Nag, A. "Heat stress of women doing manipulative work." *Am Ind Hyg Assoc J* 53, no. 12 (December 1992): 751-6. Topics discussed include: thermoregulation, environmental exposure, heart rate, oxygen consumption, skin temperature, perspiration, stress physiopathology, work.

754. Nag, P. "Effective heat load on agricultural workers during summer season." *Indian J Med Res* 72 (September 1980): 408-15. Topics discussed include: physical labor, India, work capacity, thermoregulation, work-related medicine.

755. Nayha, S. "Adjustment of blood pressure data by season." *Scand J Prim Health Care* 3, no. 2 (May 1985): 99-105. Topics discussed include: age effects, body weight, Finland, hypertension diagnosis, temperature, blood pressure.

756. Newman, M. T. "Adaptation of Man to Cold Climates." *Evolution* 10 (1956): 101-105. Topics discussed include: climate, thermoregulation, temperature, biology, physionomy, environment.

757. Noviski, N. "Exercise intensity determines and climatic conditions modify the severity of exercise induced asthma." *Am Rev Respir Dis* 136, no. 3 (September 1987): 592-4. Topics discussed include: asthma, exercise, physiopathology, cold effects, humidity, thermoregulation, physical labor.

758. Novoselov, V. "Indices of the heat state of children travelling in railroad cars without air conditioning in a dry, hot climate." *Gig Sanit no.* 10 (October 1976): 117-8. Topics discussed include: air conditioning, USSR, thermoregulation, railroads, skin temperature, tropical climates.

759. Ohara, K. "Chloride concentration in sweat; its individual, regional, seasonal and some other variations, and interrelations between them." *Jpn J Physiol* 16, no. 3 (15 June 1966): 274-90. Topics discussed include: body temperature, skin physiology.

760. Ovcharova, V. "Current status and the prospects for the development of research in the field of medical climatology, climatotherapy and climatopathology." *Vopr Kurortol Fizioter Lech Fiz Kult no.* 6 (November 1978): 7-13. Topics discussed include: anoxia, thermoregulation, cold effects, electromagnetics, homeostasis, meteorological factors, cycles, seasonality, sunlight, USSR, weather.

761. Pafnote, M. "Correlations between heart rate, body temperature and the environment in stress situations." *Fiziol Norm Patol* 18, no. 5 (1972): 425-32. Topics discussed include: body regulation, climatic effects, physical labor, heart beat.

762. Pemberton, R. "A Summary of the Effects of External Heat upon the Human Body." *Am J; Med Sci* 169, no. 637 (April 1925): 485-489. Topics discussed include: environment, temperature, biology, physiology.

763. Pickering, G. W. "Regulation of Body Temperature in Health and Disease." *Lancet* I (1958): 1-9 and 59-64. Topics discussed include: thermoregulation, health, illness.

764. Polakoff, P. "Warming up for cold weather work." *Occup Health Saf* 51, no. 10 (October 1982): 23-5. Topics discussed include: thermoregulation, the common cold, environment, exposure, frostbite, hypothermia, work-related medicine.

765. Provins, K. A. "Environmental Heat, Body Temperature, and Behavior. A Hypothesis." *Australian Journal of Psychology* 18 (1966): 118-129. Topics discussed include: environment, biology, behavior, climate.

766. Qiu, R. "Evaluation of physiological responses of bulldozer drivers in dry and hot weather." *Chung Hua Yu Fang I Hsueh Tsa Chih* 16, no. 4 (July 1982): 209-12. Topics discussed include: body temperature, perspiration, thermoregulation, heat, work-related medicine.

767. Qualey, D. A. "The influence of sodium salicylate on thermoregulation at high ambient temperatures." M. Ed. diss., Temple University, 1983. Topics discussed include: thermoregulation, heat physiology.

768. Rafalsky, D. "BTPS compensation under different climatic conditions in bodyplethysmography." *Biomed Tech Berlin* 26, no. 5 (May 1981): 102-5. Topics discussed include: microcomputers, humidity, temperature.

769. Ratner, E. "The method of physiological assessment of local climate on the basis of its effect on the thermal status of man." *Vopr Kurortol Fizioter Lech Fiz Kult* 32, no. 5 (September 1967): 446-51. Topics discussed include: thermoregulation, health resorts.

770. ---. "Study of the hygienic features of the climate of populated regions of central and southern latitudes in the USSR." *Gig Sanit* 32, no. 4 (April 1967): 17-22. Topics discussed include: environment, exposure, health, USSR, thermoregulation, weather.

771. ---. "Thermostability of the human heat state as a basis for elaborating rational work and rest regimens in a warming microclimate." *Gig Sanit no.* 9 (September 1979): 82-4. Topics discussed include: temperature, timing, body temperature, thermoregulation, climatic effects, microclimates, relaxation, work.

772. Repin, G. "Hygienic evaluation of the microclimate and the human thermal status in the performance of light work in Siberian enterprises." *Gig Sanit no.* 7 (July 1983): 77-9. Topics discussed include: hygiene, seasonality, Siberia, thermoregulation, climatic effects, microclimates, work.

773. Rublack, K. "Integrative evaluation of heat loading due to work and climate." *Z Gesamte Hyg* 27, no. 1 (1981): 12-7. Topics discussed include: body temperature, heart rate, perspiration, temperature, work.

774. Rybicka, M. "Influence of the Baltic Sea climate on the thermoregulation of child's organism in the light of selected parameters." *Pol Tyg Lek* 25, no. 40 (5 October 1970): 1491-4. Topics discussed include: appetite, body weight, Poland, timing, asthma treatment, thermoregulation, rheumatic fever treatment, thalassotherapy.

775. Sabourin, M. "Effects of light on thermal reinforcement in behaviroal thermoregulation." *Perceptual and Motor Skills* 37, no. 2 (1973): 611-614.

776. Santa, M. L. "Physiological effects of different oxygen flow rates and ambient temperatures on pressure suited subjects performing work at altitude. NADC AC 6708." *NADC MR Rep* 22 (December 1967): 1-7. Topics discussed include: thermoregulation, respiration physiology, space flight, physical labor, oxygen.

777. Sato, H. "Seasonal variation in sweating threshold of men." *Ann Physiol Anthropol* 6, no. 1 (January 1987): 33-6. Topics discussed include: thermoregulation, perspiration.

778. ---. "Sex differences in human thermoregulatory response to cold." *Journal of Human Ergology* 17, no. 1 (September 1988): 57-65. Topics discussed include: Japan, seasonal variations, thermoregulation, cold effects, body temperature.

779. Sewitch, D. "Body temperature and sleep architecture in response to a mild cold stress in women." *Physiol Behav* 36, no. 5 (1986): 951-7. Topics discussed include: adaptation, thermoregulation, reaction time, sleep stages physiology, cold effects, sleep physiology, stress physiopathology.

780. Shpangenberg, S. "Thermoregulation changes in secondary vocational and technical school students during the summer job experience." *Probl Khig* 5 (1980): 163-7. Topics discussed include: Bulgaria, temperature, timing, thermoregulation, seasonality, vocational education, work.

781. Simonovich, T. "On thermoregulation in builders working in a hot climate." *Gig Tr Prof Zabol* 10, no. 3 (March 1966): 43-5. Topics discussed include: Uzbekistan, thermoregulation, physical labor, work-related medicine, tropical climates.

782. Slobodin, A. "A method for increasing the work capacity of military personnel under hot climatic conditions." *Voen Med Zh no.* 4 (April 1988): 44-6. Topics discussed include: body temperature, breathing, exercise, endurance, timing, USSR, military personnel, tropical climates, work capacity.

783. Smolander, J. "Individual response to physical work in the heat in relation to sweating and skin blood flow." *Int Arch Occup Environ Health* 63, no. 3 (1991): 225-6. Topics discussed include: body temperature physiology, heart rate, rectum physiology, regional blood flow, exercise physiology, skin blood supply, perspiration physiology.

784. Sohar, E. "Description and forecast of summer climate in physiologically significant terms." *Int J Biometeorol* 22, no. 2 (June 1978): 75-81. Topics discussed include: thermoregulation, humidity, Israel, desert seasonality.

785. Stamford, B. "Keeping cool during hot-weather workouts." *The Physician and Sports medicine* 21 (May 1993): 144p+. Topics discussed include: body temperature, exercise.

786. Stoffer, G. R. "Effects of contingent versus yoked temperature feedback on voluntary temperature control and cold stress tolerance." *Biofeedback and Self Regulation* 4, no. 1 (March 1979): 51-61. Topics discussed include: stress, biofeedback training, thermoregulation body, pain perception.

787. Sutton, H. "The Influence of High Temperature on the Body. Especially with Regard to Heat Stroke." *J Path Bact* 3 (1908): 63-73. Topics discussed include: environment, biology, health.

788. Suzutani, T. "Studies on medico legal diagnosis in cold district. 4. An experimental study on the rise of temperature of dead body caused by the radiation of sunlight." *Hokkaido Igaku Zasshi* 54, no. 4 (July 1979): 341-5. Topics discussed include: autolysis, environmental exposure, cold climates, forensic medicine.

789. Tedder, I. "Problems of the adaptation of workers in electrolysis shops to exposure to the industrial environment in the Far North." *Gig Sanit no.* 3 (March 1977): 91-14. Topics discussed include: arctic regions, thermoregulation, environment, exposure, Russia, seasonality, physiological drug effects, cold climates, work-related medicine.

790. Thiele, F. "Relationship between skin temperature and the insensible perspiration of the human skin." *J Invest Dermatol* 47, no. 4 (October 1966): 307-12. Topics discussed include: climate, skin physiology, perspiration physiology.

791. Thornton, J. S. "Hypothermia shouldn't freeze out cold-weather athletes." *The Physician and Sports medicine* 18 (January 1990): 109-13. Topics discussed include: body temperature, cold effects, biological effects.

792. Torii, M. "Seasonal variation of sweating rate caused by exercise during thermal transient." *J Hum Ergol Tokyo* 14, no. 1 (September 1985): 53-6. Topics discussed include: body temperature, physical labor, perspiration.

793. Tromp, S. "Biological effects of simulated high altitude climate in pressurized commercial planes on passengers and flying personnel." *Aerosp Med* 43, no. 4 (April 1972): 446-9. Topics discussed include: autonomic nervous system, blood circulation, thermoregulation, circadian rhythms, drugs, infection, respiratory airflow, time factors.

794. ---. "The possible effect of meteorological stress on cancer and its importance for psychosomatic cancer research." *Experientia* 30, no. 12 (15 December 1974): 1474-8. Topics discussed include:

adaptation, altitude, thermoregulation, circadian rhythms, climatic effects, heat, physiology, neoplasm transplantation, stress, weather.

795. ---. "Preliminary report on the thermoregulatory performance of a group of 100 rheumatic patients in the Netherlands and the possible applications in balneology and climatotherapy." *Arch Phys Ther Leipz* 17, no. 1 (January 1965): 51-9. Topics discussed include: Netherlands, balneology, thermoregulation, rheumatic diseases.

796. Ty, K. T. "Characteristics of the hygienic regulation of the optimal microclimate parameters in a hot and humid climate." *Gig Sanit no.* 7 (July 1979): 60-3. Topics discussed include: thermoregulation, seasonality, temperature, humidity prevention, microtropical climate.

797. Walter, H. "Body constitution and climate. Critical reflections on the transferability of Bergmann's rule on man." *Z Morphol Anthropol* 67, no. 3 (1976): 241-63. Topics discussed include: anthropometry, body height, body weight, cold effects, racial stocks, somatotypes, temperature, body constitution, climate.

798. ---. "Body shape and temperature regulation. Studies on Indian students." *Anthropol Anz* 35, no. 1 (October 1975): 7-14. Topics discussed include: body height, body surface area, body weight, climatic effects, India, skin, body constitution, body regulation.

799. Weiss, B. "Behavioral Thermoregulation." *Science* 113 (1961): 1338-1344. Topics discussed include: behavior, temperature.

800. Wenzel, H. "Work in very hot, dry climates: evaluation of stress and procedures for reducing strain on the human body." *Z Gesamte Hyg* 36, no. 6 (June 1990): 382-8. Topics discussed include: body temperature, work schedule tolerance, heat exhaustion prevention, work-related diseases prevention, control.

801. ---. "Evaluation of tolerance limits for humans under heat stress and the problems involved." *Scand J Work Environ Health* 15 Supp, no. 1 (1989): 7-14. Topics discussed include: humidity, physical endurance, risk factors, body temperature, physical labor, stress physiopathology.

802. Westerterp, P. M. S. "Deceleration in cumulative food intake curves, changes in body temperature and diet induced thermogenesis. Special Issue: Determinants and consequences of eating and drinking." *Physiology and Behavior* 48, no. 6 (December 1990): 831-836. Topics discussed include: food intake, solid food, meals, body weight, obesity.

803. Wirz, J. A. "Seasonality in freerunning circadian rhythms in man." *Naturwissenschaften* 71, no. 6 (June 1984): 316-9. Topics discussed include: body temperature, sleep physiology, circadian rhythms, cycles.

804. Yasuda, Y. "Seasonal variation of forearm blood flow at rest and during submaximal exercise." *Nippon Seirigaku Zasshi* 45, no. 10 (1983): 640-3. Topics discussed include: body temperature, heart rate, regional blood flow, relaxation, perspiration, physical labor, forearm blood supply.

805. Young, W. J. "Effect of Exercise and Humid Heat Upon Pulse Rate, Blood Pressure, Body Temperature and Blood Concentration." *Proc Roy Soc London, B* 91 (1919): 111-126. Topics discussed include: weather, environment, biology, physionomy.

806. Yousef, M. K. "Thermoregulatory responses to desert heat: age, race and sex." *Journal of Gerontology* 39 (July 1984): 406-14. Topics discussed include: body temperature, aging, exercise physiology, perspiration, heart rate.

807. ---. *Man in stressful environments : thermal and work physiology*. 301 pages. International Symposium on Physiology of Stressful Environments (1986 : Japan). American lecture series; publication no. 1073, Springfield, Ill., U.S.A.: C.C. Thomas, 1987. Topics discussed include: thermoregulation, environment, physical labor, heat physiology, adaptation.

4

Psychological Effects: A Select Bibliography

AFFECTIVE DISORDERS

808. Arbisi, P. "Thermoregulatory response to thermal challenge in seasonal affective disorder: a preliminary report." *Psychiatry Res* 28, no. 3 (June 1989): 323-34. Topics discussed include: physiopathology, depression, phototherapy, thermoregulation, dopamine physiology, exercise.

809. Avery, D. H. "Nocturnal temperature in affective disorder." *Journal of Affective Disorders* 4, no. 1 (March 1982): 61-71. Topics discussed include: daily rhythms, drug free patients, biological rhythms.

810. Bauer, M. S. "Defining seasonal affective disorder(s). Annual Meeting of the Society for Light Treatment and Biological Rhythms, (1992, Bethesda, Maryland)." *Biological Psychiatry* 31, no. 12 (June 1992): 1185-1189. Topics discussed include: clinical definitions.

811. Bick, P. A. "Seasonal major affective disorder." *American Journal of Psychiatry* 143, no. 1 (January 1986): 90-91. Topics discussed include: affective disturbances.

812. Boule, P. "Seasonal Affective Disorder, A Review of the Literature." Ph.D. diss., Université de Bourgogne (France), 1991. Topics discussed include: affective disorders, physiology, health issues.

813. Byerley, W. F. "Treatment of seasonal affective disorder wih morning light." *Journal of Clinical Psychiatry* 48, no. 11 (November 1987): 447-448. Topics discussed include: adults, illumination, treatment.

814. Childs, A. "Seasonal affective disorder. A sad situation." *Nurs Times* 86, no. 43 (24-30 October, 1990): 46-7. Topics discussed include: nursing, psychology, counseling, phototherapy, treatment.

815. Chung, Y. "Seasonal affective disorder. Shedding light on a dark subject." *Postgrad Med* 86, no. 5 (October 1989): 309-14. Topics discussed include: circadian rhythms, depression, diagnosis, physiopathology, melatonin physiology, treatment, phototherapy.

816. Coenen, A. M. "Phototherapy in seasonal affective disorder." *Gedrag and Gezondheid Tijdschrift voor Psychologie and Gezondheid* 16, no. 1 (April 1988): 16-21. Topics discussed include: light therapy, treatment, illumination, affective disturbances.

817. Cohen, R. "Preliminary data on the metabolic brain pattern of patients with winter seasonal affective disorder." *Arch Gen Psychiatry* 49, no. 7 (July 1992): 545-52. Topics discussed include: bipolar disorders, metabolism, physiopathology.

818. Coiro, V. "Lack of seasonal variation in abnormal TSH secretion in patients with seasonal affective disorder." *Biological Psychiatry* 35, no. 1 (1 January 1994): 36-41. Topics discussed include: hormones, endogenous depression, thyroid hormone, treatment, illness, light.

819. Dahl, K. "Dim light melatonin onset and circadian temperature during a constant routine in hypersomnic winter depression." *Acta Psychiatr Scand* 88, no. 1 (July 1993): 60-6. Topics discussed include: circadian rhythms, melatonin physiology, phototherapy, treatment, light, melatonin analysis, seasonal affective disorder psychology.

820. ---. "Phase-Typing Seasonal Affective Disorder Using a Constant Routine (Light Therapy)." Ph.D. diss., University of Washington, 1992. Topics discussed include: affective disorders, health issues; available from UMI.

821. Daimon, K. "Effects of phototherapy on circadian rhythms of body temperature in affective disorders." *Jpn J Psychiatry Neurol* 46, no. 1 (March 1992): 240. Topics discussed include: seasonal affective psychology, thermoregulation, circadian rhythms, seasonal affective therapy.

822. Deltito, J. A. "Seasonal affective disorders and phototherapy." *Suicide and Life Threatening Behavior* 22 (September 1992): 386 410-11.

823. Depue, R. A. "Effect of phototherapy on spontaneous eye blink rate in subjects with seasonal affective disorder." *American Journal of Psychiatry* 145, no. 11 (November 1988): 1457-1459. Topics discussed include: eyeblink rate, illumination, reflexes.

824. Dunham, K. "Seasonal affective disorder: light makes right." *Am J Nurs* 92, no. 12 (December 1992): 44-6. Topics discussed include: nursing, physiopathology, phototherapy, treatment.

825. Eastman, C. "Natural summer and winter sunlight exposure patterns in seasonal affective disorder." *Physiol Behav* 48, no. 5 (November 1990): 611-6. Topics discussed include: depression diagnosis, circadian rhythms, psychology.

826. Eastwood, J. "Winter blues seasonal affective disorder." *Occup Health Lond* 44, no. 2 (February 1992): 44, 46. Topics discussed include: therapy, phototherapy methods, diagnosis.

827. Eastwood, M. R. "Epidemiology and Seasonal Affective Disorder." *Psychological Medicine* 18, no. 4 (November 1988): 799-806. Topics discussed include: major depression, history.

828. Elmore, S. "Body temperature and diurnal type in women with seasonal affective disorder." *Health Care Women Int* 14, no. 1 (January 1993): 17-26. Topics discussed include: phototherapy, treatment, timing, circadian rhythms, physiopathology.

829. Endo, T. "Seasonal variations of the circadian rhythms in seasonal affective disorder." *Jpn J Psychiatry Neurol* 45, no. 1 (March 1991): 178-9. Topics discussed include: thermoregulation, drug effects, physiology, blood, phototherapy, psychology, treatment, sleep, tryptophan.

830. Fluvia, X. "Seasonal affective disorder." *Revista de Psiquiatria de la Facultad de Medicina de Barcelona* 14, no. 6 (November 1987): 309-318. Topics discussed include: phototherapy, a literature review, illumination, treatment.

831. Ford, K. "A seasonal depression. Management of seasonal affective disorder." *Prof Nurse* 8, no. 2 (November 1992): 94-8. Topics discussed include: melatonin, metabolism, phototherapy, treatment.

832. Glod, C. "Seasonal affective disorder: a new light?" *J Psychosoc Nurs Ment Health Serv* 29, no. 3 (March 1991): 38-9. Topics discussed include: therapy, depressive psychology, treatment, phototherapy instrumentation.

833. Goyer, P. "Cerebral glucose metabolism in patients with summer seasonal affective disorder." *Neuropsychopharmacology* 7, no. 3 (November 1992): 233-40. Topics discussed include: tomography, glucose.

834. Graw, P. "Diurnal variation of symptoms in seasonal affective disorder." *Psychiatry Research* 37, no. 1 (April 1991): 105-111. Topics discussed include: prediction, light therapy, biological rhythms, phototherapy, relapse disorders.

835. Gupta, R. "Alternative patterns of seasonal affective disorder: Three case reports from north India." *American Journal of Psychiatry* 145, no. 4 (April 1988): 515-516. Topics discussed include: India, affective disturbances.

836. Hanada, K. "Clinical features of seasonal affective disorder in Japan: an interim report for the second year of multicenter collaborative study." *Jpn J Psychiatry Neurol* 45, no. 1 (March 1991): 182. Topics discussed include: phototherapy, psychology, treatment, diagnosis.

837. Hansen, S. "Treatment of seasonal affective disorder depression with light." *Wis Med J* 86, no. 2 (February 1987): 9-10. Topics discussed include: seasonality, treatment, phototherapy.

838. Haug, H. "On the frequency of 'diurnal variations of mood' in psychiatric patients seasonal aspects." *Prog Clin Biol Res.* 341 (1990): 105-11. Topics discussed include: depressive disorder psychology, cycles, affect physiology, circadian rhythms, mental disorders.

839. Hawkins, L. "Seasonal affective disorders: the effects of light on human behaviour." *Endeavour* 16, no. 3 (September 1992): 122-7. Topics discussed include: body temperature, blood, melatonin, epidemiology, physiopathology, circadian rhythms, light, etiology.

840. Heleniak, E. "Seasonal affective disorder: lower or raised resting metabolic rate?" *Med Hypotheses* 40, no. 6 (June 1993): 367. Topics discussed include: basal metabolism, cold effects, light, melatonin, biology.

841. Hellekson, C. "Phenomenology of seasonal affective disorder: An Alaskan perspective." *Seasonal affective disorders and phototherapy*, 33-45, ix, 386. Norman E. Rosenthal, Mary C. Blehar. New York, NY, US: Guilford Press, 1989. Topics discussed include: phenomenology, Alaska, suicide.

842. Hellekson, C. J. "Phototherapy for seasonal affective disorder in Alaska." *American Journal of Psychiatry* 143, no. 8 (August 1986): 1035-1037. Topics discussed include: artificial light, depression, Alaska, emotion, illumination, treatment.

843. Hotopf, M. "Seasonal affective-disorder, environmental hypersensitivity and somatization." *British Journal of Psychiatry* no. 164 (February 1994): 246-248.

844. Hunt, N. "Seasonal affective disorder following brain injury." *British Journal of Psychiatry* 156 (June 1990): 884-886. Topics discussed include: brain damage.

845. Isaacs, G. "Phototherapy and its mechanisms of action in seasonal affective disorder." *Journal of Affective Disorders* 14, no. 1 (January 1988): 13-19. Topics discussed include: photoperiod, bright light, dim light, light augmentation, depression, affective disturbances.

846. Jacobsen, F. "Morning versus midday phototherapy of seasonal affective disorder." *Am J Psychiatry* 144, no. 10 (October 1987): 1301-5. Topics discussed include: depressive psychology, health care, circadian rhythms, depressive therapy.

847. ---. "Seasonal affective disorder: a review of the syndrome and its public health implications." *Am J Public Health* 77, no. 1 (January 1987): 57-60. Topics discussed include: etiology, psychology, treatment, lighting.

848. Jacobsen, F. M. "The role of serotonin in seasonal affective disorder and the antidepressant response to phototherapy." *Seasonal affective disorders and phototherapy.*, pp. 333-341. eds Norman E. Rosenthal, Mary C. Blehar., New York, NY, US: Guilford Press, 1989. Topics discussed include: seasonal variations, serotonin, illumination, treatment.

849. ---. "Seasonal affective disorder and the use of light as an antidepressant." *Affective disorders. Directions in psychiatry monograph series, No. 3.*, pp. 215-229. W. W. Norton & Co, Inc, New

York, 238 pages. ed Frederic Flach. 1988. Topics discussed include: affective disturbances, depression, emotion, illumination.

850. Jain, S. "Seasonal relapses in affective disorder in the Tropics: a prospective follow up of 12 patients." *Psychopathology* 25, no. 3 (1992): 166-72. Topics discussed include: light, manic disorders, drug therapy, tranquilizers.

851. James, S. P. "Treatment of seasonal affective disorder with light in the evening." *British Journal of Psychiatry* 147 (October 1985): 424-428. Topics discussed include: bright light, dim light, illumination, biological rhythms, sleep deprivation, treatment.

852. Joerres, S. "Seasonal affective disorder in a spinal cord injury population." *J Am Paraplegia Soc* 15, no. 2 (April 1992): 66-70. Topics discussed include: pain, physiopathology, psychology.

853. Kasper, S. "Phototherapy in subsyndromal seasonal affective disorder (SSAD) and 'diagnosed' controls." *Pharmacopsychiatry* 21, no. 6 (November 1988): 428-9. Topics discussed include: psychology, treatment.

854. ---. "Recurrent brief depression and its relationship to seasonal affective disorder." *Eur Arch Psychiatry Clin Neurosci* 242, no. 1 (1992): 20-6. Topics discussed include: disorders, genetics, diagnosis, psychology.

855. Kathol, R. "Circannual rhythm and peak frequency of corticosteroid excretion: relationship to affective disorder." *Psychiatr Med* 3, no. 1 (1985): 53-63. Topics discussed include: depressive disorders, diagnosis, blood, seasonality, adrenal cortex hormone, cycles.

856. King, J. R. "Seasonal affective disorder." *British Journal of Psychiatry* 148 (April 1986): 478. Topics discussed include: light, winter depression, emotion, illumination, treatment.

857. Klein, L. "Seasonal affective disorder : January 1986 through December 1991 : 402 citations." National Library of Medicine (U.S.), Reference Section. U.S. Dept. of Health and Human Services, Public Health Service, National Institutes of Health, Supt. of Docs., U.S. G.P.O. [distributor, Bethesda, MD (8600 Rockville Pike, Bethesda 20894)] Washington, DC, 1992. Topics discussed include: depression, mental illness.

858. Koehler, K. "Season of birth and Schneider oriented diagnosis of affective disorder." *Psychiatria Clinica* 9, no. 3-4 (1976): 212-219. Topics discussed include: affective disturbances.

859. Krauchi, K. "The four seasons: food intake frequency in seasonal affective disorder in the course of a year." *Psychiatry Res* 25, no. 3 (September 1988): 323-38. Topics discussed include: body weight, drinking, food preferences, depressive psychology, eating, food habits.

860. ---. "The relationship of affective state to dietary preference: Winter depression and light therapy as a model." *Journal of Affective Disorders* 20, no. 1 (September 1990): 43-53. Topics discussed include: eating habits, winter, seasonal affective disorders, carbohydrates, seasonality.

861. Krauss, S. S. "Behavioral engagement level, variability, and diurnal rhythm as a function of bright light in bipolar II seasonal affective disorder: An exploratory study." *Psychiatry Research* 43, no. 2 (August 1992): 147-160. Topics discussed include: light, diurnal rhythms, phototherapy, biological rhythms, emotions.

862. Kripke, D. F. "Phototherapy for nonseasonal major depressive disorders." 342-356. *Seasonal Affective Disorders and Phototherapy,* eds. Norman E. Rosenthal and Mary C. Blehar, New York, Guilford Press, 1989. Topics discussed include: depression, perception, illumination, treatment.

863. Lahaie, U. "Seasonal affective disorder." *Can J Psychiatr Nurs* 31, no. 1 (January 1990): 9-13. Topics discussed include: nursing, treatment, physiopathology, phototherapy.

864. ---. "Shift workers and seasonal affective disorder." *Can Nurse* 87, no. 5 (May 1991): 33-4. Topics discussed include: etiology, treatment, phototherapy methods, work schedules, psychology, epidemiology.

865. Lahmeyer, H. "Heart rate and temperature changes during exposure to bright light in seasonal affective disorder." *Prog Neuropsychopharmacol Biol Psychiatry* 12, no. 5 (1988): 763-72. Topics discussed include: radiation effects, heart rates, light, physiopathology.

866. Lam, R. W. "Seasonal affective disorder: A Canadian sample." *Annals of Clinical Psychiatry* 1, no. 4 (December 1989): 241-245. Topics discussed include: Canada, diagnosis, treatment.

867. Lam, R. "Seasonal affective disorder presenting as chronic fatigue syndrome." *Can J Psychiatry* 36, no. 9 (November 1991): 680-2. Topics discussed include: fatigue, diagnosis, phototherapy, disorders, psychology.

868. Levendosky, A. "Core body temperature in patients with seasonal affective disorder and normal controls in summer and winter." *Biol Psychiatry* 29, no. 6 (15 March 1991): 524-34. Topics discussed include: depressive disorders, diagnosis, psychology, energy metabolism, physiology, thyrotropin, blood, thermoregulation, circadian rhythms, physiopathology.

869. Lewy, A. J. "Winter depression and the phase shift hypothesis for bright lights therapeutic effects: history, theory, and experimental evidence." *Seasonal Affective Disorders and Phototherapy*, eds. Norman E. Rosenthal and Mary C. Blehar, New York, NY. US: Guilford Press, 1989. Topics discussed include: perception, biologic rhythms, illumination, light adaptation.

870. Lewy, A. J. "The phase-shift hypothesis of seasonal affective disorder." *The American Journal of Psychiatry* 145 (August 1988): 1041-2.

871. ---. "Winter depression and the phase shift hypothesis for bright light's therapeutic effects: History, theory, and experimental evidence. Special Issue: Seasonal affective disorder: Mechanisms, treatments, and models." *Journal of Biological Rhythms* 3, no. 2 (June 1988): 121-134. Topics discussed include: light, abnormal circadian rhythms, biological rhythms, treatment, illumination.

872. Lucas, C. "Seasonal affective disorder in adolescence." *Br J Psychiatry* 159 (December 1991): 863-5. Topics discussed include: circadian rhythms, England, phototherapy, relaxation techniques, psychology, treatment, sunlight, diagnosis.

873. Martin, N. "Uncommon disorders. Summer seasonal affective disorder." *Nurs Stand* 6, no. 41 (1992 July 1-7): 32-5. Topics discussed include: diagnosis, treatment, hypothalamus physiology, hypothalamus physiopathology, cycles.

874. McIntyre, I. "Treatment of seasonal affective disorder with light: preliminary Australian experience." *Aust N Z J Psychiatry* 23, no. 3 (September 1989): 369-72. Topics discussed include: Australia, depressive psychology, treatment, phototherapy methods.

875. Mghir, R. "Phototherapy of seasonal affective disorder in an adolescent female." *J Am Acad Child Adolesc Psychiatry* 30, no. 3 (May 1991): 440-2. Topics discussed include: age effects, causes, depressive psychology, depressive therapy.

876. Mirabile, C. J. "Separation of affective disorder into seasonal and nonseasonal types using motion sickness susceptibility as a marker." *J Neuropsychiatry Clin Neurosci* 5, no. 3 (June 1993): 330-4. Topics discussed include: rehabilitation, hospitalization, patient admission, phototherapy, treatment, diagnosis.

877. Mishima, K. "Circadian rhythm of melatonin secretion and body temperature in a patient with seasonal affective disorder." *Jpn J Psychiatry Neurol* 45, no. 1 (March 1991): 161. Topics discussed include: bipolar psychology, blood, thermoregulation physiology, circadian rhythm physiology, melatonin.

878. Moreines, R. N. *Light up Your Blues: Understanding & Overcoming Seasonal Affective Disorders.* Summit, NJ: PIA Press, 1989. Topics discussed include: affective disorders, psychology.

879. Morin, G. "Seasonal affective disorder, the depression of winter: a literature review and description from a nursing perspective." *Arch Psychiatr Nurs* 4, no. 3 (June 1990): 182-7. Topics discussed include: nursing, physiopathology, treatment, circadian rhythms, phototherapy.

880. Murray, J. B. "Geophysical variables and behavior: LVII. Seasonal affective disorder and phototherapy." *Psychological Reports* 64, no. 3, Pt 1 (June 1989): 787-801. Topics discussed include: patient characteristics, a literature review, illumination, treatment, affective disturbances.

881. Nayha, S. "Mortality and seasonal change in Finland." *Duodecim* 105, no. 5 (1989): 465-70. Topics discussed include: affective disorders, cardiovascular diseases, epidemiology, Finland, respiratory tract diseases.

882. O'Brien, J. "Seasonal affective disorder." *Practitioner* 233, no. 1479 (22 November 1989): 1575-8. Topics discussed include: diagnosis, physiopathology, treatment, diagnosis.

883. Partonen, T. "Effects of light treatment on sleep structure in seasonal affective disorder." *Eur Arch Psychiatry Clin Neurosci* 242, no. 5 (1993): 310-3. Topics discussed include: ambulatory care, circadian rhythms, phototherapy, diagnosis, rehabilitation.

884. Potkin, S. "Seasonal affective disorder: prevalence varies with latitude and climate." *Clin Neuropharmacol.* 9 Supp, no. 4 (1986): 181-3. Topics discussed include: etiology, epidemiology, light, cycles.

885. Rieth, M. "Seasonal affective disorder: update 1988." *S D J Med* 41, no. 9 (September 1988): 17-20. Topics discussed include: antidepressive agents, treatment, phototherapy, depressive therapy.

886. Rosen, L. N. "Patterns of seasonal change in mood and behavior: An example from a study of military wives." *Military Medicine* 156, no. 5 (May 1991): 228-230. Topics discussed include: symptoms, Germany, army personnel, seasonal affective disorder, geography, emotion, changes in behavior.

887. Rosen, L. "Prevalence of seasonal affective disorder at four latitudes." *Psychiatry Res* 31, no. 2 (February 1990): 131-44. Topics discussed include: cross-sectional analysis, depressive psychology, risk, epidemiology, depressive epidemiology, geography.

888. Rosenthal, N. "Seasonal affective disorder in children and adolescents." *Am J Psychiatry* 143, no. 3 (March 1986): 356-8. Topics discussed include: age effects, counseling, phototherapy, student psychology, diagnosis.

889. ---. "Seasonal affective disorder. A description of the syndrome and preliminary findings with light therapy." *Arch Gen Psychiatry* 41, no. 1 (January 1984): 72-80. Topics discussed include: bipolar psychology, depression, eating disorders, hypersomnia, sleep physiology, diagnosis, phototherapy.

890. ---. "Seasonal affective disorder & light: past, present and future." *Clin Neuropharmacol.* 9 Supp, no. 4 (1986): 193-5. Topics discussed include: physiopathology, treatment, biomechanics, phototherapy, causes, light.

891. ---. "Biological effects of morning plus evening bright light treatment of seasonal affective disorder." *Psychopharmacol Bull* 23, no. 3 (1987): 364-9. Topics discussed include: psychology, timing, treatment, phototherapy.

892. ---. *Winter Blues: Seasonal Affective Disorder: What It is & How to Overcome It.* New York, NY: Guilford Press, 1993. Topics discussed include: affective disorders, psychology.

893. ---. *Seasonal Affective Disorder & Phototherapy.* New York, NY: Guilford Press, 1989. Topics discussed include: affective disorders, psychology.

894. Rosenthal, N. E. "Seasonal affective disorder and visual impairment: Two case studies." *Journal of Clinical Psychiatry* 50, no. 12 (December 1989): 469-472. Topics discussed include: cataracts, populations at risk.

895. ---. "Antidepressant effects of light in seasonal affective disorder." *American Journal of Psychiatry* 142, no. 2 (February 1985): 163-170. Topics discussed include: light treatment, winter, depression, emotion, illumination.

896. ---. "Seasonal affective disorder and phototherapy." *Annals of the New York Academy of Sciences* 453 (1985): 260-269. Topics discussed include: affective disturbances, treatment, photopic stimulation.

897. ---. "Disturbances of appetite and weight regulation in seasonal affective disorder." *Annals of the New York Academy of Sciences* 499 (June 1987): 216-230. Topics discussed include: seasonality, appetite, changes in body weight, phototherapy, reversible obesity.

898. ---. "Seasonal affective disorder: Drs. Rosenthal and James reply." *British Journal of Psychiatry* 148 (April 1986): 478-479. Topics discussed include: light, illumination, treatment, affective disturbances.

899. ---. "Effects of light treatment on core body temperature in seasonal affective disorder." *Biological Psychiatry* 27, no. 1 (January 1990): 39-50. Topics discussed include: light, illumination, treatment, physiology.

900. ---. "Seasonal affective disorders." National Institutes of Health (U.S.). U.S. Dept. of Health and Human Services, National Institutes of Health, Bethesda, MD, 1989. Topics discussed include: depression, mental illness.

901. ---. "Towards understanding the mechanism of action of light in seasonal affective disorder." *Pharmacopsychiatry* 25, no. 1 (January 1992): 56-60. Topics discussed include: pathophysiology, light therapy, phototherapy, etiology.

902. Sandyk, R. "Seasonal panic disorder: a possible variant of seasonal affective disorder." *Int J Neurosci* 62, no. 3-4 (February 1992): 263-7. Topics discussed include: anxiety, circadian rhythms, treatment, diagnosis, melatonin, blood, panic psychology, phototherapy, psychotherapy.

903. Shah, A. "Seasonal affective disorder in an elderly woman." *International Journal of Geriatric Psychiatry* 7, no. 11 (November 1992): 847. Topics discussed include: diagnosis.

904. Simonds, J. "Seasonal affective disorder: clinical aspects." *Tex Med* 84, no. 11 (November 1988): 35-9. Topics discussed include: antidepressive agents, drug therapy, psychology.

905. Skwerer, R. G. "Neurobiology of seasonal affective disorder and phototherapy. Special Issue: Seasonal affective disorder: Mechanisms, treatments, and models." *Journal of Biological Rhythms* 3, no. 2 (June 1988): 135-154. Topics discussed include: neurobiology, a literature review, biological rhythms, treatment, neurobiology, illumination.

906. ---. "Neurobiology of seasonal affective disorder and phototherapy." *Seasonal affective disorders and phototherapy*, 311-332, 386 pages. eds. Norman E. Rosenthal, and Mary C. Blehar. New York: Guilford Press, 1989. Topics discussed include: seasonality, neurobiology, illumination, treatment, neurotransmitters, sleep.

907. Smyth, A. *Seasonal Affective Disorder: Winter Depression: Who Gets It, What Causes It, How to Cure It*. San Francisco, CA: Thorsons, 1992. Topics discussed include: affective disorders, psychology.

908. Sonis, W. "The antidepressant effect of light in seasonal affective disorder of childhood and adolescence." *Psychopharmacol Bull* 23, no. 3 (1987): 360-3. Topics discussed include: psychology, treatment, phototherapy.

909. Sonis, W. A. "Seasonal affective disorder of childhood and adolescence: A review." *Seasonal affective disorders and phototherapy*, 46-54, 386 pages. eds Norman E. Rosenthal, and Mary C. Blehar. New York: Guilford Press, 1989. Topics discussed include: seasonality, adolescents, causes, phenomenology, children.

910. Spitzer, R. L. "The validity of seasonal affective disorder." *Seasonal affective disorders and phototherapy.*, 79-84, 386 pages. eds Norman E. Rosenthal, and Mary C. Blehar. New York: Guilford Press, 1989. Topics discussed include: seasonality, psychodiagnostic typologies.

911. Stewart, J. "Is seasonal affective disorder a variant of atypical depression? Differential response to light therapy." *Psychiatry Res* 33, no. 2 (August 1990): 121-8. Topics discussed include: diagnosis, psychology, psychometrics, treatment, phototherapy.

912. Stumpf, W. E. "Light, vitamin D and psychiatry: Role of 1,25 dihydroxyvitamin D sub 3 (soltriol) in etiology and therapy of seasonal affective disorder and other mental processes." *Psychopharmacology* 97, no. 3 (March 1989): 285-294. Topics discussed include: vitamins, causes, phototherapy, behavioral processes, a literature review, illumination, physiology, etiology.

913. Swade, C. "Seasonal variations in biochemical factors related to depressive illness." *J Affect Disord* 2, no. 4 (December 1980): 249-55. Topics discussed include: depressive disorders, drug therapy, folic acid, blood, serotonin.

914. Swiecicki, L. "Seasonal affective disorder and phototherapy." *Psychiatr Pol* 24, no. 5 (September 1990): 38-45. Topics discussed include: circadian rhythms, light, timing, treatment.

915. Tan, L. "Seasonal affective disorder. Winter depression." *Nurs Times* 86, no. 43 (1990 October 24-30): 44-5. Topics discussed include: etiology, treatment, phototherapy, nursing.

916. Terman, M. "Light therapy for seasonal affective disorder. A review of efficacy." *Neuropsychopharmacology* 2, no. 1 (March 1989): 1-22. Topics discussed include: circadian rhythms, multicenter studies, treatment, phototherapy.

917. ---. "On the question of mechanism in phototherapy for seasonal affective disorder: Considerations of clinical efficacy and epidemiology. Special Issue: Seasonal affective disorder: Mechanisms, treatments, and models." *Journal of Biological Rhythms* 3, no. 2 (June 1988): 155-172. Topics discussed include: clinical efficacy, treatment, illumination.

918. ---. "On the question of mechanism in phototherapy for seasonal affective disorder: Considerations of clinical efficacy and epidemiology." *Seasonal affective disorders and phototherapy*, 357-376, 386 pages. eds Norman E. Rosenthal, and Mary C. Blehar. New York, Guilford Press, 1989. Topics discussed include: seasonality, treatment, illumination.

919. Thase, M. E. "Comparison between seasonal affective disorder and other forms of recurrent depression. " *Seasonal affective disorders and phototherapy*, pp. 64-78., New York: Guilford Press, ix, 386. eds Norman E. Rosenthal, and Mary C. Blehar. 1989. Topics discussed include: phenomenology, epidemiology, family characterisitics.

920. Thompson, C. "A comparison of normal, bipolar and seasonal affective disorder subjects using the Seasonal Pattern Assessment Questionnaire." *J Affect Disord* 14, no. 3 (May 1988): 257-64. Topics discussed include: bipolar diagnosis, psychometrics, bipolar psychology, depressive psychology.

921. ---. "Seasonal affective disorder and season dependent abnormalities of melatonin suppression by light." *Lancet* 336, no. 8717 (22 September 1990): 703-6. Topics discussed include: blood, causes, trials, metabolism, timing, light, cycles, phototherapy.

922. ---. "Seasonal affective disorder a British sample: Symptomatology in relation to mode of referral and diagnostic subtype." *Journal of Affective Disorders* 14, no. 1 (January 1988): 1-11. Topics discussed include: treatment, symptoms, UK.

923. ---. "'Seasonal affective disorder a British sample: Symptomatology in relation to mode of referral and diagnostic subtype': Corrigendum." *Journal of Affective Disorders* 20, no. 4 (December 1990): 251. Topics discussed include: treatment, symptoms, UK.

924. Tian, Z. "Current research on seasonal affective disorder." *Chung Hua Shen Ching Ching Shen Ko Tsa Chih* 23, no. 2 (April 1990): 117-9. Topics discussed include: psychoses, diagnosis, treatment, phototherapy.

925. Volf, N. "Hemispheric language lateralization in seasonal affective disorder and light treatment." *Psychiatry Res* 47, no. 1 (April 1993): 99-108. Topics discussed include: physiology, motor skills, reaction time, psychology, treatment, attention, cerebral physiology, physiopathology, verbal issues.

926. Webb, M. "Response to phototherapy of an elderly patient with seasonal affective disorder." *American Journal of Psychiatry* 145, no. 12 (December 1988): 1607-1608. Topics discussed include: depression, illumination, treatment.

927. Wehr, T. A. *Seasonal Affective Disorders: A Historical Overview.* 11-32. *Seasonal Affective Disorders and Phototherapy,* eds. Norman E. Rosenthal and Mary C. Blehar, New York: Guilford Press, 1989. Topics discussed include: history, neuroses, perception, treatment.

928. ---. "Summer Depression: Description of the Syndrome and Comparison with Winter Depression." 55-63. *Seasonal Affective Disorders and Phototherapy,* eds. Norman E. Rosenthal and Mary C. Blehar, New York: Guilford Press, 1989. Topics discussed include: perception, heat effects.

929. Wehr, T. "Seasonality and affective illness." *Am J Psychiatry* 146, no. 7 (July 1989): 829-39. Topics discussed include: therapy, depression, environment, phototherapy, etiology.

930. ---. "Seasonal affective disorder with summer depression and winter hypomania." *Am J Psychiatry* 144, no. 12 (December 1987): 1602-3. Topics discussed include: epidemiology, temperature, causes.

931. Wehr, T. A. "Reverse seasonal affective disorder is identified." *American Family Physician* 36 (August 1987): 208. Topics discussed include: diseases, seasonality, psychoses.

932. Williams, R. "Frequency of seasonal affective disorder among individuals seeking treatment at a northern Canadian mental health center." *Psychiatry Res* 46, no. 1 (January 1993): 41-5. Topics discussed include: bipolar epidemiology, psychology, mental health services, diagnosis, Manitoba, risk, social environment.

933. Wirz, J. A. "Light treatment of seasonal affective disorder in Switzerland." *Acta Psychiatr Scand* 74, no. 2 (August 1986): 193-204. Topics discussed include: circadian rhythms, depressive psychology, Switzerland, depressive therapy, lighting.

934. ---. "Seasonality in biochemical determinations: a source of variance and a clue to the temporal incidence of affective illness." *Psychiatry Res* 1, no. 1 (July 1979): 53-60. Topics discussed include: biological clocks, melatonin, serotonin, tryptophan, blood, epidemiology.

935. Wolfe, M. L. "Nursing students' anticipation of experiencing symptoms of seasonal affective disorder." *Perceptual and Motor Skills* 71, no. 3, Pt 2 (December 1990): 1389-1390. Topics discussed include: symptoms, expectations.

936. Yamashita, T. "A case of reversed seasonal affective disorder." *Jpn J Psychiatry Neurol* 46, no. 1 (March 1992): 250-2. Topics discussed include: light, melatonin, blood, genetics, psychology, sensory deprivation, diagnosis.

AGGRESSION

937. Anderson, A. C. "Environmental factors and aggressive behavior." *Journal of Clinical Psychiatry* 43, no. 7 (July 1982): 280-283. Topics discussed include: pollution, drugs, alcohol, suicide, a literature review, ecology.

938. Anderson, C. A. "Temperature and aggression: Ubiquitous effects of heat on occurrence of human violence." *Psychological Bulletin* 106, no. 1 (July 1989): 74-96. Topics discussed include: heat effects, violent behavior, a literature review.

939. ---. "Temperature, aggression, and the negative affect escape model." *Psychological Bulletin* 111, no. 2 (March 1992): 347-351. Topics discussed include: curvilinear models, heat, violent behavior, a literature review.

940. Baron, R. A. "Aggression and Heat: The Mediating Role of Negative Effect." *Journal of Applied Social Psychology* 6 (1976): 18-30. Topics discussed include: violence, temperature, behavior.

941. Baron, R. "Environmental Influences on Aggression: The Facilitation of Modeling Effects by High Ambient Temperatures." *Psychonomic Science* 26 (1972): 80-82. Topics discussed include: crime, behavior, environment, weather.

942. Baron, R. A. "Aggression and Heat: The 'Long, Hot Summer' Revisited." *Advances in Environmental Psychology*, editors A. Baum, J. E. Singer, and S. Valins. Hillsdale,NJ: Erlbaum, 1979. Topics discussed include: aggression, weather, temperature, violence, psychology.

943. Baron, R. "Ambient Temperature and the Occurence of Collective Violence: The 'Long, Hot Summer' Revisited." *Journal of Personality and Social Psychology* 36 (1978): 351-360. Topics discussed include: violence, behavior, environment, weather.

944. ---. "Aggression as a function of ambient temperature and prior anger arousal." *J Pers Soc Psychol* 21, no. 2 (February 1972): 183-9. Topics discussed include: attitude, civil disorders, cold effects, electroshock, heat effects, hostility, inhibition psychology, seasonality, anger.

945. Baron, R. A. "Aggression and heat: Mediating effects of prior provocation and exposure to an aggressive model." *Journal of Personality and Social Psychology* 31, no. 5 (May 1975): 825-832. Topics discussed include: ambient temperature, behavior, emotional responses, anger.

946. ---. "Aggression and heat: The influence of ambient temperature, negative affect, and a cooling drink on physical aggression." *Journal of Personality and Social Psychology* 33, no. 3 (March 1976): 245-255. Topics discussed include: cooling drinks, emotional states, behavior.

947. Bell, P. A. "In defense of the negative affect escape model of heat and aggression." *Psychological Bulletin* 111, no. 2 (March 1992): 342-346. Topics discussed include: escape models, violence, a literature review.

948. ---. "Aggression and heat: The mediating role of negative affect." *Journal of Applied Social Psychology* 6, no. 1 (January 1976): 18-30. Topics discussed include: room temperature, attitude, anger, aggressive behavior.

949. ---. "Aggression and ambient temperature: The facilitating and inhibiting effects of hot and cold environments." *Bulletin of the Psychonomic Society* 9, no. 6 (June 1977): 443-445. Topics discussed include: cold effects, heat, aggressive behavior.

950. ---. "Heat and violence in the Dallas field data: Linearity, curvilinearity, and heteroscedasticity." *Journal of Applied Social Psychology* 19, no. 17, Pt 1 (December 1989): 1479-1482. Topics discussed include: ambient temperature, time of day, aggression, escape models, aggressive behavior.

951. Boyanowsky, E. O. "Toward a Thermoregulatory Model of Violence." *Journal of Environmental System* 1 (1981): 81-87. Topics discussed include: temperature, violence.

952. Carlsmith, J. M. "Ambient Temperature and the Occurence of Collective Violence: A New Analysis." *Journal of Personality and Social Psychology* 37 (1979): 337-344. Topics discussed include: weather, violence.

953. Goodwin, F. "Suicide, aggression, and depression. A theoretical framework for future research." *Ann N Y Acad Sci* 487 (1986): 351-5. Topics discussed include: brain metabolism, disorder genetics, risk, seasonality, psychology.

954. Gunter, B. "The Peace of a Long Cool Summer." *New Society* 45, no. 829 (25 August 1978): 394-395. Topics discussed include: weather, aggression, temperature.

955. Haertzen, C. "Seasonal changes in rule infractions among prisoners: A preliminary test of the temperature aggression hypothesis." *Psychological Reports* 72, no. 1 (February 1993): 195-200. Topics discussed include: seasonal changes, aggressive behavior.

956. Harries, K. D. "Heat and violence: New findings from Dallas field data, 1980 1981." *Journal of Applied Social Psychology* 18, no. 2 (February 1988): 129-138. Topics discussed include: heat, neighborhoods, aggravated assaults, violence, socioeconomics.

957. Hartung, J. "Light, Puberty, and Aggression: A Proximal Mechanism Hypothesis." *Human Ecology* 6 (1978): 273-297. Topics discussed include: climate, crime, biology.

958. Kenrick, D. T. "Ambient temperature and horn honking: A field study of the heat, aggression relationship." *Environment and Behavior* 18, no. 2 (March 1986): 179-191. Topics discussed include: horn honking, frustration, aggressive behavior, driving behavior.

959. Palamarek, D. L. "The effects of ambient temperature and insult on the motivation to retaliate or escape." *Motivation and Emotion* 3, no. 1 (March 1979): 83-92. Topics discussed include: anger, aggressive behavior, emotional responses, avoidance, reciprocity, heat.

960. Reifman, A. "Temper and temperature on the diamond: the heat-aggression relationship in major league baseball." *Personality and Social Psychology Bulletin* 17 (October 1991): 580-5. Topics discussed include: baseball psychology.

961. Roitman, G. "Annual rhythms of violence in hospitalized affective patients: correlation with changes in the duration of the daily photoperiod." *Acta Psychiatr Scand* 82, no. 1 (July 1990): 73-6. Topics discussed include: affective disorders, Israel, risks, schizophrenia, circadian rhythms, seasonality, social environment, violence.

962. Rule, B. G. "Environmental Stressors, Emotional Arousal and Aggression." in *Stress and Anxiety*, 87-104. editors I. Sarason, and C. Spielberger. New York: Halstead Press, 1976. Topics discussed include: environment, behavior, psychology, emotion.

963. ---. "Priming effects of heat on aggressive thoughts." *Social Cognition* 5, no. 2 (1987): 131-143. Topics discussed include: heat, stories, aggression.

964. Schwartz, D. C. "On the Ecology of Political Violence: 'The Long Hot Summer' as a Hypothesis." *The American Behavioral Scientist* 8, no. July-August (July 1968): 24-28. Topics discussed include: temperature, violence, behavior.

965. Wyndham, C. "Tolerance of very hot humid environments by highly acclimatized Bantu at rest." *Br J Ind Med* 25, no. 1 (January 1968): 22-39. Topics discussed include: aggression, body temperature, Caucasoid race, heat exhaustion, hysteria, mining, personality, South Africa, acclimation, humidity, Negroid race.

COGNITION

966. Auliciems, A. "Some observed relationships between the atmospheric environment and mental work." *Environ Res* 5, no. 2 (June 1972): 217-40. Topics discussed include: humidity, intelligence, psychology, problem solving, psychometrics, weather.

967. Bekmuratov, N. S. "On the problem of increasing the work efficiency of deaf students living in a hot climate (Uzbekistan)." *Defektologiya;* no. 3 (1989): 14-19. Topics discussed include: cognitive work, the deaf, USSR.

968. Berglund, G. "A note on intelligence and season of birth." *Br J Psychol* 58, no. 1 (May 1967): 147-51. Topics discussed include: pregnancy, Sweden, temperature, intelligence.

969. Black, F. W. "Season of birth and intelligence in a sample of learning disabled children." *Journal of Genetic Psychology* 123, no. 1 (September 1973): 31-34. Topics discussed include: children, learning disabilities.

970. Bregman, N. J. et al. "Motivation and skin temperature biofeedback: Yerkes Dodson revisited." *Psychophysiology* 19, no. 3 (May 1982): 282-285. Topics discussed include: task performance, training, skin, monetary incentives, task complexity.

971. ---. "Skin temperature biofeedback: Evaluation of positive and negative monetary incentives." *International Journal of Neuroscience* 22, no. 1-2 (1983): 119-124. Topics discussed include: monetary effects, positive and negative incentives.

972. Browne, R. A. et al. "A survey of adult patients in four hospitals for the mental handicapped." *British Journal of Mental Subnormality* 25, no. 1 (June 1979): 33-37. Topics discussed include: birth month, intelligence, seasonality.

973. Cleland, C. et al. "The influence of climate on intelligence test scores of retardates." *Am J Ment Defic* 70, no. 5 (March 1966): 726-8. Topics discussed include: seasonality, cognition, intelligence, retardation.

974. Cocchi, R. "Sensitivity to environmental temperature in Down's syndrome Subjects: A study of 432 cases." *Italian Journal of Intellective Impairment* 2, no. 2 (December 1989): 195-199. Topics discussed include: weather, Italy, atmospheric effects, children.

975. Craddick, R. "WISC and WAIS IQS as a function of season of birth." *Psychol Rep* 18, no. 1 (February 1966): 259-64. Topics discussed include: intelligence.

976. Fine, B. J. et al. "Effects of altitude and heat on complex cognitive tasks." *Human Factors* 20, no. 1 (February 1978): 115-122. Topics discussed include: stress, cognitive processes, military psychology.

977. Gopinathan, P. et al. "Role of dehydration in heat stress induced variations in mental performance." *Arch Environ Health* 43, no. 1 (January 1988): 15-7. Topics discussed include: dehydration, intelligence, memory, physiopathology, mental processes, stress.

978. Grivel, F. "The influence of ambient and body heat on human work without important physical load: III. Explanatory models of specific heat stress effects on psychomotor and mental performance." *Travail Humain* 39, no. 2 (1976): 311-328. Topics discussed include: heat, stress, mental performance, perceptual motor processes, cognitive processes.

979. Hancock, P. A. "Task categorization and the limits of human performance in extreme heat." *Aviation, Space, and Environmental Medicine* 53, no. 8 (August 1982): 778-784. Topics discussed include: ambient temperature, task performance, tracking, cognitive processes.

980. Heyns, B. "Schooling and cognitive development: Is there a season for learning? Special Issue: Schools and development." *Child Development* 58, no. 5 (October 1987): 1151-1160. Topics discussed include: summer school, cognitive development.

981. Hord, D. J. et al. *Cognitive Performance Change during a 6 Hour Hike at Low Temperature in Simulated Rain, at Controlled Walking Rates Final rept*, Performer: Naval Health Research Center, San Diego, CA., December 1983. Topics discussed include: walking, perceptions, psychology, rain, simulation, skin anatomy.

982. Jacobs, J. et al. "Support for relationship of season of birth upon intelligence." *Ment Retard* 8, no. 1 (February 1970): 12-4. Topics discussed include: pregnancy, timing, retardation.

983. Jason, G. et al. "High altitude mountaineering and brain function: neuropsychological testing of members of a Mount Everest expedition." *Aviat Space Environ Med* 60, no. 2 (February 1989): 170-3. Topics discussed include: cognition, mountaineering, neuropsychology.

984. Kennedy, R. et al. "Cognitive performance deficits in a simulated climb of Mount Everest: Operation Everest II." *Aviat Space Environ Med* 60, no. 2 (February 1989): 99-104. Topics discussed include: aerospace medicine, cognition disorders, etiology, psychometrics, task performance, altitude, anoxia complications, atmosphere exposure.

985. Kudoh, N. et al. "Effects of room environment on human cognitive activities." *Tohoku Psychologica Folia* 50 (1991): 45-54. Topics discussed include: environment, physical condition, cognitive ability, temperature effects, temperature perception, attitudes, physical comfort.

986. Minors, D. S. et al. "Circadian rhythms in deep body temperature, urinary excretion and alertness in nurses on night work." *Ergonomics* 28, no. 11 (November 1985): 1523-1530. Topics discussed include: work, biological rhythms, attention.

987. Ojha, A. et al. "A note on season of birth intelligence." *Br J Educ Psychol* 36, no. 1 (February 1966): 94-5. Topics discussed include: cognitive ability.

988. Orme, J. E. "Ability, month of birth and climatic temperature." *British Journal of Mental Subnormality* 25, no. 48 (1979): 31-32. Topics discussed include: mental retardation, environment, intelligence, etiology.

989. Regard, M. et al. "Cognitive changes at high altitude in healthy climbers and in climbers developing acute mountain sickness." *Aviat Space Environ Med* 62, no. 4 (April 1991): 291-5. Topics discussed include: cognition, memory, neurology, mountain climates.

990. Shanazarov, A. S. et al. "Psychological status and metabolism during exposure to high temperature and humidity." *Human Physiology* 15, no. 4 (July 1989): 283-287. Topics discussed include: humidity, attention, thinking, decision making, heat effects, attention.

991. Shapiro, A. "Heat, ethnic differences, and creativity in the Negev desert." *J Educ Psychol* 67, no. 2 (April 1975): 183-7. Topics discussed include: cross-cultural analysis, Europe, heart rate, intelligence, Israel, Morocco ethnology, creativeness, desert climates, ethnic groups.

992. Sharma, V. M. et al. "Influence of heat stress induced dehydration on mental functions." *Ergonomics* 29, no. 6 (June 1986): 791-799. Topics discussed include: mental functions, complexity, acclimatization, India, water deprivation, cognitive processes, exercise.

993. Sharma, V. et al. "Differential effects of hot humid and hot dry environments on mental functions." *Int Arch Occup Environ Health* 52, no. 4 (1983): 315-27. Topics discussed include: cognition, environmental exposure, heat, humidity, mental processes, physiology.

994. Shukitt, H. B. et al. "Relationships between symptoms, moods, performance, and acute mountain sickness at 4,700 meters." *Aviat Space Environ Med* 62, no. 9 Pt 1 (September 1991): 865-9. Topics discussed include: cognition, physiology, military personnel, altitude, task performance.

995. Taylor, A. J. et al. "Some cognitive effects of wintering over in the Antarctic." *New Zealand Journal of Psychology* 16, no. 2 (December 1987): 93-94. Topics discussed include: discrimination, memory, winter, cognition, cold effects, Antarctica, seasonality.

996. Triplett, G. "Research on heat and temperature in cognitive development." *Journal of Children's Mathematical Behavior* 1, no. 2 (1973): 27-43. Topics discussed include: heat, thinking.

997. Uherik, A. "Autoregulation under conditions of a cognitive load in high altitude climate simulation." *Studia Psychologica* 18, no. 3 (1976): 207-212. Topics discussed include: atmospheric pressure, memory, task performance.

998. Weihe, W. "Human adaptation to climatic changes." *Schweiz Rundsch Med Prax* 81, no. 14 (31 March 1992): 436-41. Topics discussed include: adaptation, behavior, thermoregulation, health, instinct, intelligence, homeostasis.

999. Wesensten, N. et al. "Effects of simulated high altitude exposure on long latency event related brain potentials and performance." *Aviat Space Environ Med* 64, no. 1 (January 1993): 30-6. Topics discussed include: atmosphere, timing, brain physiology, cognition, physiology, task performance.

1000. White, A. "Cognitive impairment of acute mountain sickness and acetazolamide." *Aviat Space Environ Med* 55, no. 7 (July 1984): 598-603. Topics discussed include: altitude, physiopathology, cognition, drug effects, disorders.

1001. Whorton, J. E. et al. "Season of birth and intelligence in samples of exceptional children." *Psychological Reports* 49, no. 2 (October 1981): 649-650. Topics discussed include: intelligence, learning disabilities, children.

1002. Williams, P. et al. "Season of birth and cognitive development." *Nature* 228, no. 276 (12 December 1970): 1033-6. Topics discussed include: age effects, cognition, education, United Kingdom.

1003. Wyon, D. et al. "The effects of moderate heat stress on mental performance." *Scand J Work Environ Health* 5, no. 4 (December 1979): 352-61. Topics discussed include: language, memory, psychology, skin temperature, mental processes.

MOOD & PERSONALITY

1004. Albert, P. et al. "Effect of daily variation in weather and sleep on seasonal affective disorder." *Psychiatry Res* 36, no. 1 (January 1991): 51-63. Topics discussed include: hypersomnia, psychology, personality, risk.

1005. Andersen, P. A. et al. "Changes in Latitude, Changes in Attitude: The Relationship between Climate and Interpersonal Communication Predispositions." *Communication Quarterly* 38, no. 4 (September 1990): 291-311. Topics discussed include: weather, interpersonal communication, interpersonal relations, environment.

1006. Arbisi, P. A. "Thermoregulatory Response in Seasonal Affective Disorder: Support for a Functional Dopamine Deficit in Winter Depression." Ph.D. diss., University of Minnesota, 1990. Topics discussed include: thermoregulation, affective disorder, mood, health, behavior; available from UMI.

1007. Barbato, G. et al. "Spontaneous eye blink rate in winter seasonal affective disorder." *Psychiatry Res* 47, no. 1 (April 1993): 79-85. Topics discussed include: blinking, personality, phototherapy, reaction time, physiology, physiopathology.

1008. Barnston, A. "The effect of weather on mood, productivity, and frequency of emotional crisis in a temperate continental climate." *Int J Biometeorol* 32, no. 2 (June 1988): 134-43. Topics discussed include: efficiency, emotions, affect.

1009. Bartko, J. et al. "Seasonal changes in mood and behavior: a cluster analytic approach." *Psychiatry Res* 28, no. 2 (May 1989): 227-39. Topics discussed include: depressive disorder, diagnosis, psychometrics, gender effects, space time clustering, psychology.

1010. Bauer, M. et al. "Validity of seasonal pattern as a modifier for recurrent mood disorders for DSM IV." *Compr Psychiatry* 34, no. 3 (May 1993): 159-70. Topics discussed include: body weight, phototherapy, psychology, disorder treatment, diagnosis.

1011. Bell, P. A. et al. "Environmental Influences on Attraction: Effects of Heat, Attitude Similarity, and Personal Evaluations." *Bulletin of the Psychonomic Society* 4 (1974): 479-481. Topics discussed include: environment, temperature, behavior, personality, psychology.

1012. Benoit, O. et al. "Circadian rhythms (temperature, heart rate, vigilance, mood) of short and long sleepers: Effects of sleep deprivation." *Chronobiologia* 8, no. 4 (October 1981): 341-350. Topics discussed include: circadian rhythms, biological rhythms, emotional states.

1013. Bicakova, R. A. et al. "Circadian rhythm of axillary temperature in mood illness: Prominent ultradian periodicity associated with major affective disorders. International Thematic Meetings: Unipolar and bipolar, two mood disorders: Clinical factors and treatment (1991, Bordeaux, France)." *Encephale* 18, no. 1 (January 1992): 75-76. Topics discussed include: axillary rhythms, France, diagnosis, affective disturbances, biological rhythms.

1014. Blehar, M. C. et al. "Seasonal mood disorders: Consensus and controversy." *Psychopharmacology Bulletin* 26, no. 4 (1990): 465-494. Topics discussed include: epidemiology, light treatment, a literature review, phototherapy.

1015. Boubekri, M. et al. "Impact of window size and sunlight penetration on office workers' mood and satisfaction: a novel way of assessing sunlight." *Environment and Behavior* 23 (July 1991): 474-93. Topics discussed include: sunshine, windows, work, light, psychology.

1016. Boyce, P. et al. "Seasonal affective disorder in the southern hemisphere." *Am J Psychiatry* 145, no. 1 (January 1988): 96-9. Topics discussed include: Australia, circadian rhythms, depression, etiology, psychology, carbohydrates, eating disorders, light, sleep disorders, temperature.

1017. Brandstatter, H. et al. "Effects of weather and air pollution on mood: An individual difference approach." *Environmental social psychology NATO ASI series Series D, Behavioural and social sciences, No 45*, pp. 149-159. Kluwer Academic Publishers, Dordrecht, Netherlands; xii, 330. Eds Jorge Correia Jesuino, Luis Soczka, Geoffrey M. Stephenson, David Canter. 1988. Topics discussed include: atmospheric conditions, pollution, emotion, individual differences.

1018. Carskadon, M. et al. "Parental reports of seasonal mood and behavior changes in children." *J Am Acad Child Adolesc Psychiatry* 32, no. 2 (March 1993): 264-9. Topics discussed include: age effects, epidemiology, psychology, gender effects, parents psychology, personality, diagnosis.

1019. Cerbus, G. et al. "Seasonal variation of personality of college students as measured by the MMPI." *Psychological Reports* 33, no. 2 (October 1973): 665-666. Topics discussed include: traits.

1020. Cervinka, R. "Psychosocial aspects of fitness for employment in the tropics." *Offentl Gesundheitswes* 45, no. 4 (April 1983): 180-2. Topics discussed include: Germany, ethnology, Nigeria, timing, personality, personnel management, personnel selection.

1021. Christie, M. J. et al. "Deep body temperature: Diurnal variation, sex, and personality." *Journal of Psychosomatic Research* 21, no. 3 (1977): 207-211. Topics discussed include: biological rhythms, gender differences, eating, young adults.

1022. Colquhoun, W. P. "Effects of personality on body temperature and mental efficiency following transmeridian flight." *Aviation, Space, and Environmental Medicine* 55, no. 6 (June 1984): 493-496. Topics discussed include: biological rhythms, young males, biological rhythms, aviation.

1023. Colquhoun, W. P. et al. "Personality differences in body temperature rhythm, and their relation to its adjustment to night work." *Ergonomics* 21, no. 10 (October 1978): 811-817. Topics discussed include: body rhythms, night work, introversion, neuroticism, extraversion, workday shifts.

1024. Crowley, J. et al. "Effect of high terrestrial altitude and supplemental oxygen on human performance and mood." *Aviat Space Environ Med* 63, no. 8 (August 1992): 696-701. Topics discussed include: oxygen inhalation treatment, task performance, affect, sickness psychology, cognition.

1025. Cunningham, M. R. "Weather, mood, and helping behavior: Quasi experiments with the sunshine samaritan." *Journal of Personality and Social Psychology* 37, no. 11 (November 1979): 1947-1956. Topics discussed include: helping behavior, environmental effects, emotion, assistance behavior, social behavior.

1026. Depue, R. et al. "Seasonal and mood independence of low basal prolactin secretion in premenopausal women with seasonal affective disorder." *Am J Psychiatry* 146, no. 8 (August 1989): 989-95. Topics discussed include: circadian rhythms, depressive disorder, physiopathology, psychology, personality, blood, menstrual cycle.

1027. Depue, R. A. et al. "A two dimensional threshold model of seasonal bipolar affective disorder." *Psychopathology: An interactional perspective Personality, psychopathology, and psychotherapy (David Magnusson, Arne Ohman, Eds,* pp. 95-123. Academic Press, Inc., Orlando, FL; xix, 408. 1987. Topics discussed include: seasonality, affective disturbances, manic depression.

1028. Diebold, K. "The Problem of the Relations between Month of Birth and Psychic Disorders." *Fortschritte der Neurologie, Psychiatrie und ihrer Grenzgebiete* 43 (1975): 71-81. Topics discussed include: seasonality, mental disorders, illness, mood.

1029. Dirlich, G. et al. "Temporal coordination of rest activity cycle, body temperature, urinary free cortisol, and mood in a patient with 48 hour unipolar depressive cycles in clinical and time cue free environments." *Biol Psychiatry* 16, no. 2 (February 1981): 163-79. Topics discussed include: activity cycles, cues, hydrocortisone urine, social isolation, time, depressive disorder physiopathology, cycles.

1030. Edgerton, R. "Cultural vs. Ecological Factors in the Expression of Values, Attitudes, and Personality Characteristics." *American Anthropologist* 67 (1965): 442-447. Topics discussed include: culture, climatic effects, ecology, environment, behaviour, psychology.

1031. Ekvall, R. B. "Tibetan Nomadic Pastoralists: Environments, Personality, and Ethos." *Proceedings of the American Philosophical Society* 118, no. 6 (December 1974): 519-537. Topics discussed include: children, lifestyle.

1032. Elmore, S. K. "Seasonal Affective Disorder and Performance: A Test of the Phase-Delay Hypothesis." Ph.D. diss., University of Washington, 1990. Topics discussed include: affective disorders, mood, health, thermoregulation; available from UMI.

1033. Endo, T. et al. "Seasonal variations of the circadian rhythms in seasonal affective disorder." *Jpn J Psychiatry Neurol* 46, no. 1 (March 1992): 253-5. Topics discussed include: personality, psychology, wakefulness, circadian rhythms, diagnosis, sleep stages.

1034. Faedda, G. et al. "Seasonal mood disorders. Patterns of seasonal recurrence in mania and depression." *Arch Gen Psychiatry* 50, no. 1 (January 1993): 17-23. Topics discussed include: age effects, bipolar disorderdiagnosis, epidemiology, Italy, gender effects.

1035. Faravelli, C. et al. "Circadian rhythm in primary affective disorder." *Comprehensive Psychiatry* 26, no. 4 (July 1985): 364-369. Topics discussed include: oral temperature, blood pressure, heart rate, performance, mood, antidepressant drug treatment, biological rhythms, body temperature.

1036. Farley, F. "Season of birth, intelligence and personality." *Br J Psychol* 59, no. 3 (August 1968): 281-3. Topics discussed include: age effects, emotions, England, extraversion, intelligence tests, introversion, mental retardation, socioeconomics, delivery.

1037. Fossey, E. et al. "Seasonality in personality: A review." *Canadian Journal of Psychiatry* 37, no. 5 (June 1992): 299-308. Topics discussed include: psychiatric illness, mortality, a literature review, mental disorders, death, dying.

1038. Gaist, P. et al. "Effects of bright light on resting metabolic rate in patients with seasonal affective disorder and control subjects." *Biol Psychiatry* 28, no. 11 (1 December 1990): 989-96. Topics discussed include: body weight, caloric intake, depressive disorder, metabolism, psychology, personality, treatment, energy metabolism, phototherapy.

1039. Genhart, M. J. "Seasonality, Depression, and the Psychological Effects of Phototherapy in an Elderly Sample." Ph.D. diss., University of Maryland College Park, 1989. Topics discussed include: affective disorder, physiology, health, mood, behavior; available from UMI.

1040. Genhart, M. et al. "Effects of bright light on mood in normal elderly women." *Psychiatry Res* 47, no. 1 (April 1993): 87-97. Topics discussed include: personality, reference values, seasonal affective disorder psychology, disorder treatment, seasonality, aging psychology, phototherapy.

1041. Goldstein, K. "Weather, mood, and internal external control." *Percept Mot Skills* 35, no. 3 (December 1972): 786. Topics discussed include: reinforcement psychology, self esteem, gender effects, affect.

1042. Griffitt, W. "Environmental Effects on Interpersonal Affective Behavior: Ambient Effective Temperature and Attraction." *Journal of Personality and Social Psychology* 15 (1970): 240-244. Topics discussed include: environment, weather, behavior.

1043. Griffitt, W. et al. "Hot and Crowded: Influences of Population Density and Temperature on Interpersonal Affective Behavior." *Journal of Personality and Social Psychology* 17 (1971): 92-99. Topics discussed include: demography, behavior, emotion.

1044. Gupta, S. "Season of birth in relation to personality and blood groups." *Personality and Individual Differences* 13, no. 5 (May 1992): 631-633. Topics discussed include: extraversion, neuroticism, impulsiveness, venturesomeness, empathy, male doctors, blood groups, traits, extraversion.

1045. Haertzen, C. "Geophysical variables and behavior: LXV. Seasonal changes in mood in opioid addicts on the Addiction Research Center Inventory." *Psychol Rep* 68, no. 2 (April 1991): 360-2. Topics discussed include: anxiety disorders, diagnosis, depressive disorder, narcotic dependence rehabilitation, prisoners, psychology, psychometrics, personality.

1046. Haggag, A. et al. "Seasonal mood variation: an epidemiological study in northern Norway." *Acta Psychiatr Scand* 81, no. 2 (February 1990): 141-5. Topics discussed include: cross-sectional analysis, depressive disorder, psychology, Norway, epidemiology, personality.

1047. Hare, E. H. "The Season of Birth of Siblings of Psychiatric Patients." *British Journal of Psychiatry* 129 (1976): 49-54. Topics discussed include: mental disorders, illness, mood.

1048. ---. "Variations in the Seasonal Distribution of Births of Psychotic Patients in England and Wales." *British Journal of Psychiatry* 132 (1978): 155-158. Topics discussed include: mental disorders, illness, mood.

1049. Hentschel, U. et al. "Season of birth and personality: Another instance of noncorrespondence." *Journal of Social Psychology* 125, no. 5 (October 1985): 577-585. Topics discussed include: women.

1050. Herbert, D. T. "Urban Deprivation: Definition, Measurement and Spatial Qualities." *Geographical Journal* 141 (1975): 362-372. Topics discussed include: behavior, psychology, mood, geography, environment.

1051. Hill, M. A. "Light, circadian rhythms, and mood disorders: A review." *Annals of Clinical Psychiatry* 4, no. 2 (June 1992): 131-146. Topics discussed include: circadian rhythms, seasonal affective disorder, nonseasonal depression, a literature review, illumination, biological rhythms.

1052. Howarth, E. et al. "A multidimensional approach to the relationship between mood and weather." *British Journal of Psychology* 75, no. 1 (February 1984): 15-23. Topics discussed include: performance, atmospheric effects, emotion, work performance.

1053. Huntington, E. "Geographical Environment and Japanese Character." *J Race Development* II (1912): 256-281. Topics discussed include: environment, geography, society, anthropolgy, personality.

1054. Ito, A. et al. "Prevalence of seasonal mood changes in low latitude area: Seasonal Pattern Assessment Questionnaire score of Quezon City workers." *Jpn J Psychiatry Neurol* 46, no. 1 (March 1992): 249. Topics discussed include: cross-sectional analysis, Philippines, epidemiology, seasonal affective disorder, diagnosis, psychology, cross-cultural analysis.

1055. Jasnoski, M. B. "The Physical Environment Affects Quality of Life Based upon Environmental Sensitivity." *Journal of Applied Developmental Psychology* 13, no. 2 (April 1992): 139-142. Topics discussed include: cognition, performance, economics, life satisfaction, thinking, mood.

1056. Jobe, J. et al. "Effects of dexamethasone and high terrestrial altitude on cognitive performance and affect." *Aviat Space Environ Med* 62, no. 8 (August 1991): 727-32. Topics discussed include: personality, drug effects, sickness, drug treatment, cognition.

1057. Joffe, R. et al. "Seasonal variation of peripheral thyroid hormone levels in major depression." *J Affect Disord* 23, no. 1 (September 1991): 49-51. Topics discussed include: diagnosis, psychology, thyroid tests, blood.

1058. Joseph, V. J. et al. "Seasonal variation in behavioral responses to m CPP in patients with seasonal affective disorder and controls." *Biol Psychiatry* 33, no. 7 (1 April 1993): 496-504. Topics discussed include: circadian rhythms, drug effects, physiology, euphoria, infusions, intravenous, neural transmission, personality, physiopathology.

1059. Kamo, T. et al. "Fifty young women's seasonal changes in mood and behavior in Tokyo." *Jpn J Psychiatry Neurol* 46, no. 1 (March 1992): 246-8. Topics discussed include: cross-sectional analysis, Hong Kong, seasonal affective disorder, diagnosis, psychology, cross-cultural analysis, epidemiology.

1060. Kanekar, S. et al. "Intelligence, extraversion, and neuroticism in relation to season of birth." *J Soc Psychol* 86, no. 2 (April 1972): 309-10. Topics discussed include: cross-cultural analysis, extraversion psychology, India, intelligence, neurotic disorders, personality, delivery, personality.

1061. Kasper, S. et al. "Psychological effects of light therapy in normals." 260-270. *Seasonal Affective Disorders and Phototherapy,* eds. Norman E. Rosenthal and Mary C. Blehar, New York: Guilford Press, 1989. Topics discussed include: personality, psychology.

1062. Kasper, S. et al. "Phototherapy in individuals with and without subsyndromal seasonal affective disorder." *Arch Gen Psychiatry* 46, no. 9 (September 1989): 837-44. Topics discussed include: circadian rhythms, depressive diagnosis, psychology, health care, personality disorders, weather, treatment.

1063. ---. "Immunological correlates of seasonal fluctuations in mood and behavior and their relationship to phototherapy." *Psychiatry Res* 36, no. 3 (March 1991): 253-64. Topics discussed include: depressive disorder, psychology, dose response relationship, radiation, natural immunology, personality, immunology.

1064. ---. "Epidemiological findings of seasonal changes in mood and behavior. A telephone survey of Montgomery County, Maryland." *Arch Gen Psychiatry* 46, no. 9 (September 1989): 823-33. Topics discussed include: cross-sectional analysis, depressive disorder, diagnosis, psychology, health, Maryland, weather.

1065. Kastl, O. "Occupational and environment analysis of railway employees with myocardial infarct." *Med Klin* 64, no. 42 (17 October 1969): 1911-7. Topics discussed include: age effects, physical labor, Germany, myocardial infarction, personality, climatic effects, epidemiology, occupations, railroads.

1066. Lacoste, V. et al. "Seasonal variation in normal subjects: an update of variables current in depression research." 167-229, *Seasonal Affective Disorders and Phototherapy,* eds. Norman E. Rosenthal and Mary C. Blehar, New York: Guilford Press, 1989. Topics discussed include: biochemistry, metabolism, perception, personality, sleep, dreams, central nervous system, thermoregulation, Switzerland.

1067. Lacoste, V. et al. "Seasonality in personality dimensions." *Psychiatry Research* 21, no. 2 (June 1987): 181-183. Topics discussed include: seasonality in personality traits, diurnal rhythms, healthy adults, biological rhythms, affective disturbances.

1068. Lam, R. et al. "Seasonal mood symptoms in bulimia nervosa and seasonal affective disorder." *Compr Psychiatry* 32, no. 6 (November 1991): 552-8. Topics discussed include: feeding behavior, hypersomnia diagnosis, hypersomnia psychology, psychometrics, bulimia diagnosis, bulimia psychology.

1069. Lewy, A. et al. "Treating phase typed chronobiologic sleep and mood disorders using appropriately timed bright artificial light." *Psychopharmacol Bull* 21, no. 3 (1985): 368-72. Topics discussed include:

affective disorders, blood, circadian rhythms, seasonality, treatment, melatonin, phototherapy, sleep disorders treatment.

1070. ---. "The use of plasma melatonin levels and light in the assessment and treatment of chronobiologic sleep and mood disorders." *J Neural Transm Suppl* 21 (1986): 311-22. Topics discussed include: affective disorders, blood, biological clocks, intraocular pressure drug effects, seasonality, treatment, phototherapy.

1071. Lewy, A. J. "Treating chronobiologic sleep and mood disorders with bright light." *Psychiatric Annals* 17, no. 10 (October 1987): 664-669. Topics discussed include: illumination, biological rhythms, affective disturbances, sleep disorders, seasonality, treatment.

1072. Lewy, A. J. et al. "Bright artificial light treatment of a manic depressive patient with a seasonal mood cycle." *American Journal of Psychiatry* 139, no. 11 (November 1982): 1496-1498. Topics discussed include: winter depression, biological rhythms, illumination, manic depression.

1073. Madden, P. A. "Seasonal Change in Mood and Behavior: A Twin Study." Ph.D. diss., University of Pittsburgh, 1992. Topics discussed include: behavior, psychology; available from UMI.

1074. Maes, M. et al. "Seasonal-variation and meteotropism in various self-rated psychological and physiological features of a normal couple." *International Journal of Biometeorology* 36, no. 4 (October 1992): 195-200. Topics discussed include: biometeorology, chronopsy, meteotropism, affective disorders, biopsychosociometeorology, weather, treatment, mood.

1075. Marshall, P. et al. "Effects of allergy season on mood and cognitive function." *Ann Allergy* 71, no. 3 (September 1993): 251-8. Topics discussed include: psychomotor performance, reaction time, affect physiology, cognition physiology, hypersensitivity, immediate psychology.

1076. Mascie, T. C. G. "Season of birth, IQ components, and personality traits." *Journal of Genetic Psychology* 137, no. 1 (September 1980): 151-152. Topics discussed include: traits, intelligence.

1077. Meesters, Y. "Test retest reliability of personality questionnaires for people with winter depression complaints." *Nederlands Tijdschrift voor de Psychologie en haar Grensgebieden* 47, no. 3 (June 1992): 143-147. Topics discussed include: Netherlands, traits, seasonality.

1078. Monk, T. H. et al. "Diurnal variation in mood and performance in a time isolated environment." *Chronobiology International* 2, no. 3 (1985): 185-193. Topics discussed include: rectal temperature, task performance, biological rhythms, emotion, body temperature, performance, time disorientation.

1079. Nagayama, H. "Phototherapy of seasonal affective disorder." *Jpn J Psychiatry Neurol* 45, no. 1 (March 1991): 183-4. Topics discussed include: personality, psychology, treatment.

1080. Nelson, M. "Psychological testing at high altitudes." *Aviat Space Environ Med* 53, no. 2 (February 1982): 122-6. Topics discussed include: anger, anoxia, compulsive behavior, expeditions, hostility, obsessive behavior, paranoid behavior, personality disorders, mountaineering, personality.

1081. Nelson, T. M. et al. "Interaction of temperature, illuminance and apparent time on sedentary work fatigue." *Ergonomics* 27, no. 1 (January 1984): 89-101. Topics discussed include: air, illumination, mood, time perception, fatigue, emotion, employee.

1082. Oren, D. A. et al. "Environmental light and mood." *Environment and psychopathology (Abdu'l Missagh A Ghadirian, Heinz Edgar Lehmann, Eds*, pp. 3-20. New York: Springer Publishing Co., Inc.; xv, 188. 1993. Topics discussed include: seasonal affective disorder.

1083. Papolos, D. F. "Serotonin, seasonality, and mood disorders." *The role of serotonin in psychiatric disorders Clinical and experimental psychiatry monograph, No 4 (Serena Lynn Brown, Herman M van Praag, Eds*, pp. 260-283. New York: Brunner, Mazel, Inc.; xiii, 349. 1991. Topics discussed include: serotonin, etiology, melatonin, seasonal affective disorder, biological rhythms, phototherapy.

1084. Persinger, M. A. et al. "Geophysical variables and behavior: XII. The weather matrix accommodates large portions of variance of measured daily mood." *Perceptual and Motor Skills* 57, no. 3, Pt 1 (December 1983): 868-870. Topics discussed include: daily cycles, environment, emotion, atmospheric effects.

1085. Persinger, M. "Geophysical variables and behavior: LV. Predicting the details of visitor experiences and the personality of experients: the temporal lobe factor." *Percept Mot Skills* 68, no. 1 (February 1989): 55-65. Topics discussed include: electric stimulation, hippocampus physiology, extraterrestrial environment, parapsychology physiology, temporal lobe physiology.

1086. ---. "Lag responses in mood reports to changes in the weather matrix." *Int J Biometeorol* 19, no. 2 (June 1975): 108-14. Topics discussed include: adaptation, self assessment psychology, emotions, reaction time.

1087. Roseman, C. J. "Seasonal Affective Disorder and Medication Errors with Light Intervention in an Alaskan Medical Center." Ph.D. diss., University of Alaska Anchorage, 1992. Topics discussed include: affective disorder, mood, health; available from UMI.

1088. Rosen, L. N. et al. "Patterns of seasonal change in mood and behavior: An example from a study of military wives." *Military Medicine* 156, no. 5 (May 1991): 228-230. Topics discussed include: symptoms, Germany, army personnel, seasonal affective disorder, geography, emotion, changes in behavior.

1089. ---. "Seasonal variations in mood and behavior in the general population: A factor analytic approach." *Psychiatry Research* 38, no. 3 (September 1991): 271-283. Topics discussed include: socializing, appetite, weight gain or loss, sleep length, implications for winter depression, emotion, behavior.

1090. Rosen, S. *Weathering: How the Atmosphere Conditions Your Body, Your Mind, Your Moods and Your Health*. New York: Evans, 1979. Topics discussed include: weather, biology, physiology, behavior, cognition, mood, health.

1091. Rosenthal, N. et al. "No mood altering effects found after treatment of normal subjects with bright light in the morning." *Psychiatry Res* 22, no. 1 (September 1987): 1-9. Topics discussed include: arousal, feeding behavior, psychological tests, psychometrics, sleep stages, affect, circadian rhythms, phototherapy methods, seasonality.

1092. Rubin, Z. "Seasonal rhythms in behavior." *Psychology Today* 13, no. 7 (1979): 12-16. Topics discussed include: environment, personality, psychophysiology.

1093. Rudorfer, M. et al. "Biogenic amines in seasonal affective disorder: effects of light therapy." *Psychiatry Res* 46, no. 1 (January 1993): 19-28. Topics discussed include: bipolar, depressive psychology, treatment, blood, personality, cerebrospinal fluid, psychology, phototherapy.

1094. Sakamoto, K. et al. "A nationwide survey of seasonal affective disorder at 53 outpatient university clinics in Japan." *Acta Psychiatr Scand* 87, no. 4 (April 1993): 258-65. Topics discussed include: cross-sectional analysis, depressive psychology, Japan epidemiology, personality, diagnosis, psychology, sunlight.

1095. Sanders, J. L. et al. "Relationships between weather and mood." *Journal of General Psychology* 107, no. 1 (July 1982): 155-156. Topics discussed include: humidity, temperature, barometric pressure, emotion, atmospheric effects.

1096. Schlosberg, A. et al. "Psychiatric Admissions and Climatic Factors in the Coastal Regions of Israel." *Israel Annals of Psychiatry and Related Disciplines* 11 (1973): 23-32. Topics discussed include: climate, mental disorders, health, mood.

1097. Schneider, F. W. et al. "Helping Behavior in Hot, Comfortable, and Cold Temperatures: A Field Study." *Environment and Behavior* 12, no. 2 (June 1980): 231-240. Topics discussed include: helping others, behavior, adults.

1098. Schulz, P. et al. "Personality as a dimension of summer and winter depression." *Psychopharmacol Bull* 24, no. 3 (1988): 476-83. Topics discussed include: seasonality, disorder psychology.

1099. Sedivec, V. "Effect of seasons of the year on the development of pathological mood phases of affective melancholia." *Ceskoslovenska Psychiatrie* 72, no. 2 (April 1976): 98-103. Topics discussed include: seasonality, depression.

1100. Shafii, M. E. et al. "Biological rhythms, mood disorders, light therapy, and the pineal gland." xix, 213 pages. Progress in psychiatry., Washington, DC: American Psychiatric Press, Inc., 1990. Topics discussed include: biological rhythms, affective disturbances, seasonality, organic therapies, premenstrual tension, melatonin.

1101. Shukitt, B. et al. "Mood states at 1600 and 4300 meters terrestrial altitude." *Aviat Space Environ Med* 59, no. 6 (June 1988): 530-2. Topics discussed include: timing, affect.

1102. Shukitt, H. B. et al. "Altitude symptomatology and mood states during a climb to 3,630 meters." *Aviat Space Environ Med* 61, no. 3 (March 1990): 225-8. Topics discussed include: affect, affective disorders, etiology, barometric pressure, mountaineering.

1103. Smithers, A. G. et al. "Personality and season of birth." *Journal of Social Psychology* 105, no. 2 (August 1978): 237-241. Topics discussed include: extraversion, neuroticism.

1104. Sonis, W. A. "Chronobiology of seasonal mood disorders." *Clinical guide to depression in children and adolescents*, 89-114. eds Sharon Lee Shafii Mohammad Shafii. xv, 304 pages.Washington, DC, American Psychiatric Press, Inc., 1992. Topics discussed include: seasonal affective disorder, biological rhythms, children, adolescents, adults.

1105. Spoont, M. R. "The Effects of Full Spectrum Ligth on the Circadian Rhythms of Individuals With Seasonal Affective Disorder and Normal Controls." Ph.D. diss., University of Minnesota, 1990. Topics discussed include: affective disorder, mood, health, temperature; available from UMI.

1106. Spoont, M. et al. "Dimensional measurement of seasonal variation in mood and behavior." *Psychiatry Res* 39, no. 3 (December 1991): 269-84. Topics discussed include: depressive disorder, psychometrics, seasonal affective disorder, personality.

1107. Suter, E. et al. "Effects of jogging on mental well being and seasonal mood variations: a randomized study with healthy women and men." *Schweiz Med Wochenschr* 121, no. 35 (31 August 1991): 1254-63. Topics discussed include: endurance, physical work, gender effects, affect, jogging, mental health.

1108. Szabo, C. P. "Mood disorders and season of presentation - a preliminary study of an inpatient sample at Baragwanath Hospital." *South African Medical Journal* 84, no. 1 (January 1994): 35-37. Topics discussed include: mania, depression.

1109. Tichner. "The Psycho-Physics of Climate." *Am J Psychol* 20 (1909): 1-14. Topics discussed include: physiology, behavior, mood.

1110. Van, H. C. et al. "The use of light in the treatment of season connected mood disorders." *Ned Tijdschr Geneeskd* 131, no. 48 (28 November 1987): 2178-9. Topics discussed include: depressive etiology, phototherapy methods, depressive treatment, phototherapy.

1111. Walter, O. "Personality fitness for the tropics." *Med Klin* 63, no. 34 (23 August 1968): 1337-40. Topics discussed include: Caucasoid race, competitive behavior, exploratory behavior, mental health, motivation, medicine, tropical climates.

1112. Wendt, H. W. "Season of birth, introversion, and astrology: A chronobiological alternative." *Journal of Social Psychology* 105, no. 2 (August 1978): 243-247. Topics discussed include: introversion, astrology, personality traits.

1113. ---. "Seasons of preverbal learning and personality development: An ecological model." *Psychologische Rundschau* 28, no. 4 (October 1977): 266-293. Topics discussed include: seasonality, cognition, cognitive development.

1114. White, D. M. "An Investigation of the Clinical Features of Seasonal Affective Disorder." Ph.D. diss., The University of Alabama, 1986. Topics discussed include: affective disorder, mood, health; available from UMI.

1115. Whitton, J. L. et al. "Weather and infradian rhythms in self reports of health, sleep and mood measures." *Journal of Psychosomatic Research* 26, no. 2 (1982): 231-235. Topics discussed include: infradian rhythms, health, sleep, cognition, biological rhythms, seasonality, health, emotion.

1116. Winton, F. et al. "Effects of light treatment upon mood and melatonin in patients with seasonal affective disorder." *Psychol Med* 19, no. 3 (August 1989): 585-90. Topics discussed include: depression, psychology, treatment, melatonin, blood, phototherapy.

1117. Wirz, J. A. et al. "Mood follow up over 6 years in a patient with season dependent form of depression (SAD)." *Fortschr Neurol Psychiatr* 59, no. 11 (November 1991): 453-5. Topics discussed include: symptoms, diagnosis, genetics, personality, risks, seasonal affective disorder, psychology.

SCHIZOPHRENIA & OTHER MENTAL ILLNESS

1118. Abe, K. "Seasonal Fluctuation of Psychiatric Admissions, based on the Data for Seven Prefectures of Japan for a Seven-Year Period 1955-1961." *Folia Psychiatr Neurol Jpn* 17 (1963): 101. Topics discussed include: health, mental illness, disease, a literature review.

1119. ---. "Seasonal patterns of recurrence in schizophreniform psychosis." *Schizophrenia Research* 7, no. 1 (April 1992): 43-47. Topics discussed include: seasonality, catatonia, atypical psychosis, relapse disorders, onset disorders.

1120. Bagedahl, S. M. "Does the season of conception influence the frequency of post partum mental illness?" *Acta Psychiatr Scand* 67, no. 3 (March 1983): 159-62. Topics discussed include: psychology, pregnancy, puerperal disorders, Sweden, epidemiology.

1121. Baron, M. "Risk factors in schizophrenia. Season of birth and family history." *Br J Psychiatry* 152 (April 1988): 460-5. Topics discussed include: age effects, risk, genetics, etiology.

1122. Blouin, A. "Seasonal patterns of bulimia nervosa." *The American Journal of Psychiatry* 149 (January 1992): 73-81. Topics discussed include: seasonality, mental illness, bulimia.

1123. Bojanovsky, J. "Seasonal Birthdates of Psychiatric Patients." *Ceslovenska Psychiatrie* 57 (1961): 47-49. Topics discussed include: birth, mental illness, health.

1124. Bolton P. "Season of Birth - Issues, Approaches and Findings for Autism." *Journal of Child Psychology and Psychiatry and Allied Disciplines* 33, no. 3 (March 1992): 509-530. Topics discussed include: infection, genetics, infantile autism, southern hemisphere, children, schizophrenia, epidemiology, infection, psychosis, disorder, exposure, diseases.

1125. Bourgeois, M. "Survey on the seasonal changes in the birth of patients with schizophrenia." *Ann Med Psychol Paris* 148, no. 8 (October 1990): 698-708. Topics discussed include: delivery, France, infants, risk, causes, epidemiology.

1126. Boyd, J. "Season of birth: schizophrenia and bipolar disorder." *Schizophr Bull* 12, no. 2 (1986): 173-86. Topics discussed include: age effects, brain damage, chronic complications, fertilization, prognosis, risk, social class, etiology.

1127. Bradbury, T. N. "Racial factors and season of birth in schizophrenia: A reinterpretation of Gallagher, McFalls, and Jones." *Journal of Abnormal Psychology* 95, no. 2 (May 1986): 187-188. Topics discussed include: methodology, ethnic differences.

1128. ---. "Season of birth in schizophrenia: A review of evidence, methodology, and etiology." *Psychological Bulletin* 98, no. 3 (November 1985): 569-594. Topics discussed include: a literature review.

1129. Bradley, M. "Seasonal Variations in the Incidence of Severely Crippling Mental Disorders." *American Journal of Psychiatry* 115 (1958): 343-345. Topics discussed include: mental illness, health.

1130. Brewerton, T. D. "Serotonin function in depression: effects of seasonality?" *The American Journal of Psychiatry* 149 (September 1992): 1277. Topics discussed include: seasonality, mental illness, tryptophan, prolactin, somatotropin, metabolism.

1131. Brown, C. "Symptomatology, Life Events, and Coping Strategies in Seasonal Affective Disorder." Ph.D. diss., The American University, 1992. Topics discussed include: mental illness, stress; available from UMI.

1132. Brown, P. J. "Seasonal variations in prolactin levels in schizophrenia." *Psychiatry Research* 25, no. 2 (August 1988): 157-162. Topics discussed include: cycles, prolactin.

1133. Buck, C. "Season of birth among the sibs of schizophrenics." *British Journal of Psychiatry* 133 (October 1978): 358-360. Topics discussed include: siblings.

1134. Cerbus, G. "Seasonal Differences of Depression in Mental Hospital Admissions as Measured by the MMPI." *Psychological Reports* 36 (1975): 737-738. Topics discussed include: mental illness, health.

1135. Cohen, P. "Seasonal patterns of bulimia nervosa." *The American Journal of Psychiatry* 150 (February 1993): 357. Topics discussed include: menarche, mental illness, bulimia.

1136. Corgiat, M. D. "Seasonality of schizophrenic births and age of onset." *Journal of Orthomolecular Psychiatry* 12, no. 4 (1983): 268-269. Topics discussed include: season of birth, psychiatric hospitals.

1137. Cortes, B. H. "Seasonality of birth in schizophrenia and affective disease in Mexico: Season of birth in endogenous psychosis." *New Trends in Experimental and Clinical Psychiatry* 6, no. 2 (April 1990): 79-84. Topics discussed include: mania, neurotic depression, Mexico.

1138. Crow, T. "Mutation and psychosis: a suggested explanation of seasonality of birth." *Psychol Med* 17, no. 4 (November 1987): 821-8. Topics discussed include: risk, schizophrenia, genetics, mutation, psychotic disorders.

1139. Csernansky, J. G. "Seasonal changes of CSF monoamine metabolites in psychiatric patients: What is the source?" *Psychiatry Research* 25, no. 3 (September 1988): 361-363. Topics discussed include: metabolites, affective disorder, biological rhythms, schizophrenia, major depression, hypomania, personality.

1140. Dalen, P. *Season of Birth: A Study of Schizophrenia and Other Mental Disorders*. Amsterdam: North-Holland, 1975. Topics discussed include: mental disorders, illness, health.

1141. ---. "Does age incidence explain all seasons of birth effects in the literature?" *Schizophrenia Bulletin* 16, no. 1 (1990): 11-12. Topics discussed include: age groups, etiology.

1142. ---. "Schizophrenia, season of birth, and maternal age." *British Journal of Psychiatry* 153 (December 1988): 727-733. Topics discussed include: seasonality, a literature review, age differences, mothers.

1143. Danneel, R. "Schizophrenia, mental deficiencies and season of birth." *Archiv fur Psychiatrie und Nervenkrankheiten* 217, no. 1 (1973): 37-41. Topics discussed include: precocious imbecility, mental retardation, seasonality.

1144. Dassa, D. "Season of birth and bipolar disorder." *The American Journal of Psychiatry* 150 (March 1993): 526-7. Topics discussed include: manic depressive psychoses, seasonality, mental illness, schizophrenia, affective disorders, mental illness classification.

1145. Degreef, G. "Season of birth and CT scan findings in schizophrenic patients." *Biol Psychiatry* 24, no. 4 (August 1988): 461-4. Topics discussed include: brain damage, chronic pathology, cerebral ventricles pathology, chronic disease, brain pathology, organic mental disorders, psychotic pathology.

1146. Del, M. V. J. "Seasonal worsening of bulimia nervosa." *The American Journal of Psychiatry* 148 (December 1991): 1753. Topics discussed include: mental illness, depression.

1147. Di, M. V. "Minimal brain damage and seasonality of birth in schizophrenic disorders." *European Psychiatry* 7, no. 2 (1992): 91-92. Topics discussed include: brain damage, Italy.

1148. Dill, D. B. *Life, heat, and altitude; physiological effects of hot climates and great heights.* 211 pages. Cambridge: Harvard University Press, 1938. Topics discussed include: heat, man, weather, mental effects, physiology.

1149. Ede, A. "The seasonality of schizophrenic births in Alberta." *American Journal of Orthopsychiatry* 55, no. 3 (July 1985): 451-453. Topics discussed include: Canada, birth rates.

1150. ---. "'Seasonality of schizophrenic births': Reply." *American Journal of Orthopsychiatry* 56, no. 1 (January 1986): 169-171. Topics discussed include: Canada.

1151. Fananas, L. "Seasonality of birth in schizophrenia: An insufficient stratification of control population?" *Social Psychiatry and Psychiatric Epidemiology* 24, no. 5 (September 1989): 266-270. Topics discussed include: Spain.

1152. Faustman, W. "Season of birth and neuropsychological impairment in schizophrenia." *J Nerv Ment Dis* 180, no. 10 (October 1992): 644-8. Topics discussed include: memory, epidemiology, diagnosis, psychology.

1153. Felice, A. "Temperature and admission rates to Pennsylvania state mental hospitals." *J Clin Psychol* 25, no. 3 (July 1969): 329-31. Topics discussed include: Pennsylvania, seasonality, socioeconomics, hospitals.

1154. Fickova, E. "Effect of high altitude climatotherapy on the functional state of the nervous system in asthmatics." *Studia Psychologica* 21, no. 3 (1979): 196-208. Topics discussed include: climatotreatment, psychic load, asthma, skin, physiology, arousal, atmospheric conditions, physical treatment methods.

1155. Franzek, E. "Schizophrenia and birth seasonality contrary results in relation to genetic risk." *Fortschr Neurol Psychiatr* 61, no. 1 (January 1993): 22-6. Topics discussed include: Berlin, epidemiology, genotype, pregnancy, risk, causes, genetics, psychology.

1156. ---. "Birth seasonality in schizophrenia: Different results with respect to the genetic loading." *Fortschritte der Neurologie, Psychiatrie* 61, no. 1 (January 1993): 22-26. Topics discussed include: Germany, genetics.

1157. ---. "Season of birth effect reveals the existence of etiologically different groups of schizophrenia." *Biological Psychiatry* 32, no. 4 (August 1992): 375-378. Topics discussed include: cycloid psychosis, genetics.

1158. Freeman, H. "Relationship of schizophrenia to the environment. Special Issue: Schizophrenia as a systems disorder: The relevance of mediating processes for theory and therapy." *British Journal of Psychiatry* 155, no. Suppl 5 (July 1989): 90-99. Topics discussed include: etiology, social factors, genetics.

1159. Gallagher, B. J. "Racial factors in birth seasonality among schizophrenics: A preliminary analysis." *Journal of Abnormal Psychology* 92, no. 4 (November 1983): 524-527. Topics discussed include: ethnic differences.

1160. Garralda, M. E. "Seasonal variations in the child and adolescent psychoses." *Journal of Adolescence* 12 (September 1989): 315-21. Topics discussed include: mental illness, adolescent psychiatry.

1161. Gomez, F. P. I. "Seasonal nature of birth of schizophrenic patients." *Actas Luso Esp Neurol Psiquiatr Cienc Afines* 18, no. 2 (March 1990): 125-31. Topics discussed include: paranoid etiology, gender differences, diseases, delivery.

1162. Hafner, H. "Abnormal seasonality of schizophrenic births. A specific finding?" *Eur Arch Psychiatry Neurol Sci* 236, no. 6 (1987): 333-42. Topics discussed include: physiopathology, Germany, mental retardation, neurotic disorders, personality, gender differences.

1163. Hare, E. H. "Seasonal Variations in Psychiatric Illness." *Trends in Neurosciences* 3 (1980): 295-298. Topics discussed include: mental disorders, illness.

1164. Hare, E. "A relation between seasonal temperature and the birth rate of schizophrenic patients." *Acta Psychiatr Scand* 63, no. 4 (April 1981): 396-405. Topics discussed include: age effects, England, Wales, epidemiology.

1165. Hare, E. H. "Mental Disorder and Season of Birth: Comparison of Psychose with Neuroses." *British Journal of Psychiatry* 115 (1968): 533-540. Topics discussed include: illness.

1166. ---. "Mental Disorder and Season of Birth: A National Sample Compared with the General Population." *British Journal of Psychiatry* 124 (1974): 81-86. Topics discussed include: illness.

1167. ---. "Season of birth in schizophrenia and neurosis." *American Journal of Psychiatry* 132, no. 11 (November 1975): 1168-1171. Topics discussed include: neurotic patients, England, Wales.

1168. Hare, E. "Season of birth and birth order in schizophrenia." *Vestn Akad Med Nauk SSSR* 26, no. 5 (1971): 39-42. Topics discussed include: family composition, London, neurotic disorders, epidemiology.

1169. ---. "Schizophrenia and season of birth." *Br J Psychiatry* 120, no. 554 (January 1972): 124-5. Topics discussed include: bipolar disorder, epidemiology, England, Wales.

1170. Haug, H. "On the frequency of 'diurnal variations of mood' in psychiatric patients seasonal aspects." *Prog Clin Biol Res.* 341 (1990): 105-11. Topics discussed include: depressive disorder psychology, cycles, affect physiology, circadian rhythms, mental disorders.

1171. Hsieh, H. H. "Seasons of birth and sub types of schizophrenia." *British Journal of Psychiatry* 149 (July 1986): 125. Topics discussed include: gender differences, paranoia, viral disorders, causes.

1172. ---. "Seasons of birth and subtypes of schizophrenia." *Acta Psychiatrica Scandinavica* 75, no. 4 (April 1987): 373-376. Topics discussed include: gender differences, paranoia, subtypes.

1173. Iannuzzo, G. "The hypothesis of season of birth as a risk factor in schizophrenia. Recapitulation and critical review." *Minerva Psichiatr* 31, no. 2 (April 1990): 63-75. Topics discussed include: risk, epidemiology, etiology.

1174. James, R. "Seasonal Admission Rates in Texas Mental Hospitals." *Journal of Clinical Psychology* 24 (1968): 190. Topics discussed include: health, mental illness, statistics.

1175. James, W. "Schizophrenia and season of birth." *Br J Psychiatry* 119, no. 549 (August 1971): 229-30. Topics discussed include: epidemiology.

1176. Jones, I. H. "Seasonal births in schizophrenia: A Southern Hemisphere study using matched pairs." *Acta Psychiatrica Scandinavica* 59, no. 2 (February 1979): 164-172. Topics discussed include: differential resistance, infection, Australia, birth.

1177. Kellett, J. "Seasonality in Schizophrenia." *Lancet* 1 (1978): 664. Topics discussed include: mental disorders, illness.

1178. Kendler, K. "The season of birth of schizophrenic, neurotic and psychiatrically normal twins." *Br J Psychiatry* 141 (August 1982): 186-90. Topics discussed include: diseases, twins, neurotic disorders, epidemiology.

1179. Knobloch, H. "Seasonal Variation in the Births of the Mentally Deficient." *Am J Public Health* 48, no. 9 (1958): 1201-1208. Topics discussed include: climate, birth, mental illness, health.

1180. Kocur, J. "Season related incidence of schizophrenia and affective psychoses." *Psychiatr Pol* 16, no. 4 (1982): 261-6. Topics discussed include: Poland, gender effects, urban population, affective disorders, psychotic epidemiology.

1181. Koehler, K. "Season of birth and Schneider oriented diagnosis of schizophrenia, neurosis and psychopathy." *Archiv fur Psychiatrie und Nervenkrankheiten* 223, no. 1 (1976): 69-75. Topics discussed include: diagnosis, psychodiagnosis.

1182. Krupinski, J. "Season of birth in schizophrenia: An Australian study." *Australian and New Zealand Journal of Psychiatry* 10, no. 4 (December 1976): 311-314. Topics discussed include: southern versus northern hemisphere, environment, seasonality.

1183. Lewis, M. S. "An explanation for the season of birth effect in schizophrenia and certain other diseases." *Psychological Bulletin* 89, no. 3 (May 1981): 589-596. Topics discussed include: causes, etiology.

1184. Lewis, M. "Age incidence and schizophrenia: Part I. The season of birth controversy." *Schizophr Bull* 15, no. 1 (1989): 59-73. Topics discussed include: age effects, risk, epidemiology, psychology.

1185. Losonczy, M. "Seasonal variations of human lumbar CSF neurotransmitter metabolite concentrations." *Psychiatry Res* 12, no. 1 (May 1984): 79-87. Topics discussed include: alzheimer's disease, cerebrospinal fluid, neuroregulators, schizophrenia.

1186. Machon, R. A. "Seasonality, birth complications and schizophrenia in a high risk sample." *British Journal of Psychiatry* 151 (July 1987): 122-124. Topics discussed include: pregnancy, development, winter, populations at risk, obstetrical complications, urban areas.

1187. ---. "The interaction of seasonality, place of birth, genetic risk and subsequent schizophrenia in a high risk sample." *British Journal of Psychiatry* 143 (October 1983): 383-388. Topics discussed include: urban areas, genetics, risk, viral infections, Denmark, children.

1188. Mann, S. "Psychotropic drugs, summer heat and humidity, and hyperpyrexia: a danger restated." *Am J Psychiatry* 135, no. 9 (September 1978): 1097-1100. Topics discussed include: thermoregulation, adverse effects, acute etiology, mental disorders, muscular diseases, tranquilizing agents, tropical climates.

1189. McNeil, T. "Season of birth among siblings of schizophrenics. A test of the parental conception habits interpretation." *Acta Psychiatr Scand* 54, no. 4 (October 1976): 267-74. Topics discussed include: Sweden, epidemiology.

1190. Mills, C. A. "Climate and Metabolic Stress." *Am J Hyg* 29, no. 3 (1939): 147-164. Topics discussed include: weather, health issues, mental illness, climate.

1191. ---. "Depression, Weather and Health." *Human Biology* 10, no. 3 (1938): 388-399. Topics discussed include: health issues, mental illness, climate.

1192. Miura, T. "Seasonal distribution of births of schizophrenic patients." *Nippon Eiseigaku Zasshi* 30, no. 1 (April 1975): 254. Topics discussed include: Japan, birth rate, epidemiology.

1193. Morgan, R. "Does light rather than social factors synchronize the temperature rhythm of psychiatric patients?" *Chronobiologia* 7, no. 3 (July 1980): 331-5. Topics discussed include: seasonality, circadian rhythms, light, mental disorders, physiopathology, social environment.

1194. Moskowitz, R. A. "Seasonality in Schizophrenia." *Lancet* 1 (1978): 664. Topics discussed include: mental disorders, illness.

1195. Muller, H. G. "A hypothesis on the abnormal seasonality of schizophrenic births." *European Archives of Psychiatry and Neurological Sciences* 239, no. 5 (April 1990): 331-334. Topics discussed include: seasonality, conception, premature births, children, Germany, pregnancy.

1196. Nasrallah, H. A. "Seasonality of birth in subtypes of chronic schizophrenia." *Acta Psychiatrica Scandinavica* 69, no. 4 (April 1984): 292-295. Topics discussed include: seasonality, gender differences, paranoia, predisposition.

1197. Norris, A. S. "Season of Birth and Mental Illness: A Critical Examination." *Archives of General Psychiatry* 7 (1962): 207-212. Topics discussed include: fertility, mental illness.

1198. O'Callaghan, E. "Season of birth in schizophrenia: Evidence for confinement of an excess of winter births to patients without a family history of mental disorder." *British Journal of Psychiatry* 158 (June 1991): 764-769. Topics discussed include: family history, psychiatric disorders, Ireland, mental disorders, genetics.

1199. Odegard, O. "Season of Birth in the General Population and in Patients with Mental Disorder in Norway." *British Journal of Psychiatry* 125 (1974): 397-405. Topics discussed include: mental disorders, illness.

1200. O'Hare, A. "Seasonality of schizophrenic births in Ireland." *British Journal of Psychiatry* 137 (July 1980): 74-77. Topics discussed include: seasonality of births, Ireland, demography.

1201. Ohlund, L. S. "Electrodermal orienting response, maternal age, and season of birth in schizophrenia." *Psychiatry Research* 36, no. 2 (February 1991): 223-232. Topics discussed include: inpatients, age effects, mothers.

1202. Ohlund, L. "Season of birth and electrodermal unresponsiveness in male schizophrenics." *Biol Psychiatry* 27, no. 3 (1 February 1990): 328-40. Topics discussed include: mental disorders, prognosis, risk, arousal, diagnosis, psychology.

1203. Pantleo, P. "Climatic Temperature and Psychiatric Admissions." *Journal of Clinical Psychology* 26 (1970): 308-310. Topics discussed include: climate, mental illness, health.

1204. Parker, G. "Mental Disorders and Season of Birth." *British Journal of Psychiatry* 125 (1974): 329-330. Topics discussed include: mental disorders, illness.

1205. ---. "Schizophrenia and season of birth: Further southern hemisphere studies." *Australian and New Zealand Journal of Psychiatry* 12, no. 1 (March 1978): 65-67. Topics discussed include: southern hemisphere, New Zealand.

1206. ---. "Season of birth and schizophrenia: An equatorial study." *Acta Psychiatrica Scandinavica* 56, no. 2 (August 1977): 143-146. Topics discussed include: frequency, Philippines.

1207. Parry, B. L. "Treatment of a patient with seasonal premenstrual syndrome." *The American Journal of Psychiatry* 144 (June 1987): 762-6. Topics discussed include: depression, seasonality, mental illness, melatonin, light, psychology.

1208. Pasamanick, B. "Seasonality of schizophrenic births." *American Journal of Orthopsychiatry* 56, no. 1 (January 1986): 168-169. Topics discussed include: seasonality, births, Canada.

1209. Perez, R. H. "Seasonal predominance in the date of birth in schizophrenia and affective disease." *Gac Med Mex* 127, no. 1 (January 1991): 7-11; discussion 11-2. Topics discussed include: Mexico, nervous system, timing, affective disorders, epidemiology.

1210. Petersen, W. F. *The Patient and the Weather: Vol III Mental and Nervous Diseases.* Ann Arbor, MI: Edwards Brothers, 1934. Topics discussed include: health, mental illness, climate.

1211. Pulver, A. "Risk factors in schizophrenia. Season of birth, gender, and familial risk." *Br J Psychiatry* 160, (January 1992): 65-71. Topics discussed include: age effects, chronic disease, risk, diagnosis, genetics, gender identity, causes, psychology.

1212. ---. "The association between season of birth and the risk for schizophrenia." *Am J Epidemiol* 114, no. 5 (November 1981): 735-49. Topics discussed include: theory, New York, risk, epidemiology, etiology.

1213. ---. "Risk factors in schizophrenia: season birth in Maryland, USA." *Br J Psychiatry* 143 (October 1983): 389-96. Topics discussed include: hospitalization, Maryland, risk, etiology.

1214. Pulver, A. E. "Season of birth of siblings of schizophrenic patients." *British Journal of Psychiatry* 160 (January 1992): 71-75. Topics discussed include: winter, spring, siblings.

1215. ---. "Age incidence artifacts do not account for the season of birth effect in schizophrenia." *Schizophrenia Bulletin* 16, no. 1 (1990): 13-15. Topics discussed include: risk, age groups, causes, populations at risk.

1216. Rodrigo, G. "Season of birth of schizophrenics in Mississippi, USA." *Acta Psychiatr Scand* 86, no. 5 (November 1992): 327-31. Topics discussed include: age effects, Mississippi, pregnancy, labor, epidemiology.

1217. Rybakowski, J. "Seasonal variations of the dexamethasone suppression test in depression compared with schizophrenia: a gender effect." *J Affect Disord* 24, no. 2 (February 1992): 87-91. Topics discussed include: blood, seasonal affective disorder, diagnosis.

1218. Sacchetti, E. "The brain damage hypothesis of the seasonality of births in schizophrenia and major affective disorders: evidence from computerised tomography." *Br J Psychiatry* 160 (March 1992): 390-7. Topics discussed include: diagnosis, atrophy, Italy, epidemiology, risk, brain damage, etiology.

1219. Sankar, D. V. S. "Demographic studies on childhood schizophrenia: preliminary report on season of birth, diagnosis, sex, I.Q. and race distribution." *In: Sankar, D, Schizophrenia: current concepts and research Hicksville, N Y, PJD Publications, 1969 944 p* (1969): 450-469. Topics discussed include: children, intelligence, cross-cultural analysis, neurosciences, demography.

1220. Satel, S. L. "Seasonal cocaine abuse." *The American Journal of Psychiatry* 146 (April 1989): 534-5. Topics discussed include: seasonality, mental illness, drug abuse, cocaine, depression.

1221. Shimura, M. "Season of birth of schizophrenics in Tokyo, Japan." *Acta Psychiatrica Scandinavica* 55, no. 3 (March 1977): 225-232. Topics discussed include: monthly cycles, Tokyo, Japan, epidemiology.

1222. Shur, E. "Season of birth in high and low genetic risk schizophrenics." *British Journal of Psychiatry* 140 (April 1982): 410-415. Topics discussed include: seasonality, genetic risk, predisposition.

1223. ---. "Age prevalence and the season of birth effect in schizophrenia: A response to Lewis and Griffin." *Psychological Bulletin* 93, no. 2 (March 1983): 373-377. Topics discussed include: age prevalence.

1224. Sothern, R. B. "Circannual and menstrual rhythm characteristics in manic episodes and body temperature." *Biol Psychiatry* 33, no. 3 (1 February 1993): 194-203. Topics discussed include: manic disorder, psychophysiology, psychotic disorders psychology, thermoregulation, menstrual cycle physiology, cycles, seasonality.

1225. Stecher, L. I. "The Effect of Humidity on Nervousness and on General Efficiency." *Archiv Physiol New York* 38 (1916): 1-94. Topics discussed include: environment, weather, biology, behavior, mental illness, labor.

1226. Stiles, W. B. "Lack of synchronized seasonal variation in the intensity of psychological problems." *Journal of Abnormal Psychology* 102 (August 1993): 388-94. Topics discussed include: mental illness.

1227. Templer, D. I. "Comment on the Lewis and Griffin explanation for season of birth effect in schizophrenia." *Journal of Orthomolecular Psychiatry* 11, no. 3 (1982): 219-221.

1228. ---. "Confirmation of relationship between temperature and the conception and birth of schizophrenics." *Journal of Orthomolecular Psychiatry* 9, no. 3 (1980): 220-222. Topics discussed include: conception, warm months, cold months, seasonality.

1229. ---. "Multiple sclerosis, schizophrenia, temperature and latitude." *Journal of Orthomolecular Medicine* 5, no. 3 (1990): 125-128. Topics discussed include: multiple sclerosis, latitude, geography, Italy.

1230. ---. "Month of conception and birth of schizophrenics as related to the temperature." *Journal of Orthomolecular Psychiatry (Regina)* 7, no. 4 (1978): 31-235. Topics discussed include: environment, genetics, causes, research.

1231. ---. "Schizophrenia prevalence: Wheat, milk and temperature." *Journal of Orthomolecular Psychiatry* 9, no. 4 (1980): 284-286. Topics discussed include: milk, wheat consumption, prevalence, epidemiology, diets.

1232. Torrey, E. F. "Headache in schizophrenia and seasonality of births." *Biological Psychiatry* 26, no. 8 (December 1989): 852-853. Topics discussed include: headaches, season of birth, childhood illnesses, family effects, predisposition.

1233. ---. "'Seasonality of schizophrenic births': Comment." *American Journal of Orthopsychiatry* 56, no. 1 (January 1986): 169. Topics discussed include: births, Canada, seasonality.

1234. ---. "An influenza epidemic and the seasonality of schizophrenic births." *Psychiatry and biological factors*, 109-116. ed Edouard Kurstak. New York: Plenum Medical Book Co., Plenum Press, 1991. Topics discussed include: influenza, epidemiology, congenital disorders, demography, etiology.

1235. ---. "The seasonality of schizophrenic births: A reply to Marc S. Lewis." *Schizophrenia Bulletin* 16, no. 1 (1990): 1-3. Topics discussed include: theory, age effects, season of birth, causes, epidemiology.

1236. ---. "Seasonality of schizophrenic births in the United States." *Archives of General Psychiatry* 34, no. 9 (September 1977): 1065-1070. Topics discussed include: etiology, regional differences, seasonality of births.

1237. ---. "A shifting seasonality of schizophrenic births." *British Journal of Psychiatry* 134 (February 1979): 183-186. Topics discussed include: seasonality of births, temperature effects.

1238. Videbech, T. H. "Endogenous Psychoses and Season of Birth." *Acta Psychiatrica Scandinavia* 50 (1974): 202-218. Topics discussed include: mental disorders, illness.

1239. Vieweg, W. "Seasonal stability of water balance among schizophrenic patients subject to water intoxication." *Prog Neuropsychopharmacol Biol Psychiatry* 14, no. 2 (1990): 215-22. Topics discussed include: blood pressure, circadian rhythms, weight gain, physiopathology, water electrolyte balance.

1240. Watson, C. "Schizophrenic birth seasonality and the age incidence artifact." *Schizophr Bull* 16, no. 1 (1990): 5-10; discussion 17-28. Topics discussed include: age effects, epidemiology, risk, causes, psychology.

1241. ---. "Schizophrenic birth seasonality in relation to the incidence of infectious diseases and temperature extremes." *Arch Gen Psychiatry* 41, no. 1 (January 1984): 85-90. Topics discussed include: climate, marriage, Minnesota, causes, epidemiology.

1242. Watson, C. G. "The relationships of anhedonia and the process reactive dimension to season of birth and infectious disease incidence in schizophrenia." *Journal of Nervous and Mental Disease* 175, no. 1 (January 1987): 34-40. Topics discussed include: symptoms, infectious disorders, causes, prenatal development, epidemiology.

1243. ---. "Season of birth and schizophrenia: A response to the Lewis and Griffin critique." *Journal of Abnormal Psychology* 91, no. 2 (April 1982): 120-125. Topics discussed include: hospitals, epidemiology.

1244. Welham J.L. "Implications of an irregular seasonal effect in schizaphrenia birth rates." Schizaphrenia Research, 101. 1994.

1245. Witz-Justice, A. "Seasonality in Biochemical Determinations: A Source of Variance and a Clue to Temporal Incidence of Affective Illness." *Psychiatry Res* 1 (1979): 53-60. Topics discussed include: biology, mental illness, health.

1246. Woodard, P. "Seasonal birthrates and schizophrenia." *American Journal of Psychiatry* 147, no. 4 (April 1990): 537. Topics discussed include: winter, spring, birth rates.

1247. Woodruff, R. J. "Psychiatric illness and season of birth." *Am J Psychiatry* 131, no. 8 (August 1974): 925-6. Topics discussed include: astrology, bipolar disorder, epidemiology, Missouri, neurotic disorders, schizophrenia, mental disorders.

1248. Yoney, T. H. "Seasonal variation in obsessive-compulsive disorder: preliminary experience with light treatment." *The American Journal of Psychiatry* 148 (December 1991): 1727-9. Topics discussed include: light, psychology, mental illness.

1249. Zipursky, R. B. "Seasonality of birth and CT findings in schizophrenia." *Biological Psychiatry* 22, no. 10 (October 1987): 1288-1292. Topics discussed include: family history, populations at risk.

STRESS (PSYCHOLOGICAL, PHYSIOLOGICAL)

1250. Berglund, B. "Quantitative approaches in environmental studies." *International Journal of Psychology* 12, no. 2 (1977): 111-123. Topics discussed include: stress, methodologies, pollution, attitudes.

1251. Fraser, J. et al. "Receptor reaction to odor stress. A paradigm and methodological exploration." *Atmos Environ* 10, no. 2 (1976): 151-7. Topics discussed include: atmospheric effects, meteorology, socioeconomics, wind, odors, stress physiopathology.

1252. Gonzalez, R. R. et al. *Effect of Atropine on Local Skin Wettedness and Sensible Heat Loss*, Performer: Army Research Inst. of Environmental Medicine, Natick, MA., December 1984. Topics discussed include: exercise, atropine, perspiration, biology, skin, humidity, heart rate, pharmacology, medicine, stress.

1253. Greene, T. C. et al. "Environmental stress." *Sex differences in human performance Wiley series on studies in human performance*, pp. 81-106. John Wiley & Sons, Chichester, England; 202 pages. ed Mary Anne Baker. 1987. Topics discussed include: stress, gender differences, psychophysiology, noise effects, temperature effects, personal space, endocrine system.

1254. Hazard, C. J. et al. "Perception of heat stress in urban microclimates." *Perceptual and Motor Skills* 46, no. 3, Pt 2 (June 1978): 1146. Topics discussed include: psychology, environmental heat, temperature effects, physiological stress, endurance.

1255. Henschel, A. et al. "Assessment of industrial heat stress." *Am Ind Hyg Assoc J* 27, no. 1 (January 1966): 13-6. Topics discussed include: climate, work-related medicine, environmental health, physical labor.

1256. Kalkstein, L. S. et al. "An evaluation of winter weather severity in the United States using the weather stress index." *Bulletin of the American Meteorological Society* 68 (December 1987): 1535-40. Topics discussed include: winter, climate, physiological effects.

1257. Kerslake, D. M. *The stress of hot environments.* 316 pages. Monographs of the Physiological Society, no. 29, Cambridge Eng.: University Press, 1972. Topics discussed include: heat, stress, physiology.

1258. Livingstone, S. D. et al. *Effect of Cold Exposure on Various Sites of Core Temperature Measurements*, Performer: Defence and Civil Inst. of Environmental Medicine, Downsview (Ontario)., 1983. Topics discussed include: stress, physiology, exposure, medicine.

1259. McCorriston, J. et al. "The ecology of seasonal stress and the origins of agriculture in the Near East." *American Anthropologist* 93 (March 1991): 46-69. Topics discussed include: prehistory, domestication, the ecosystem, archeology.

1260. Modrzejewski, J. "The effect of tropical climate on the state of emotional stress." *Biul Inst Med Morsk Gdansk* 21, no. 1 (1970): 37-42. Topics discussed include: blood, emotions, fear, heat exhaustion, humidity, naval medicine, stress, tropical climates.

1261. Planalp, J. M. *Heat stress and culture in North India.* 557 pages. United States. Army Medical Research and Development Command. United States Army Medical Research and Development Commmand Special technical report, Washington: U.S. Army Medical Research and Development Command, 1971. Topics discussed include: India, ethnology, physiology.

1262. Rim, Y. "Psychological test performance during climatic heat stress from dessert winds." *Int J Biometeorol* 19, no. 1 (March 1975): 37-40. Topics discussed include: desert climates, air movements, psychology, stress, wind.

1263. Russell, R. L. *Plasma Catecholamines and Stress Assessment in Men Exposed to Moderate Altitudes Master's thesis*, Performer: Air Force Inst. of Tech., Wright-Patterson AFB, OH., 1982. Topics discussed include: stress, physiology, sickness, exposure, oxygen, blood.

1264. Ryd, H. et al. *Methods of evaluating human stress due to climate.* 42 pages. National Swedish building research. Document D6, 1970, Stockholm: Svensk byggtjanst (distr.), 1970. Topics discussed include: weather, physiology, stress physiology.

1265. Sharma, V. M. et al. "Ethnic variations in psychological performance under altitude stress." *Aviation, Space, and Environmental Medicine* 47 (March 1976): 248-251. Topics discussed include: stress, ethnic groups, variations, depression, anxiety, concentration, eye coordination, India.

5

Sociological Effects: A Select Bibliography

ACCIDENTS & INJURIES

1266. Axelson, O. "Working environment and health in small firms." *Lakartidningen* 68, no. 42 (13 October 1971): 4741-51. Topics discussed include: accidents, air pollution, climatic effects, environment, exposure, light, noise, state medicine, work-related diseases, work-related medicine.

1267. Bellis, P. "Study on the relationship between meteorological seasonal factors and acute vascular accidents." *Minerva Med* 68, no. 25 (19 May 1977): 1735-54. Topics discussed include: acute disease, coronary disease, mortality, hypertension, Italy, temperature, cardiovascular diseases, cerebrovascular disorders.

1268. Bernett, P. "Relationship between sport injuries and weather." *Fortschr Med* 93, no. 3 (23 January 1975): 99-103. Topics discussed include: adaptation, cold climates, skiing, temperature, athletic injuries etiology.

1269. Biener, K. "The weather and skiing accidents." *MMW Munch Med Wochenschr* 121, no. 12 (23 March 1979): 427-30. Topics discussed include: atmospheric pressure, safety, Switzerland, athletic injuries epidemiology.

1270. Borek, C. "The Role of Meteorological and Individual Factors in Occurrence of Work Accidents; ." *Dissertation Abstracts International C: European Abstracts* 48, no. 3 (September 1987): 477. Topics discussed include: injuries, Poland, work-related safety, health issues, weather, the workplace.

1271. Chilaiditis, G. "Meteorological factors and acute cerebral vascular accidents." *Sem Hop Paris* 44, no. 17 (8 April 1968): 1152-7. Topics discussed include: seasonality, atmospheric effects, cerebrovascular disorders epidemiology, temperature, weather.

1272. Danis, D. "Warm weather injuries." *J Emerg Nurs* 19, no. 4 (August 1993): 351-2. Topics discussed include: accidents, epidemiology, seasonality, swimming, wounds.

1273. De, L. B. P. "Acute accidents caused by heat." *Cah Anesthesiol* 17, no. 3 (April 1969): 239-65. Topics discussed include: environment, tropical climates, dehydration etiology, exhaustion.

1274. Delahaye, R. "Acute accidents due to heat. Clinical study of 122 cases observed in the northern Sahara." *Rev Corps Sante Armees Terre Mer Air* 5, no. 1 (February 1964): 69-104. Topics discussed include: Africa, convulsions, dehydration, fever, exhaustion, tropical climates.

1275. Dreyfuss, I. "Desert Shield: military wins battle against heat injury." *The Physician and Sports Medicine* 19 (June 1991): 141-5. Topics discussed include: heat, body temperature, medicine, military.

1276. Duclos, P. "Lightning related mortality and morbidity in Florida." *Public Health Rep* 105, no. 3 (May 1990): 276-82. Topics discussed include: age effects, death, employment, epidemiology, work-related diseases, risk, seasonality, gender effects, electric injuries.

1277. Gonzalez, A. "Meteorology and acute cerebrovascular accidents." *Acta Neurol Latinoam* 18, no. 1 (1972): 15-9. Topics discussed include: acute disease, Argentina, humidity, seasonality, atmospheric effects, epidemiology, temperature, weather.

1278. Harlfinger, O. "Industrial accidents and the weather." *MMW Munch Med Wochenschr* 118, no. 3 (16 January 1976): 69-72. Topics discussed include: Germany, transients, migrants.

1279. Helzer-Julin, M. "Sun, heat, and cold injuries in cyclists." *Clin Sports Med* 13, no. 1 (January 1994): 219-34. Topics discussed include: weather, cold climates, therapy, exhaustion, hypothermia, sunlight.

1280. Henry, C. "Heat stress and its effects on illness and injury rates." *Mil Med* 150, no. 6 (June 1985): 326-9. Topics discussed include: desert climates, exhaustion, epidemiology, military medicine.

1281. Hernandez, Q. J. "Effect of meteorological changes on the occurrence of cerebrovascular accidents." *Rev Clin Esp* 162, no. 3-4 (15-31 August 1981): 127-30. Topics discussed include: atmospheric effects, cerebrovascular disorders epidemiology, etiology, risk, seasonality, Spain, temperature, weather.

1282. Hubert, E. "Attempt to correlate accident and weather conditions." *Hefte Unfallheilkd* no. 130 (1978): 217-24. Topics discussed include: climate.

1283. Kiwerski, J. "Incidence of spinal injuries in different seasons of the year." *Chir Narzadow Ruchu Ortop Pol* 48, no. 5-6 (1983): 417-21. Topics discussed include: accidents, Poland.

1284. Koska, W. "Seasonal incidence of accidents in the countryside." *Pol Tyg Lek* 25, no. 32 (10 August 1970): 1224-6. Topics discussed include: agriculture, Poland.

1285. Kossaia, A. "Dependence of injuries on the weather." *Ortop Travmatol Protez* no. 5 (May 1975): 33-7. Topics discussed include: seasonality, Ukraine, urban areas, accidents, traffic, wounds, injuries.

1286. Langworthy, J. "Survey of occupational injuries on a large pineapple plantation during the harvesting season." *Hawaii Med J* 46, no. 2 (February 1987): 52, 55. Topics discussed include: Hawaii, accidents, agriculture, fruit, wounds, injuries etiology.

1287. Little, R. "Body temperature after accidental injury." *Br J Surg* 68, no. 4 (April 1981): 221-4. Topics discussed include: body regulation, seasonality, skin, wounds, injuries physiopathology.

1288. Luisto, M. "Tetanus caused by occupational accidents." *Scand J Work Environ Health* 18, no. 5 (October 1992): 323-6. Topics discussed include: agricultural workers, Finland, forestry, seasonality.

1289. Marcinkowski, J. "The meteorological factors influencing the number of man hours lost by workers in the engineering industry." *Zentralbl Bakteriol Orig B* 166, no. 1 (January 1978): 37-44. Topics discussed include: accidents, Poland, respiratory infections, seasonality, weather, absenteeism, work-related medicine.

1290. Martinez, B. F. "Geographic distribution of heat-related deaths among elderly persons: use of county-level dot maps for injury surveillance and epidemiologic research." *The Journal of the American Medical Association* 262 (27 October 1989): 2246-50. Topics discussed include: mortality.

1291. Masotti, G. "On the influence of meteorological and cosmic factors on the frequency of traffic accidents. Statistical study." *Minerva Medicoleg* 85, no. 6 (November 1965): 207-14. Topics discussed include: statistics, weather.

1292. Masterson, E. "Victims of our climate." *Injury* 24, no. 4 (April 1993): 247-8. Topics discussed include: age effects, fractures, epidemiology, gender effects, seasonality, weather.

1293. Mita, S. "Meteorological influences on occurence of foreign body of the cornea in an industrial factory." *Nippon Ganka Gakkai Zasshi* 68, no. 13 (December 1964): 1893-903. Topics discussed include: Japan, accidents, foreign bodies, weather.

1294. Mobed, K. "Occupational health problems among migrant and seasonal farm workers." *West J Med* 157, no. 3 (September 1992): 367-73. Topics discussed include: accidents, agricultural workers, California, family health, health issues services, transients.

1295. Mullner, R. "Seasonal variations of accidental poisonings in children: Illinois, 1972-73." *Ill Med J* 148, no. 2 (August 1975): 139-40. Topics discussed include: Illinois, poisoning epidemiology.

1296. Niskanen, T. "Assessing the safety environment in work organization of road maintenance jobs." *Accident Analysis and Prevention* 26, no. 1 (February 1994): 27-39. Topics discussed include: motivation, climate.

1297. Novak, R. "Seasonal and circadian trends in occupational injuries." *Prog Clin Biol Res* 341B (1990;): 355-362. Topics discussed include: accidents, chronobiology, work schedule tolerance, circadian rhythm.

1298. Oehmisch, W. "Seasonal rhythm of mortality. Mortality according to age groups and seasons." *Dtsch Gesundheitsw* 24, no. 22 (29 May 1969): 1046-52. Topics discussed include: accidents, arteriosclerosis, diabetes mellitus, Germany, lung diseases, neoplasms, cycles, urologic diseases.

1299. Ono, Y. "Biometeorologic studies on cerebrovascular diseases. II. Seasonal observation on effects of meteorologic factors on the death from cerebrovascular accident." *Jpn Circ J* 33, no. 11 (November 1969): 1299-308. Topics discussed include: atmospheric pressure, mortality, humidity, infarction mortality, Japan, temperature, cerebrovascular disorders mortality, weather.

1300. ---. "Biometeorologic studies on cerebrovascular diseases. I. Effects of meteorologic factors on the death from cerebrovascular accident." *Jpn Circ J* 33, no. 11 (November 1969): 1285-98. Topics discussed include: age effects, atmospheric effects, mortality, humidity, infarction, Japan, seasonality, gender effects, temperature, weather.

1301. ---. "Biometeorologic studies on cerebrovascular diseases. III. Effects by the combination of meteorologic changeson the death from cerebrovascular accident." *Jpn Circ J* 33, no. 11 (November 1969): 1309-14. Topics discussed include: Japan, atmospheric effects, cerebrovascular disorders mortality, humidity, temperature.

1302. ---. "Biometeorologic studies on cerebrovascular diseases. V. A multivariate analysis of meteorologic effects on cerebrovascular accident." *Jpn Circ J* 38, no. 3 (March 1974): 195-208. Topics discussed include: atmospheric pressure, cerebral embolism, thrombosis etiology, mortality, humidity, Japan, seasonality, temperature, weather.

1303. Palao, S. A. "Ischemic cerebrovascular accident and climatologic factors." *Arch Neurobiol Madr* 52, no. 3 (May 1989): 161-5. Topics discussed include: atmospheric effects, humidity, Spain, temperature, cerebrovascular disorders etiology, meteorology.

1304. Pawlowska, H. "The influence of meteorological factors on the incidence of maxillofacial injuries." *Czas Stomatol* 24, no. 10 (October 1971): 1143-8. Topics discussed include: Poland, epidemiology, weather.

1305. Persinger, M. A. "Geophysical variables and behavior: XX. Weekly numbers of mining accidents and the weather matrix: The importance of geomagnetic variation and barometric pressure." *Perceptual and Motor Skills* 59, no. 3 (December 1984): 719-722. Topics discussed include: geomagnetic variation, barometric pressure, frequency of mining, atmospheric effects, magnetism.

1306. Pfaff, G. "More accidents due to daylight saving time? A comparative study on the distribution of accidents at different times of day prior to and following the introduction of Central European Summer Time (CEST)." *Int Arch Occup Environ Health* 49, no. 3-4 (February 1982): 315-23. Topics discussed include: circadian rhythms, Germany, light, time.

1307. Roberts, K. "The automobile and heat stress." *Pediatrics* 58, no. 1 (July 1976): 101-4. Topics discussed include: accident prevention, body temperature regulation, infant care, sunlight, ventilation, weather, fever etiology.

1308. Sampson, J. "Injury and illness during cold weather training." *Mil Med* 148, no. 4 (April 1983): 324-30. Topics discussed include: New York, gender effects, cold climates, military medicine, morbidity, wounds, injuries epidemiology.

1309. Schweiger, I. "Comparative analysis of seasonal incidence of accidents and temporary disability at the Szczecin Shipyard in 1976." *Med Pr* 32, no. 4 (1981): 303-6. Topics discussed include: Poland, timing, work-related, disability evaluation, naval medicine.

1310. ---. "Seasons and the incidence of occupational and non occupational accidents at the Szczecin shipyard in 1976." *Med Pr* 32, no. 3 (1981): 233-6. Topics discussed include: Poland.

1311. Shapiro, S. "A study of racial difference, seasonal variation, and temporal variation in hip fracture related mortality, Cuyahoga County, Ohio 1958 1962, 1968 1972." *J Chronic Dis* 30, no. 3 (March 1977): 185-92. Topics discussed include: fractures, hip injuries.

1312. Shults, R. A. "Investigation of a heat stress-related death of a fire fighter." *Fire Technology* 28 (November 1992): 317-31. Topics discussed include: firefighters health, hygiene, injuries, fatalities.

1313. Sinks, T. "Surveillance of work related cold injuries using workers' compensation claims." *J Occup Med* 29, no. 6 (June 1987): 504-9. Topics discussed include: absenteeism, work-related diseases, economics, Ohio, seasonality, temperature, wind, cold.

1314. Skrinjaric, I. "Seasonal variations of dental injuries." *Stomatol Vjesn* 14, no. 3-4 (1985): 121-7. Topics discussed include: tooth injuries.

1315. Sotaniemi, E. "Effect of environmental temperature on hospital admissions for cerebrovascular accidents." *Ann Clin Res* 4, no. 4 (August 1972): 233-5. Topics discussed include: seasonality, cerebral hemorrhages, cerebrovascular disorders, environment.

1316. Steinberg, H. "The weather and psychopathology." *Hillside Journal of Clinical Psychiatry* 1, no. 2 (1979): 231-243. Topics discussed include: environment, cold climates, injuries.

1317. Takala, E. "Seasonal variation in neck and shoulder symptoms." *Scand J Work Environ Health* 18, no. 4 (August 1992): 257-61. Topics discussed include: age effects, multivariate analysis, work-related diseases epidemiology, repetition strain injury, etiology, shoulder injuries.

1318. Taylor, M. "Cold weather injuries during peacetime military training." *Mil Med* 157, no. 11 (November 1992): 602-4. Topics discussed include: risk, trench foot epidemiology, cold climates, frostbite epidemiology, military personnel.

1319. ---. "Frostbite injuries during winter maneuvers: a long term disability." *Mil Med* 154, no. 8 (August 1989): 411-2. Topics discussed include: acute disease, weather, frostbite complications, military personnel.

1320. Turk, H. "Effect of meteorologic environmental factors on the incidence of accidents." *Z Gesamte Hyg* 32, no. 12 (December 1986): 704-7. Topics discussed include: cross-sectional analysis, Germany, work, seasonality, weather.

1321. Vernon, H. M. "Industrial Accidents in Relation to Temperature." *Eng and Ind Man,* London (December 1922): 28. Topics discussed include: industry, economy, environment, labor.

1322. ---. *The Relation of Atmospheric Conditions to the Working Capacity and the Accident Rate of Coal Miners*, Indus. Fatigue Res. Board (British), 1927. Topics discussed include: environment, labor, health issues.

1323. Will, D. P. *Prediction of Police Incidents and Accidents by Meteorological Variables*, Group Psychology Branch, Texas Christian University, Fort Worth, TX, 1969. Topics discussed include: weather, behavior, health issues.

CIVILIZATION, CULTURE & HISTORY

1324. Allen, R. O. "Impact of the environment on Egyptian civilization before the Pharaohs." *Analytical Chemistry* 65 (January 1993): 32A-34A. Topics discussed include: rivers, sediments, the Nile, Egypt, civilization, archeology, climatic changes.

1325. Allen, W. D. "Music history in five five-hundred-year cycles (Greek-Early Christian-Gregorian-Medieval-Modern); conditioned by the parallel climatic-cultural cycles." 41 pages. Weather Science Foundation. *Journal of Human Ecology,* 1 no. 13 (1951); Crystal Lake, Illinois: Published for the Weather Science Foundation by Weather Forecasts, 1951. Topics discussed include: history, physiology.

1326. Anderson, E. *Latvia-past and present ... 1918-1968. Historical past in pictures, climate, fauna and flora, the country and its cities, people and their culture.* 2d. rev. ed Waverly, IA: Latvju Gramata, 1969. Topics discussed include: Latvia.

1327. Barnett, G. A. "Seasonality in Television Viewing: A Mathematical Model of Cultural Processes." *Communication Research* 18, no. 6 (December 1991): 755-772. Topics discussed include: sociocultural factors, television ratings.

1328. Barrell, J. "Probable Relations of Climatic Changes to the Origin of the Tertiary Ape-Man." *Scientific Monthly* (1917). Topics discussed include: climate, history, anthropology, archeology.

1329. Basu, A. "High Altitude Hypoxia, Culture, and Human Fecundity, Fertility." *American Anthropologist* 86 (December 1984): 994-6.

1330. Bates, C. C. "America's weather warriors, 1814-1985." *Technology and Culture* 29 (January 1986): 161-2.

1331. Beals, K. L. "Brain size, cranial morphology, climate, and time machines." *Current Anthropology* 25, no. 3 (June 1984): 301-330. Topics discussed include: head measures, ethnic groups, environment, morphology, evolution.

1332. Bishop, C. W. "The Geographical Factor in the Development of Chinese Civilisation." *Geographical Review* (1922): 7 Topics discussed include: geography, civilisation, history, culture.

1333. Boyden, S. "Cultural Adaptation to Ecological Maladjustment." in *The Impact of Civilization on the Biology of Man*, 190-218. editor S. Boyden. Canberra: Australian National University Press, 1970. Topics discussed include: culture, environment, society, behavior, biology.

1334. Brown, N. "Climate-change and human history - some indications from Europe, AD400-1400." *Environmental Pollution* 83, no. 1-2 (1994): 37-43. Topics discussed include: history, science.

1335. Bryson, R. A. "Climatic Change and the Mill Creek Culture of Iowa." *Journal of the Iowa Archeological Society* 15-16 (1968): 1-358. Topics discussed include: history.

1336. ---. "Some cultural and economic consequences of climate change." University of Wisconsin, Madison. Center for Climatic Research. Institute for Environmental Studies, 1975. Topics discussed include: crops, weather, mental illness, physiology, man.

1337. Buchner, A. P. "Cultural responses to Altithermal (Atlantic) climate along the eastern margins of the North American grasslands, 5500 to 3000 B.C." Ph.D. diss., University of Calgary, National Museums of Canada, 1980. Topics discussed include: paleoclimatology, North America, Indians, antiquities, paleoecology.

1338. Budyko, M. I. "Anthropogenic Climatic Changes." *Climate and development: Climatic change and variability and the resulting social, economic and technological implications*, 270-84. eds H. J. Otten, D. Trinidade, S. C. Karpe. New York; Berlin; London and Tokyo: Springer in cooperation with the United Nations, 1990. Topics discussed include: natural resources, climatic effects.

1339. Bullen, A. K. ed. "Development of high civilizations in hot climates." 101-149. Florida Academy of Science. *Florida anthropologist* 6, no. 4 (December 1953) Gainesville, FL: Florida Anthropological Society, 1953.

1340. Butzer, K. "Environment, culture, and human evolution." *Am Sci* 65, no. 5 (September 1977): 572-84. Topics discussed include: climate, fossils, history, medicine, ancient civilizations, culture, evolution.

1341. Carter, G. F. *Man and the Land: A Cultural Geography*. New York: Holt, Rinehart and Winston, 1964. Topics discussed include: culture, geography, behavior.

1342. Chaunu, P. "Le Climat et l'Histoire, à Propos d'un Livre Récent." *Rev Historique* (October 1967). Topics discussed include: climate, history, civilization.

1343. Claiborne, R. *Climate, man and history*. 444 pages. London: Angus & Robertson, 1973. Topics discussed include: civilization, culture.

1344. Condon, R. G. *Inuit behavior and seasonal change in the Canadian Arctic*. 228 pages. Studies in cultural anthropology; no. 2, Ann Arbor, Mich.: UMI Research Press, 1983. Topics discussed include: Eskimos, physical measures, psychology, arctic medicine, disease, seasonality, Indians of North America, man.

1345. ---. "Seasonal Variation and Interpersonal Conflict in the Central Canadian Arctic." *Ethnology* 21, no. 2 (April 1982): 151-164. Topics discussed include: sociology, Canada.

1346. Custer, J. F. "Cultural ecology and archaeology in the Middle Atlantic: Are people really more than what they eat: Part I: A short history of studies of past environments in Middle Atlantic archaeology." *Pennsylvania Archaeologist* 57, no. 1 (1987): 47-51. Topics discussed include: Indians, culture, prehistory, ecology, archaeology.

1347. de Vries, J. "Measuring the Impact of Climate on History: The Search for Appropriate Methodologies." *Journal of Interdisciplinary History* 10 (1980): 599-630. Topics discussed include: history.

1348. Edgerton, R. "Cultural vs. Ecological Factors in the Expression of Values, Attitudes, and Personality Characteristics." *American Anthropologist* 67 (1965): 442-447. Topics discussed include: culture, climatic effects, ecology, environment, behaviour, psychology.

1349. Ekiguchi, K. *Japanese Craft & Customs: A Seasonal Approach*. New York: Kodansha America, 1993. Topics discussed include: sociology, anthropology, archeology, social life, customs, arts.

1350. Feldman, D. A. "The history of the relationship between environment and culture in ethnological thought: An overview." *J of the Hist of the Behavioral Sci* 11, no. 1 (1975): 67-81. Topics discussed include: ethnology, cultural ecology.

1351. Fischer, D. H. "Climate and History: Priorities for Research." *Journal of Interdisciplinary History* 10 (1980): 821-831. Topics discussed include: history.

1352. Friedman, R. M. "Appropriating the weather." *Technology and Culture* 31 no. 31 (October 1990): 251 888-9.

1353. Gates, D. M. *Man and his environment: climate*. 175 pages. Man and his environment series, New York: Harper & Row, 1971. Topics discussed include: climatology, environment, policy.

1354. George, H. B. *The Relations of Geography and History*. Oxford University Press: 1901. Topics discussed include: climate, culture, geography, history.

1355. Gourou, P. "Civilisations et Malchance Geographique." *Annales Economies Societes Civilisations* 4 (1949): 445-450. Topics discussed include: climate, geography, history, economy, society, culture.

1356. Gribbin, J. R. *Climate and mankind*. 56 pages. World Climate Conference (Geneva, Switzerland). London: Earthscan, 1979. Topics discussed include: climatic changes.

1357. Griffin, J. B. "Some Correlations of Climatic and Cultural Change in Eastern American Prehistory." *Annals of the New York Academy of Sciences* 95, no. 1 (1961): 710-717. Topics discussed include: climate, culture, history, archeology.

1358. Guterbock, T. M. "The Effect of Snow on Urban Density Patterns in the United States." *Environment and Behavior* 22, no. 3 (May 1990): 358-386. Topics discussed include: weather, population density.

1359. Hadlow, L. *Climate, vegetation and man*. 288 pages. Fundamental geography, book 3, London: University of London Press, 1964. Topics discussed include: bioclimatology, crops, agricultural zones.

1360. Hardy, J. D. "Measurement of the radiant heat load on man in summer and winter Alaskan climates." 43 pages. United States. Arctic Aeromedical Laboratory, Ladd Air Force, Alaska Special report, Ladd Air Force Base, Alaska: 1953. Topics discussed include: Alaska, climatic effects, temperature, physiology.

1361. Harris, M. *Culture, Man and Nature*. New York: Crowell, 1971. Topics discussed include: culture, civilization, environment.

1362. Hippocrates. *Airs, Waters and Places*. 460 B.C. Topics discussed include: weather, climatic effects, mortality, health.

1363. Hoch, I. "Wages, Climate and the Quality of Life." *Journal of Environmental Economics and Management* 1 (1974): 268-295. Topics discussed include: environment, economy, society, culture.

1364. Huntington, E. "Changes of Climate and History." *Am Hist Rev* 18 (1913): 213-232. Topics discussed include: history, society, culture.

1365. ---. *Mainsprings of Civilization*. New York: Wiley, 1945. Topics discussed include: theory, culture, history, society, geography, climate.

1366. ---. "Maya Civilization and Climatic Changes." Proc 19th Intern Congr Americanists, 150-164. Washington, 1917. Topics discussed include: climate, civilization, culture, history, environment.

1367. ---. "Physical Environment as a Factor in the Present Conditions of Turkey." *J Race Development* I (1911): 460-481. Topics discussed include: environment, geography, society, history, anthropolgy.

1368. ---. *The Character of Races As Influenced by Physical Environment, Natural Selection and Historical Development*. New York: Charles Scribner's Sons, 1924. Topics discussed include: climate, geography, biology, history, society, culture.

1369. ---. *Civilization and climate*. 453 pages. 3d ed., New Haven: Yale University Press, 1924. Topics discussed include: man, geography.

1370. ---. "Climate and the Evolution of Civilization." in *The Evolution of Earth and its Inhabitants*, editor R. S. Lull. New Haven: Yale University Press, 1919. Topics discussed include: civilization, culture.

1371. ---. "Death Valley and our future climate." Harper's Monthly Magazine (May 1916). Topics discussed include: climate.

1372. Jones, A. M. "Exploiting a Marginal European Environment: Population Control and Resource Management under the Ancien Regime." *Journal of Family History* 16, no. 4 (October 1991): 363-379. Topics discussed include: resources, demography, France, Malthus.

1373. Junior League of McAllen, Inc. Staff. *Some Like It Hot: The Climate, Culture & Cuisine of South Texas.* McAllen, TX: Junior League of McAllen, 1992. Topics discussed include: social life, customs, cookery, Texas.

1374. Lamb, H. H. "Climate and history: Report on the international conference on 'Climate and history: Climatic research unit, University of East Anglia, Norwich, England, 8-14 July 1979." *Past & Present* 88 (1980): 136-141. Topics discussed include: methodology, history.

1375. ---. *Climatic History & the Modern World.* New York, NY: Routledge, Chapman & Hall, 1982. Topics discussed include: weather, meteorology.

1376. ---. *Climatic History & the Future.* Princeton, NJ: Princeton University Press, 1985. Topics discussed include: weather, meteorology.

1377. Laudien, H. *Temperature relations in animals and man.* 229 pages. Biona report; 4. Schriftenreihe der Akademie der Wissenschaften und der Literatur, Mainz, Kommission fur Biologische Kybernetik, Mainz, Stuttgart ; New York: Akademie der Wissenschaften und der Literatur ; G. Fischer, 1986. Topics discussed include: animals, heat.

1378. Le Roy Ladurie, E. "Histoire et Climat." *Annales: Economies, Societes, Civilisations* 14 (1959): 3-34. Topics discussed include: climate, economy, society, weather, culture.

1379. Lee, D. "Culture and climatology." *Int J Biometeorol* 12, no. 4 (December 1968): 317-9. Topics discussed include: acclimation, culture.

1380. Leff, D. N. "Biology, Behavior and Environment in Human Evolution." *Mosaic* 10, no. 2 (March 1979): 2-12. Topics discussed include: evolution, biology, behavior, history.

1381. Ludlum, D. M. *The Weather Factor: Fascinating Accounts of How Weather Shaped America's Battles, Sports, & History.* Boston, MA: Houghton Mifflin Company, 1984. Topics discussed include: history, sports.

1382. Mackey, W. C. "A Statistical Analysis of Whiting's 'Effects of Climate on Certain Cultural Practices' with Special Reference to Galton's Problem." *Southeastern Review* 1, no. 1 (December 1973): 64-87. Topics discussed include: climates, culture, anthropology, Africa, ecology, diffusion.

1383. Meggers, B. "Environmental Limitations on the Development of Culture." *American Anthropologist* 56 (1954): 801-824. Topics discussed include: environment, geography, culture, economy, history, society.

1384. Mills, C. A. *Climate makes the man.* 320 pages. [1st ed.] ed. New York, London: Harper & Brothers, 1942. Topics discussed include: weather, mental illness, physiology, climatology, medicine.

1385. Milov, L. V. "Natural and Climatic Factors and Peculiarities of Russian Historic Process." *Historical Social Research, Historische Sozialforschung* 16, no. 2(58) (1991): 40-59. Topics discussed include: USSR, history, Russia, nature, climatic effects.

1386. Monge M. *Acclimatization in the Andes; historical confirmations of climatic aggression in the development of Andean man*, 130 pages. Baltimore: Johns Hopkins Press, 1948. Topics discussed include: altitude, mountain sickness.

1387. Montesquieu, C. d. S. d. *The Spirit of Laws.* Anne M. Cohler, Basia Carolyn Miller and Harold Samuel Stone, translators, Cambridge; New York: Cambridge University Press, 1989; original 1748.

Topics discussed include: climate, culture, civilization, mortality, fertility, alcohol, labor, laws, economics.

1388. Nelles, H. V. "Commerce in a Cold Climate: Bliss on Canadian Business History." *Business History Review* 62, no. 2 (June 1988): 310-16. Topics discussed include: economics, history, Canada.

1389. Newman, R. W. "The Relation of Climate and Body Size in U. S. males." *Am J Phys Anthropol N S* 13 (1955): 1-17. Topics discussed include: biology, physionomy.

1390. Numata, K. "Fudo, or climate and culture." *Shakaigaku Hyoron, Japanese Sociological Review* 25, no. 3 (December 1974): 99, 113-119. Topics discussed include: culture, climate, national character.

1391. Nuttonson, M. Y. *Rice culture and rice-climate relationships with special reference to the United States rice areas and their latitudinal and thermal analogues in other countries.* 555 pages. American Institute of Crop Ecology, and United States. Weather Bureau. Washington: American Institute of Crop Ecology, 1965. Topics discussed include: rice, climatic effects.

1392. Ogum, G. E. O. "Seasonality of Births in South-Eastern Nigeria." *Journal of Biosocial Science* 11, no. 2 (April 1979): 209-217. Topics discussed include: Nigeria, birth, season, culture, birth.

1393. Pain, S. "Rigid cultures caught out by climate-change." *New Scientist* 141, no. 1915 (5 March 1994): 13.

1394. *Global dermatology: diagnosis and management according to geography, climate, and culture*, eds Lawrence Charles Parish, Larry E. Millikan, and Mohamed Amer. New York: Springer-Verlag, 1994. Topics discussed include: medicine, geography, skin disease, epidemiology, dermatology, cross-cultural analysis, diagnosis, treatment, anthropology.

1395. Pearce, F. "Climate and man." *New Scientist* 125 (13 January 1990): 67.

1396. Peattie, R. *Geography in Human Destiny.* New York: George W. Stewart, 1940. Topics discussed include: geography, history, behaviour, biology.

1397. Petersen, W. F. *Man, weather, sun.* 463 pages. [1st ed.] ed. Springfield, IL.: C. C. Thomas, 1947. Topics discussed include: environment, climatology, medicine.

1398. Pfister, C. "Climate and Economy in Eighteenth-Century Switzerland." *Journal of Interdisciplinary History* 9, no. 2 (1978): 223-243. Topics discussed include: Switzerland, history, climatic effects, economy, demography, eighteenth century.

1399. Post, J. D. "The Impact of Climate on Political, Social, and Economic Change: A Comment." *Journal of Interdisciplinary History* 10, no. 4 (March 1980): 719-723. Topics discussed include: demography, economics, cycles, climatic variation.

1400. ---. "A Study in Meteorological and Trade Cycle History: The Economic Crisis Following the Napoleonic Wars." *Journal of Economic History* 34, no. 2 (June 1974): 315-49. Topics discussed include: history, fluctuations.

1401. "Proceedings of the World Climate Conference : a conference of experts on climate and mankind, Geneva, 12-23 February 1979." World Climate Conference (1979 : Geneva, Switzerland), and World Meteorological Organization. Secretariat of the World Meteorological Organization, Geneva, 1979. Topics discussed include: climatology, climatic changes.

1402. Rentsch, J. R. "Climate and culture: interaction and qualitative differences in organizational meanings." *Journal of Applied Psychology* 75 (December 1990): 668-81. Topics discussed include: sociology, interpersonal interaction, psychology.

1403. Robbins, M. C. "Climate and Behavior: A Biocultural Study." *Journal of Cross Cultural Psychology* 3 (1972): 331-344. Topics discussed include: behavior, biology, culture.

1404. ---. "Explorations in Psychocultural Bioclimatology." Ph.D. diss., University of Minnesota, 1966. Topics discussed include: anthropology, bioclimatology; available from UMI.

1405. Roberts, D. F. "Body Weight, Race and Climate." *Am JPhys Anthropol* II (1953): 533-558. Topics discussed include: biology, physionomy, anthropology.

1406. ---. *Climate and human variability*. 123 pages. 2d ed ed. Cummings modular program in anthropology, Menlo Park, CA.: Cummings Pub. Co., 1978.

1407. Ross, A. *Strange weather : culture, science, and technology in the age of limits,* 1956 Andrew Ross. London; New York: Verso, 1991. Topics discussed include: science, philosophy, sociology, technology, change in society.

1408. Rotberg, R. L. *Climate & History*. 1981. Topics discussed include: history.

1409. Semple, E. Ch. *American History and Its Geographic Conditions*. 1903. Topics discussed include: history, geography, environment, climate.

1410. ---. *Influences of Geographic Environment*. New York: Holt, Rinehart & Winston, 1911. Topics discussed include: history, geography, climate, economy.

1411. Shapley, H. *Climate and Civilization*. Cambridge: Harvard University Press, 1953. Topics discussed include: culture, society, history.

1412. Silverberg, R. *The challenge of climate: man and his environment.* 326 pages. [1st ed.] ed. New York: Meredith Press, 1969.

1413. Stahl, P. "On climate and occupation of the Santa Elena Peninsula: implications of documents for Andean prehistory." *Current Anthropology* 25 (June 1984): 351-2. Topics discussed include: geography, Ecuador, history, social forces.

1414. Upham, S. "Smallpox and Climate in the American Southwest." *American Anthropologist* 88, no. 1 (March 1986): 115-128. Topics discussed include: culture, interpersonal contact, epidemics, diseases, weather, American Indians, Spanish exploration, climatic effects.

1415. Valentine, J. H. "Human Crises and the Physical Environment." *Man Environment Systems* 5, no. 1 (January 1975): 23-28. Topics discussed include: weather, crisis, weather, news, pollution, a literature review.

1416. Walker, E. G. *The Gowen Sites: Cultural Responses to Climatic Warming on the Northern Plains, 7500 500 BP*. Chicago, IL: University of Chicago Press, 1993. Topics discussed include: sociology, anthropology, archeology, history, Canada, British Columbia.

1417. Walker, T. S. *Village and household economies in India's semi-arid tropics.* 394 pages. Baltimore and London: Johns Hopkins University Press, 1990. Topics discussed include: rural areas, economics.

1418. Wallis, W. "Geographical Environment and Culture." *Social Forces* 4 (1926): 702-708. Topics discussed include: geography, culture, society.

1419. Watsuji, T. *Climate and culture : a philosophical study*. 235 pages. Classics of modern Japanese thought and culture, New York: Greenwood Press, 1988. Topics discussed include: monsoons, ecology, Japan, philosophy.

1420. Whitebeck, R. H. "The Influence of Geographical Environment upon Religious Belief." *The Geographical Review* 5 (1918): 316-324. Topics discussed include: geography, behavior, religion, culture.

1421. Whiting, J. "Effects of Climate on Certain Cultural Practices." *Explorations in Cultural Anthropology*, W. Goodenough. New York: McGraw-Hill, 1964.

1422. Wigley, T. M. L. *Climate and history : studies in past climates and their impact on man.* 530 pages. Cambridge ; New York: Cambridge University Press, 1981. Topics discussed include: climatic change, sociology, paleoclimatology.

1423. Wulsin, F. R. "Adaptations to Climate Among NON-European Peoples." in *Physiology of Heat Regulation*, editor L. N. Newburgh. Philadelphia: Saunders, 1949. Topics discussed include: anthropology, biology, environment, temperature.

CRIME & HOMICIDE

1424. Anderson, C. A. "Temperature and aggression: Effects on quarterly, yearly, and city rates of violent and nonviolent crime." *Journal of Personality and Social Psychology* 52, no. 6 (June 1987): 1161-1173. Topics discussed include: cities, atmospheric effects, heat effects, epidemiology.

1425. ---. "Ambient temperature and violent crime: Tests of the linear and curvilinear hypotheses." *Journal of Personality and Social Psychology* 46, no. 1 (January 1984): 91-97. Topics discussed include: theory.

1426. Bagley, C. "Social Structure and the Ecological Distribution of Mental Illness, Suicide and Delinquency." *Psychological Medicine* 3 (1973): 177-187. Topics discussed include: environment, mental disorders, crime, society.

1427. Bell, P. A. "Linear and curvilinear relationships between temperature, affect, and violence: Reply to Cotton." *Journal of Applied Social Psychology* 16, no. 9 (1986): 802-807. Topics discussed include: aggression, heat, crime.

1428. Bicakova, R. A. "Seasonal variations in socially and legally unacceptable sexual behaviour." *Chronobiol Int* 2, no. 3 (1985): 203-8. Topics discussed include: France, rape, Texas, circadian rhythms, sex offenses.

1429. Chang, D. H. "Environmental Influences on Criminal Activity in Korea." *Criminology* 10, no. 3 (November 1972): 338-352. Topics discussed include: crime, Korea, physical factors.

1430. Cheatwood, D. "Is there a season for homicide?" *Criminology* 26 (May 1988): 287-306. Topics discussed include: crime, homicide.

1431. Cohen, J. "The Geography of Crime." *The Annals of the American Academy of Political and Social Science* 217 (1941): 29-37. Topics discussed include: geography, aggression.

1432. Cohn, E. G. "Weather and violent crime: A reply to Perry and Simpson, 1987." *Environment and Behavior* 22, no. 2 (March 1990): 280-294. Topics discussed include: population, unemployment rates, murder, rape, aggravated assault, seasonality, homicide, attacks.

1433. ---. "Weather and Crime." *British Journal of Criminology* 30, no. 1 (December 1990): 51-64. Topics discussed include: criminology, environment, criminal behavior, climatic influences, a literature review.

1434. Cotton, J. L. "Ambient temperature and violent crime." *Journal of Applied Social Psychology* 16, no. 9 (1986): 786-801. Topics discussed include: violence, heat effects.

1435. Curry, M. "The Relationship of Weather Conditions, Facial Characteristics and Crime." *Journal of Criminal Law and Criminology* 39 (August 1948): 253-261. Topics discussed include: climate, aggression.

1436. DeFronzo, J. "Climate and crime: Tests of an FBI assumption." *Environment and Behavior* 16, no. 2 (March 1984): 185-210. Topics discussed include: temperature, precipitation, climatic conditions, violence, urban areas, atmospheric effects.

1437. Dodge, R. W. "The seasonality of crime victimization." United States. Bureau of Justice Statistics. U.S. Dept. of Justice, Bureau of Justice Statistics, Washington, D.C., 1988. Topics discussed include: seasonality.

1438. ---. "Crime and seasonality." United States. Bureau of the Census. Crime Statistics Analysis Staff. U.S. Dept. of Justice, Bureau of Justice Statistics, Washington, 1980. Topics discussed include: weather statistics, criminal statistics.

1439. Ellis, F. "Mortality during heat waves in New York City July, 1972 and August and September, 1973." *Environ Res* 10, no. 1 (August 1975): 1-13. Topics discussed include: age effects, cardiovascular diseases, homicide, kidney diseases, weather.

1440. Falk, G. J. "The Influence of the Seasons on the Crime Rate." *The Journal of Criminal Law, Criminology and Police Science* 43 (1952): 199-213. Topics discussed include: aggression.

1441. Field, S. "The Effect of Temperature on Crime." *British Journal of Criminology* 32, no. 3 (June 1992): 340-351. Topics discussed include: England, Wales, weather, seasonality.

1442. Ganjavi, O. "Geophysical variables and behavior: impact of atmospheric conditions on occurrences of individual violence among Canadian penitentiary populations." *Perceptual and Motor Skills* 61 (August 1985): 259-75. Topics discussed include: psychology, aggression, seasonality, crime, criminals.

1443. Granger, A. *A Season for Murder*. New York, NY: Avon Books, 1993. Topics discussed include: crime, mortality.

1444. Harries, K. D. *Crime and the Environment*. Springfield, IL: Charles C. Thomas, 1980. Topics discussed include: environment.

1445. ---. "Spatial and Temporal Dimensions of Assaults Against Children in Dallas, Texas, 1980-1981." *Journal of Family Violence* 3, no. 4 (December 1988): 327-338. Topics discussed include: environment, socioeconomics, Texas, assault, climate.

1446. ---. "Determinism revisited: Assault and heat stress in Dallas, 1980." *Environment and Behavior* 15, no. 2 (March 1983): 235-256. Topics discussed include: thermal discomfort, aggravated assault, temperature effects, aggressive behavior, violence.

1447. ---. "Seasonality and assault: explorations in inter-neighborhood variation, Dallas 1980." *Annals of the Association of American Geographers* 74 (December 1984): 590-604. Topics discussed include: crime, criminals, Dallas Texas, assault, deviant behavior, heat psychology.

1448. Hartung, J. "Light, Puberty, and Aggression: A Proximal Mechanism Hypothesis." *Human Ecology* 6 (1978): 273-297. Topics discussed include: climate, crime, biology.

1449. Heller, N. "A Climatological Model for Forecasting the Demand for Police Service." *Journal of Research in Crime and Deliquency* 7 (1970): 167-176. Topics discussed include: climate, seasonality, society, behavior.

1450. Kaplan, S. J. "Climatic Factors and Crime." *The Professonal Geographer* 12, no. 6 (November 1960): 1-5. Topics discussed include: climate, aggression.

1451. ---. "The Geography of Crime." *Sociology of Crime*, 166-167. J. S. ed Roucek. New York: Greenwood Press, 1969. Topics discussed include: geography, aggression.

1452. Kevan, S. "Weather climate criminality." *ZFA Stuttgart* 52, no. 5 (20 February 1976): 257-64. Topics discussed include: accidents, suicide, crime.

1453. Kop, P. P. "Human aggressive behaviour: The period of perpetration in relation to month of birth of male offenders." *Journal of Biosocial Science* 9, no. 3 (July 1977): 309-316. Topics discussed include: seasonality, crime, aggressive behavior, male criminals, age differences.

1454. Lab, S. P. "Predicting crime from the weather." *The Futurist* 21 (March 1987): 30-2. Topics discussed include: seasonality, criminals, forecasting.

1455. Landau, S. F. "The Seasonality of Violent Crime: The Case of Robbery and Homicide in Israel." *Journal of Research in Crime and Delinquency* 30, no. 2 (May 1993): 163-191. Topics discussed include: Israel, robbery, homicide, seasons.

1456. LaRoche, E. "Weather and Crime in Tallahassee during 1954." *Journal of Criminal Law, Criminology and Police Science* 47 (1956): 218-219. Topics discussed include: aggression, climate.

1457. Leffingwell, A. *Illegitimacy and the Influence of the Seasons upon Conduct.* New York: Scribners, 1892. Topics discussed include: climate, behavior, crime.

1458. ---. *Illegitimacy & the Influence of Seasons Upon Conduct: Two Studies in Demography.* Salem, NH: Ayer Company Publishers, 1976. Topics discussed include: anthropology, archeology, sociology, crime, United Kingdom.

1459. Lester, D. "Variation in homicide rate with latitude and longitude in the United States." *Percept Mot Skills* 36, no. 2 (April 1973): 532. Topics discussed include: geography, suicide.

1460. Lewis, L. T. "The Influence of Season on Assault." *Professional Geographer* 28 (1975): 214-217. Topics discussed include: violence, aggression.

1461. Lucero, R. "Weather, Crime, and Mental Illness." *Journal of the Minnesota Academy of Science* 32 (1965): 223-226. Topics discussed include: seasonality, behavior, health, mental illness.

1462. Michael, R. P. "Annual rhythms in human violence and sexual aggression in the United States and the role of temperature." *Social Biology* 30, no. 3 (September 1983): 263-278. Topics discussed include: seasonality, violent crimes, violence, aggressive behavior, sex offenses, crime, rape.

1463. ---. "'Rape and seasonal variations': Drs. Michael and Zumpe reply." *American Journal of Psychiatry* 141, no. 8 (August 1984): 1015-1016. Topics discussed include: sexual violence, rape, sex offenses.

1464. ---. "Sexual violence in the United States and the role of season." *American Journal of Psychiatry* 140, no. 7 (July 1983): 883-886. Topics discussed include: rape, assault, robbery, murder, epidemiology, theft, homicide, environment.

1465. Michael, R. "An annual rhythm in the battering of women." *Am J Psychiatry* 143, no. 5 (May 1986): 637-40. Topics discussed include: rape, seasons, gender effects, temperature, violence, aggression psychology.

1466. Nadelson, C. C. "Rape and seasonal variations." *American Journal of Psychiatry* 141, no. 8 (August 1984): 1015. Topics discussed include: sexual violence, rape, sex offenses.

1467. Perry, J. D. "Violent crimes in a city: Environmental determinants." *Environment and Behavior* 19, no. 1 (January 1987): 77-90. Topics discussed include: murder, rape, seasonality, unemployment, homicide, violence, attack behavior, urban areas.

1468. Rotton, J. "Atmospheric and temporal correlates of sex crimes: endogenous factors do not explain seasonal differences in rape." *Environment and Behavior* 25 (September 1993): 625-42. Topics discussed include: seasonality, criminals, sexual crimes, weather, psychology.

1469. ---. "Geophysical variables and behavior: ubiquitous errors: a reanalysis of Anderson's (1987) 'temperature and aggression'." *Psychological Reports* 73 (August 1993): 259-71. Topics discussed include: violence, seasonality, criminals, aggression, psychology, heat.

1470. ---. "Air pollution, weather, and violent crimes: Concomitant time series analysis of archival data." *Journal of Personality and Social Psychology* 49, no. 5 (November 1985): 1207-1220. Topics discussed include: family violence, crime, time series analysis.

1471. Schmid, C. F. "A Study of Homocides in Seattle 1914 to 1924." *Social Forces* 4 (1926): 745-756. Topics discussed include: crime, aggression.

1472. Sells, S. B. *Accidents, Police Incidents, and Weather: A Further Study of the City of Fort Worth, Texas, 1968*, Group Psychology Branch, Office of Naval Research and Behavioral Research, Texas Christian University, Fort Worth, 1971. Topics discussed include: crime, behavior, health.

1473. Simpson, M. "Crime and climate: A reconsideration." *Environment and Behavior* 22, no. 2 (March 1990): 295-300. Topics discussed include: murder, rape, assault, urban areas, violence, homicide.

1474. U. S. Department of Justice. *Crime and Seasonality*. Washington, D.C.: Government Printing Office, 1980. Topics discussed include: seasonality.

FERTILITY (BIRTH RATES)

1475. Abelson, A. "Altitude and fertility." *Anthropological studies of human fertility*, 83-91. ed B. A. Kaplan. Detroit: Wayne State Univ Press, 1976. Topics discussed include: anoxia, culture, Peru.

1476. ---. "Altitude and fertility." *Hum Biol* 48, no. 1 (February 1976): 83-91. Topics discussed include: age effects, Peru, population growth, socioeconomics.

1477. Abelson, A. et al. "Altitude, migration, and fertility in the Andes." *Soc Biol* 21, no. 1 (March 1974): 12-27. Topics discussed include: Caucasoid race, ethnic groups, Indians, South America, transients, migrants.

1478. Bangham, C. et al. "Fertility of Nepalese Sherpas at moderate altitudes: comparison with high altitude data." *Ann Hum Biol* 7, no. 4 (July 1980): 323-30. Topics discussed include: China ethnology, marriage, menarche, Nepal, sex ratio.

1479. Bantje, H. "Seasonality of births and birthweights in Tanzania." *Soc Sci Med* 24, no. 9 (1987): 733-9. Topics discussed include: diet, fertility, food supply, infants, malaria physiopathology, pregnancy, rain, birth rate.

1480. Bantje, H. F. W. "Female Stress and Birth Seasonality in Tanzania." *Journal of Biosocial Science* 20, no. 2 (April 1988): 195-202. Topics discussed include: fertility, diseases, conception rate, malarial infection, hospitals.

1481. Barry, H. et al. "Season of Birth: An Epidemiological Study in Psychiatry." *Archives of General Psychiatry* 5 (1961): 292-300. Topics discussed include: fertility, climate.

1482. Batschelet, E. et al. "Anguliar Linear Correlation Co-Efficient for Rhythmometry and Circannual Changes in Births at Different Latitudes." *International Journal of Chronobiology* 1 (1973): 183. Topics discussed include: seasonality, birth, fertility, geography.

1483. Becker, S. "Seasonality of Fertility in Matlab, Bangladesh." *Journal of Biosocial Science* 13, no. 1 (January 1981): 97-105. Topics discussed include: rates, seasonal patterns.

1484. ---. "Seasonal Patterns of Fertility Measures: Theory and Data." *Journal of the American Statistical Association* 76, no. 374 (June 1981): 249-59. Topics discussed include: demography, economics.

1485. Becker, S. et al. "Seasonal Patterns of Reproduction in Matlab, Bangladesh." *Population Studies* 40, no. 3 (November 1986): 457-472. Topics discussed include: fertility, fecundity, Bangladesh, rural areas.

1486. Bell, P. "Bio-meteorological control of human fertility." *J R Coll Gen Pract* 21, no. 113 (December 1971): 705-9. Topics discussed include: fertilization, pregnancy, sunlight, climate.

1487. Bernard, R. P. et al. "Seasonality of Birth in India." *J Biosoc Sci* 10 (1978): 409-421. Topics discussed include: fertility, demography.

1488. Bloch, A. "Sudden infant death syndrome in the Ashkelon district. A 10 year survey." *Isr J Med Sci* 9, no. 4 (April 1973): 452-8. Topics discussed include: age effects, aspiration, autopsy, birth rate, mortality, Israel, pericarditis, pneumonia, respiratory tract infections, seasonality, gender effects.

1489. Bongaarts, J. et al. "Fertility effect of seasonal migration and seasonal variation in fecundability: test of a useful approximation under more general conditions." *Demography* 16, no. 3 (August 1979): 475-9. Topics discussed include: birth rate, transients, migrants.

1490. Brackenridge, C. "The seasonal variation of births of offspring from couples heterozygous for cystic fibrosis." *Ann Hum Genet* 42, no. 2 (October 1978): 197-201. Topics discussed include: Australia, birth rate, fertility, genes, infants, temperature.

1491. Caldwell, J. C. et al. "Periodic high risk as a cause of fertility decline in a changing rural environment: survival strategies in the 1980-1983 south Indian drought." *Economic Development and Cultural Change* 34 (July 1986): 677-701. Topics discussed include: India, rural areas, food supply.

1492. Chang, K. S. F. et al. "Climate and Conception Rates in Hong Kong." *Human Biology* 35 (September 1963): 366-376. Topics discussed include: fertility, demography.

1493. Chaudhury, R. H. "Socioeconomic and Seasonal Variations in Birth: A Replication." *Soc Biol* 19 (1972): 65-68. Topics discussed include: society, economy, fertility.

1494. Clark, S. J. et al. "Seasonal distribution of live births in a rural community in the southern United States." *Human Ecology* 15 (September 1987): 289-300. Topics discussed include: seasonality, fertility, childbearing age, family size.

1495. Clegg, E. et al. "Reproduction in human high altitude populations." *Hormones* 2, no. 1 (1971): 13-25. Topics discussed include: endocrine glands physiology, fertility physiology, gametogenesis physiology, infants, pregnancy, reproduction physiology.

1496. Cohen, I. C. et al. "Monthly Variation in Conceptions Leading to Induced Abortion." *Soc Biol* 24 (1977): 245-250. Topics discussed include: biology, fertility, seasonality, mortality, health.

1497. Condon, R. G. et al. "The ecology of human birth seasonality." *Human Ecology* 10 (December 1982): 495-511. Topics discussed include: Papua New Guinea, culture, customs, Eskimos, sociology, sex customs, primitive cultures, fertility.

1498. Condon, R. G. "Birth Seasonality, Photoperiod, and Social Change in the Central Canadian Arctic." *Human Ecology* 19, no. 3 (September 1991): 287-321. Topics discussed include: fertility, traditional social systems, arctic regions, seasonality, light, ethnography.

1499. Cooperstock, M. et al. "Seasonality of preterm birth in the Collaborative Perinatal Project: demographic factors." *Am J Epidemiol* 124, no. 2 (August 1986): 234-41. Topics discussed include: epidemiology, infants, marriage, childbearing age, pregnancy, socioeconomics.

1500. Cowgill, U. M. "Historical Study of the Season of Birth in the City of York, England." *Nature* 209 (12 March 1966): 1067-1070. Topics discussed include: demography, fertility, geography.

1501. ---. "Recent Variations in the Season of Birth in Puerto Rico." *Proc Natl Acad Sci USA* 52 (1964): 1149-1151. Topics discussed include: demography, fertility.

1502. ---. "Season of Birth in Man: Contemporary Situation with Special Reference to Europe and the Southern Hemisphere." *Ecology* 47 (1966): 614-623. Topics discussed include: demography, geography, fertility.

1503. Czeizel, E. et al. "Seasonal changes in the frequency of fetal damages and fertility." *Gynaecologia* 164, no. 2 (1967): 89-95. Topics discussed include: Hungary, cycles, pregnancy, demography, abortion, epidemiology, birth rate, fetal death, epidemiology.

1504. Drugan, A. et al. "Seasonal variation in conception does not appear to influence the rate of prenatal diagnosis of nondisjunction." *Fetal Ther* 4, no. 4 (1989): 195-9. Topics discussed include: fertilization, ovulation, physiology, pregnancy, risk.

1505. Dutt, J. "Altitude and fertility: the confounding effect of childhood mortality a Bolivian example." *Soc Biol* 27, no. 2 (June 1980): 101-13. Topics discussed include: anoxia, infants, female etiology, pregnancy, socioeconomics.

1506. Eckes, L. "Altitude adaptation. IV. Fertility and reproduction at high altitudes." *Gegenbaurs Morphol Jahrb* 122, no. 5 (1976): 761-70. Topics discussed include: birth weight, fetal anoxia physiopathology, infant mortality, infants, physiology, acclimation, pregnancy.

1507. Ehrenkranz, J. "Seasonal breeding in humans: birth records of the Labrador Eskimo." *Fertil Steril* 40, no. 4 (October 1983): 485-9. Topics discussed include: Newfoundland, pregnancy, coitus, Eskimos.

1508. Erhardt, C. L. et al. "Seasonal patterns of conception in New York City." *Am J Public Health* 61, no. 11 (November 1971): 2246-58. Topics discussed include: birth, death, ethnic groups, illegitimacy, income, maternal age, menstruation, pregnancy, Puerto Rico, gender effects, timing, fertilization.

1509. Goldstein, M. C. et al. "High altitude hypoxia, culture, and human fecundity, fertility: a comparative study." *American Anthropologist* 85 (March 1983): 28-49. Topics discussed include: anoxemia, Ladakh India, anthropometry.

1510. Gunn, P. A. "Seasonality of births in nineteenth-century urban Tasmania." *Human Biology* 64 (February 1992): 51-65. Topics discussed include: birth rate, population, Tasmania Australia, biology.

1511. Gupta, H. D. "Climate and Conception Rates in Punjab India." *Ind J publ Hlth* 19 (1975): 122-125. Topics discussed include: fertility, birth.

1512. Hajek, E. et al. "Seasonality of conception in human populations in Chile." *Int J Biometeorol* 25, no. 4 (December 1981): 281-91. Topics discussed include: birth rate, Chile, climate, pregnancy, temperature, fertilization.

1513. Halli, S. S. "The Seasonality of Births in Canada." *Journal of Biosocial Science* 21, no. 3 (July 1989): 321-327. Topics discussed include: Canada, pregnancy, birth, seasonality, women.

1514. Hart, N. "Famine, Maternal Nutrition and Infant Mortality: A Re-examination of the Dutch Hunger Winter." *Population Studies* 47, no. 1 (March 1993): 27-46. Topics discussed include: economics, medical subsidy programs, disability, fertility, child care, children.

1515. Himes, J. H. et al. "Seasonality of Births in a Rural Zapotec Municipio, 1945-1970." *Human Biology* 49 (1977): 2. Topics discussed include: birth, fertility.

1516. Hoffmann, F. "Seasonal fluctuations in the sex ratio of new borns and the influence of the declining birth rate on the sex ratio." *Geburtshilfe Frauenheilkd* 39, no. 3 (March 1979): 243-7. Topics discussed include: orral contraceptives, hormones, Germany, sex ratio.

1517. Hoffmann, F. et al. "Seasonal variations in the birth rate and conception rate within the last 200 years." *Geburtshilfe Frauenheilkd* 36, no. 9 (September 1976): 780-5. Topics discussed include: family planning, Germany, pregnancy, fertilization.

1518. Holland, B. "Seasonality of Births: Stability and Change in a Developing Country." *Human Biology* 61, no. 4 (August 1989): 591-598. Topics discussed include: Malaysia, fertility, urban population, religious cultural groups, demographic characteristics, Muslims, residential patterns, socioeconomic determinants.

1519. Houdaille, J. "Seasonal changes in birth rates in rural France from 1640 to 1869." *Population* 40, no. 2 (1985): 360-362. Topics discussed include: rural areas, France.

1520. Hrushesky, W. "The multifrequency (circadian, fertility cycle, and season) balance between host and cancer." *Ann N Y Acad Sci* 618, (1991): 228-56. Topics discussed include: breast neoplasms, surgery, carcinoma, drug therapy, floxuridine, immune system physiology, kidney neoplasms, physiopathology.

1521. Hum, D. et al. "Plantations, Staple Exports and the Seasonality of Births in Jamaica: 1880-1938." *Social and Economic Studies* 26, no. 1 (March 1977): 63-85. Topics discussed include: agriculture, birth rate, crops, production methods, Jamaica, rural areas.

1522. Huntington, E. *Season of Birth: its Relation to Human Abilities*. New York: John Wiley and Sons, 1938. Topics discussed include: fertility, demography.

1523. Hurtado, A. M. et al. "Seasonality in a Foraging Society: Variation in Diet, Work Effort, Fertility, and Sexual Division of Labor among the Hiwi of Venezuela." *Journal of Anthropological Research* 46, no. 3 (September 1990): 293-346. Topics discussed include: gender differences, labor force, hunting, gathering, diet, work, foragers.

1524. James, W. H. "Social Class and Season of Birth." *J BiosocSci* 3 (1971): 309-320. Topics discussed include: fertility, society.

1525. James, W. "Coital rate, sex ratio, and season of birth." *Lancet* 2, no. 716 (17 July 1971): 159. Topics discussed include: England, infants, marriage, pregnancy, demography, Wales, coitus, delivery.

1526. ---. "Seasonality in twin and triplet births." *Ann Hum Biol* 7, no. 2 (March 1980): 163-75. Topics discussed include: abortion, birth rate, fertilization, ovulation, pregnancy, sex behavior, triplets, twins.

1527. Johnson, J. T. et al. "Seasonality of Births for West Malaysia's Two Main Racial Groups." *Human Biology* 47.(1975): 3. Topics discussed include: birth, fertility, biology.

1528. Jongbloet, P. H. "Menses and mon phases, ovulation and seasons, vitality and month of birth." *Developmental Medicine and Child Neurology (London)* 25, no. 4 (August 1983): 527-531. Topics discussed include: mental retardation, correlations, lunar synodic cycle, menstruation, Netherlands, ovulation, endurance, biostatistics.

1529. Jongbloet, P. "Discrimination between two competing hypotheses on seasonality of birth in subfecundability traits." *Med Hypotheses* 40, no. 4 (April 1993): 211-6. Topics discussed include: pregnancy, fertility, cycles.

1530. Kallan, J. et al. "Demographic components of seasonality of pregnancy." *J Biosoc Sci* 21, no. 1 (January 1989): 101-8. Topics discussed include: coitus, fertility, pregnancy.

1531. Kestenbaum, B. "Seasonality of birth: two findings from the decennial census." *Social Biology* 34 (September 1987): 244-8. Topics discussed include: seasonality, fertility, sociology, population, history.

1532. Kosambi, D. D. et al. "Seasonal Variation in the Indian Birth Rate." *Ann Eugenics* 16 (1951): 165-192. Topics discussed include: fertility, demography, anthropology.

1533. Lam, D. et al. "Seasonality of births in human populations." *Social Biology* 38 (March 1991): 51-78. Topics discussed include: seasonality, fertility, cross-cultural analysis.

1534. Land, K. C. et al. "Arima models of seasonal variation in U.S. birth and death rates." *Demography* 20 (November 1983): 541-68. Topics discussed include: fertility, mortality.

1535. Laurenson, I. et al. "Fertility at low and high altitude in Central Nepal." *Soc Biol* 32, no. 1-2 (March 1985): 65-70. Topics discussed include: newborns, Nepal, pregnancy.

1536. Lee, R. D. "Short-Term Variation: Vital Rates, Prices, and Weather." in *The Population History of England, 1541-1871: A Reconstruction*, 356-401. editors E. A. Wrigley, and R. S. Schofield. Cambridge: Harvard University Press, 1981. Topics discussed include: mortality, demography, fertility, economy.

1537. Leridon, H. "Biometric Aspects of Human Fecundity and Birth-Rates, Seasons and Economic Conjuncture. Presentation of Two I.N.E.D. Essays; Aspects Biometriques de la Fecondite Humaine et Natalite Saisons et Conjoncture Economique. Presentation de Deux Cahiers de I.N.E.D." *Population* 28, no. 2 (March 1973): 285-289. Topics discussed include: fecundity, economy, demography, seasonality.

1538. Leslie, P. et al. "Extreme seasonality of births among nomadic Turkana pastoralists." *Am J Phys Anthropol* 79, no. 1 (May 1989): 103-15. Topics discussed include: birth rate, coitus, diet, food, Kenya, lactation, nutrition, pregnancy, temperature, weaning, fertility.

1539. Levine, R. J. "Male fertility in hot environment." *The Journal of the American Medical Association* 252 (21 December 1984): 3250-1. Topics discussed include: heat, biology.

1540. Levine, R. et al. "Deterioration of semen quality during summer in New Orleans." *Fertil Steril* 49, no. 5 (May 1988): 900-7. Topics discussed include: seasonality, sperm count, sperm mortality, heat, infertility, semen analysis.

1541. ---. "Air conditioned environments do not prevent deterioration of human semen quality during the summer." *Fertil Steril* 57, no. 5 (May 1992): 1075-83. Topics discussed include: employment, spermatozoa, temperature, seasonality, semen physiology.

1542. Levy, V. "Seasonal fertility cycles in rural Egypt: behavioral and biological linkages." *Demography* 23, no. 1 (February 1986): 13-30. Topics discussed include: birth rate, Egypt, employment, family planning, income, infant mortality, rural areas, population.

1543. Llewellyn, J. "Light and Sexual Periodicity." *Nature* 129 (1932): 868. Topics discussed include: climate, behaviour, fertility.

1544. Lyster, W. R. "New Seasonal Distribution of Births in New South Wales." *Med J Australia* 2 (1979): 150. Topics discussed include: birth, fertility.

1545. Lyster, W. "Three patterns of seasonality in American births." *Am J Obstet Gynecol* 110, no. 7 (1 August 1971): 1025-8. Topics discussed include: fertility, pregnancy, sex ratio, birth rate.

1546. Macfarlane, W. V. "Seasonality of Conception of Human Population." *Biometeorology* (1970): 4. Topics discussed include: demography, birth, fertility, biology.

1547. Macfarlane, W. V. et al. "Seasonal Conception Rates in Australia." *Med J Australia* 47 (1960): 121-124. Topics discussed include: demography, fertility.

1548. Malina, R. et al. "Seasonality of births in a rural Zapotec municipio, 1945 1970." *Hum Biol* 49, no. 2 (May 1977): 125-37. Topics discussed include: agriculture, Mexico, weather, birth rate.

1549. Massey, D. et al. "A demonstration of the effect of seasonal migration on fertility." *Demography* 21, no. 4 (November 1984): 501-17. Topics discussed include: preschool, marriage, Mexico, rural areas, population, socioeconomics, timing, ethnology, transients, migrants.

1550. Menken, J. "Seasonal migration and seasonal variation in fecundability: effects on birth rates and birth intervals." *Demography* 16, no. 1 (February 1979): 103-19. Topics discussed include: statistics, fertility, theory.

1551. Mills, C. A. et al. "Does Climate Affect the Human Conception Rate?" *Arch Internal Med* 46 (1930): 921-929. Topics discussed include: demography, biology, fertility.

1552. Mitchell, R. J. et al. "Social class and seasonality of birth in the midlands of Tasmania during the nineteenth century." *Human Biology* 57 (May 1985): 213-28. Topics discussed include: population, Tasmania Australia, socioeconomics, biology.

1553. Miura, T. et al. "The relation between seasonal birth variation and the season of the mother's birth." *Arch Gynecol* 229, no. 2 (1980): 115-22. Topics discussed include: Japan, mothers.

1554. Miura, T. *Seasonal effects on reproduction, infection and psychoses*, edited by T. Miura. 223 pages. Progress in biometeorology; v. 5, The Hague: SPB Academic, 1987. Topics discussed include: biological rhythms, diseases, seasonality, photoperiodism, medicine, climatology.

1555. Mosher, S. W. "Birth Seasonality among Peasant Cultivators: The Interrelationship of Workload, Diet, and Fertility." *Human Ecology* 7, no. 2 (June 1979): 151-181. Topics discussed include: Taiwan, nutrition, fishermen, nutritional cycles, labor, fishing, culture.

1556. Mukherjee, A. et al. "Wages and Involuntary Unemployment in the Slack Season of a Village Economy." *Journal of Development Economics* 37, no. 1-2 (November 1991): 227-64. Topics discussed include: education, demography, labor, mobility, migration, economic development, resources, income distribution, health, fertility.

1557. Naeye, R. L. "Seasonal Variations in Coitus and Other Risk Factors, and the Outcome of Pregnancy." *Ear Hum Dev* 4 (1980): 61-68. Topics discussed include: sex, behavior, biology, fertility.

1558. National Office of Vital Statistics "Seasonal Variation of Birth in the United States 1948-1954", *Vital Statistics - Special Report* 47, no.4 (1959). Topics discussed include: demography, fertility, statistics.

1559. Nelson, R. J. et al. "Mechanisms of seasonal cycles of behavior." *Annual Review of Psychology* 41 (1990): 81-108. Topics discussed include: seasonality, physiology, fertility.

1560. Nonaka, K. et al. "Effects of maternal birth season on birth seasonality in the Canadian population during the seventeenth and eighteenth centuries." *Hum Biol* 62, no. 5 (October 1990): 701-17. Topics discussed include: Canada, fertility, France, ethnology, history, medicine, marriage, childbearing age.

1561. Otto, W. "Seasonal reciprocity relation between mortality and reproduction figures." *Z Arztl Fortbild Jena* 65, no. 12 (15 June 1971): 621-4. Topics discussed include: biological clocks, cybernetics, Germany, birth rate.

1562. Panter, B. C. et al. "Seasonality of reproductive function and weight loss in rural Nepali women." *Hum Reprod* 8, no. 5 (May 1993): 684-90. Topics discussed include: aging, body mass, energy metabolism, fertility, lactation, Nepal, nutrition, reproduction, physiology, weight loss.

1563. Parker, G. "New Seasonal Distribution of Births in New South Wales." *Med J Australia* 2 (1979): 424. Topics discussed include: birth, fertility.

1564. ---. "Season of Birth in New South Wales." *Med J Australia* 2 (1978): 563-566. Topics discussed include: fertility.

1565. Parkes, A. S. "Seasonal Changes in Sexual Activity and Birth Rate." in *Patterns of Sexual Equality and Reproduction*, London: Oxford University Press, 1976. Topics discussed include: sex, fertility, behavior.

1566. Parkes, A. "Environmental influences on human fertility." *J Biosoc Sci Suppl* no. 3 (June 1971): 13-28. Topics discussed include: altitude, birth rate, climate, coitus, culture, infants, light, marriage, population growth, pregnancy, seasonality, temperature, twins.

1567. ---. "Seasonal variation in human sexual activity." *Eugen Soc Symp* 4 (1968): 128-45. Topics discussed include: abortion, epidemiology, ovulation, birth rate, coitus, emotion, geography, illegitimacy, libido, pregnancy, sexual behavior.

1568. Pasamanick, B. et al. "Geographic and Seasonal Variations in Births." *Public Health Reports* 74 (April 1959): 285-288. Topics discussed include: geography, demography, fertility, birth.

1569. Pasternak, B. "Seasonality in Childbirth and Marriage: A Chinese Case." *Bulletin of the Institute of Ethnology, Academia Sinica* 41 (1976): 25-45. Topics discussed include: birth, fertility, behavior, society, demography.

1570. Paxson, C. H. "Consumption and Income Seasonality in Thailand." *Journal of Political Economy* 101, no. 1 (February 1993): 39-72. Topics discussed include: microeconomics, economic development, health, education, fertility, consumer behavior.

1571. Petersen, D. J. et al. "Seasonal variation in adolescent conceptions, induced abortions, and late initiation of prenatal care." *Public Health Reports* 107 (November 1992): 701-6. Topics discussed include: fertility, teenage pregnancy, abortion, prenatal care.

1572. Prasad, B. G. et al. "A Study of the Seasonal Variation of Births at Maternity Hospitals of Some Medical Colleges in ndia." *Indian Journal of Medical Research* 57 (1969): 4. Topics discussed include: birth, fertility, health.

1573. Randall, W. "An eleven year cycle in human births." *Int J Biometeorol* 35, no. 1 (June 1991): 33-8. Topics discussed include: newborns, magnetics, cycles, pregnancy, sunlight, birth rate, meteorology.

1574. ---. "Photoperiod, temperature, and regional patterns of conceptions in the USA." *Int J Biometeorol* 37, no. 1 (February 1993): 52-60. Topics discussed include: birth rate, light, meteorology, pregnancy, fertility, physiology, cycles.

1575. ---. "The solar wind and human birth rate: a possible relationship due to magnetic disturbances." *Int J Biometeorol* 34, no. 1 (May 1990): 42-8. Topics discussed include: Finland, magnetics, melatonin, blood, cycles, wind, meteorology.

1576. Richards, T. "Weather, nutrition, and the economy: short run fluctuations in births, deaths, and marriages, France 1740-1909." *Demography* 20, no. 2 (May 1983): 197-212. Topics discussed include: France, history, medicine, birth rate, economics, marriage, mortality.

1577. Rodgers, J. L. et al. "The season-of-birth paradox." *Social Biology* 35 (September 1988): 171-85. Topics discussed include: seasonality, fertility.

1578. Rojansky, N. et al. "Seasonality in human reproduction: an update." *Hum Reprod* 7, no. 6 (July 1992): 735-45. Topics discussed include: physiology, melatonin, pregnancy, reproduction.

1579. Ronnike, F. "Seasonality of human conception associated with births in countries bordering the Norwegian Sea with special reference to northern Norway." *Tidsskr Nor Laegeforen* 101, no. 2 (20 January 1981): 151-5. Topics discussed include: the arctic, history, medicine, Norway, pregnancy, birth rates, fertilization.

1580. Rosenhouse Persson, S. "The Effects of Seasonal Migration on Fertility: Taking into Account Migration's Effects on the Proximate Determinants of Fertility." *Dissertation Abstracts International, A: The Humanities and Social Sciences* 46, no. 12(1) (June 1986): 3871-A. Topics discussed include: labor migration, rural areas, seasonality, migration.

1581. Sandahl, B. "Seasonal Birth Pattern in Sweden in Relation to Birth Order and Maternal Age." *Acta Obstet Gynecol Scand* 57 (1978): 393-396. Topics discussed include: fertility.

1582. Scaglion, R. "Seasonal births in a Western Abelam village, Papua New Guinea." *Hum Biol* 50, no. 3 (September 1978): 313-23. Topics discussed include: agriculture, ethnic groups, New Guinea, birth rate.

1583. Seiver, D. "Trend and variation in the seasonality of U.S. fertility, 1947-1976." *Demography* 22, no. 1 (February 1985): 89-100. Topics discussed include: heat effects.

1584. Seiver, D. A. "Seasonality of Fertility: New Evidence." *Population and Environment* 10, no. 4 (June 1989): 245-257. Topics discussed include: demography.

1585. Sherrets, S. "Weather and human conception." *Psychol Rep* 44, no. 3 Pt 2 (June 1979): 1241-2. Topics discussed include: birth rate, pregnancy, seasonality, fertilization.

1586. Shimura, M. J. et al. "Geographical and Secular Chages in the Seasonal Distribution of Birth." *Social Science and Medicine* 15D (1981): 103-109. Topics discussed include: geography, fertility.

1587. Slatis, H. M. et al. "Seasonal Variation in Stillbirth Frequencies." *Human Biology* 39 (1967): 284-294. Topics discussed include: demography, fertility, birth.

1588. Spencer, B. et al. "Spectral Analysis and the Study of Seasonal Fluctuations in Historical Demography." *J Europ Econ Hist* 5 (1976): 171-190. Topics discussed include: demography, fertility, mortality.

1589. ---. "Workload and Seasonal Variation in Birthrates Reconsidered." *Amer Anthropol* 79 (1977): 911-912. Topics discussed include: demography, fertility, birth.

1590. Stoeckel, J. et al. "Seasonal Variation in Births in Rural East Pakistan." *Journal of Biosocial Science* 4 (1972): 107-116. Topics discussed include: birth, fertility, demography.

1591. Stutvoet, H. J. "Seasonal Birth Frequencies in Parameters." *Acta Genet* 2 (1951): 177-192. Topics discussed include: fertility, demography.

1592. Suarez, M. R. "Various aspects of fertility and fecundity in high altitudes." *Ginecol Obstet Mex* 22, no. 133 (November 1967): 1527-36. Topics discussed include: Bolivia.

1593. Tembon, A. "Seasonality of births in the North West Province, Cameroon: implications for family planning programme." *Cent Afr J Med* 36, no. 4 (April 1990): 90-3. Topics discussed include: Cameroon, culture, infants, birth rates, family planning.

1594. Thompson, R. W. et al. "Seasonal Variation in Conception in Rural Uganda and Mexico." *American Anthropologist* 75 (1973): 676-686. Topics discussed include: fertility, birth, biology, demography.

1595. Timonen, S. et al. "Seasonal Changes in Obstetrical Phenomena." *Acta Obstet Gynec Scand* 44 (1965): 507-533. Topics discussed include: fertility, birth.

1596. Udry, J. R. et al. "Seasonality of Coitus and Seasonality of Birth." *Demography* 4 (1967): 673-79. Topics discussed include: sex, fertility, demography.

1597. U.S. National Center for Health Statistics. "Seasonal Variation of Births, United States, 1933-1963." *Vital and Health Statistics* series 21 (1966): 9. Topics discussed include: demography, fertility, birth.

1598. Valentine, T. R. "Drought, Transfer Entitlements, and Income Distribution: The Botswana Experience." *World Development* 21, no. 1 (January 1993): 109-26. Topics discussed include: education, labor, mobility, development, income distribution, migration nutrition, health, education, fertility, households.

1599. ---. "Mineral-Led Economic Growth, Drought Relief, and Incomes Policy: Income Distribution in Botswana Reconsidered." *American Journal of Economics and Sociology* 52, no. 1 (January 1993): 31-49. Topics discussed include: labor, mobility, international migration, development, wealth, income distribution, health, education, fertility.

1600. Warren, C. W. et al. "Social Status and Season of Birth: A Study of a Metropolitan Area in the Southeastern United States." *Soc Biol* 26 (1981): 275-287. Topics discussed include: society, fertility.

1601. Warren, C. W. "Seasonal Variation in Fertility: The Extent to Which Month-of-Birth as It is Related to Age-at-Entry into School Affects the Academic Performance of Children in the United States." Ph.D. diss., Emory University, 1978. Topics discussed include: sociology, demography, cognition; available from UMI.

1602. Warren, C. W. et al. "Seasonal variation in conception and various pregnancy outcomes." *Social Biology* 33 (March 1986): 116-26. Topics discussed include: fertility, abortion, miscarriages.

1603. Wickens, C. H. "The Vitality of White Races in Low Latitudes." *Economic Record and Proc Roy Soc Victoria* 40 (May 1927). Topics discussed include: latitude, fertility, mortality, climate.

1604. Zelnik, M. "Socioeconomic and Seasonal Variations in Birth: A Replication." *Milbank Mem Fund Quarterly* 47, (1969): 159-165. Topics discussed include: society, fertility.

INFANT MORTALITY

1605. Al'Bitskii, V. et al. "Statistical analysis of seasonal fluctuations in infant mortality." *Sov Zdravookhr* no. 5 (1977): 45-9.

1606. Anderson, S. et al. "Effect of ambient temperature on SIDS rate." *Med J Aust* 158, no. 10 (17 May 1993): 703-4. Topics discussed include: cold effects, infants, seasonality, urban areas, population, sudden infant death, epidemiology.

1607. Artemov, V. et al. "Infectious morbidity of children up to 7 years of age in relation to the season of their birth." *Zh Mikrobiol Epidemiol Immunobiol no.* 3 (March 1985): 34-7. Topics discussed include: preschool, infants, risk, Russia, timing, urban population, communicable diseases, epidemiology.

1608. Aubenque, M. et al. "Seasonal variations and chronological series in causes of death in France from 1900 to 1972." *Cah Sociol Demogr Med* 19, no. 1 (January 1979): 17-22. Topics discussed include: age effects, France, infants, mortality.

1609. Auliciems, A. et al. "Sudden infant deaths and clear weather in a subtropical environment." *Soc Sci Med* 24, no. 1 (1987): 51-6. Topics discussed include: humidity, etiology, temperature, wind, epidemiology, tropical climates.

1610. Beal, S. et al. "Sudden infant death syndrome related to climate." *Acta Paediatr Scand* 80, no. 3 (March 1991): 278-87. Topics discussed include: Australia, cold effects, infants, epidemiology, etiology.

1611. Bonser, R. et al. "Sudden infant death syndrome in Cardiff, association with epidemic influenza and with temperature 1955-1974." *Int J Epidemiol* 7, no. 4 (December 1978): 335-40. Topics discussed include: social classes, Wales, cold climates, diseases, propagation, influenza, etiology.

1612. Bouvier, C. M. et al. "Sudden infant death and immunization: an extensive epidemiological approach to the problem in France winter 1986." *Int J Epidemiol* 18, no. 1 (March 1989): 121-6. Topics discussed include: autopsy, causes, epidemiology, hospitalization, seasonality, etiology.

1613. ---. "Sudden infant death in France during the winter of 1986." *Arch Fr Pediatr* 45, no. 1 (January 1988): 21-6. Topics discussed include: newborns, seasonality, etiology, epidemiology.

1614. Breschi, M. et al. "The effect of season and climate on the survival of children in Italy during the 19th century." *Population* 41, no. 1 (1986): 9-35. Topics discussed include: infants, Italy, mortality.

1615. Bridges, W. C. et al. "The Australian general practice morbidity and prescribing survey, 1969 to 1974. 3. Workload and morbidity." *Med J Aust* 2, no. 1 Suppl (2 October 1976): 7-11. Topics discussed include: age effects, Australia, infants, seasonality, gender effects, urban areas, population.

1616. Buve, A. et al. "Sudden infant death syndrome: does winter affect poor and rich babies equally?" *J Epidemiol Community Health* 46, no. 5 (October 1992): 485-8. Topics discussed include: birth weight, England, risks, epidemiology, Wales, seasonality, social classes, etiology.

1617. Campbell, M. "Sudden infant death syndrome and environmental temperature: further evidence for a time lagged relationship." *Med J Aust* 151, no. 7 (2 October 1989): 365-7. Topics discussed include: age effects, body regulation, culture, risks, epidemiology, timing, etiology.

1618. Campbell, M. et al. "Sudden infant deaths and cold weather: was the rise in infant mortality in 1986 in England and Wales due to the weather?" *Paediatr Perinat Epidemiol* 5, no. 1 (January 1991): 93-100. Topics discussed include: epidemiology, seasonality, Wales, cold climates, etiology.

1619. Carpenter, R. et al. "Environmental findings and sudden infant death syndrome." *Lung* suppl (1990): 358-367. Topics discussed include: cross-sectional analyses, England, risks, Wales, air pollution, causes, diseases, mortality, seasonality.

1620. Chukanin, N. et al. "Role of hardening in lowering morbidity of pre school children in hot climate conditions." *Pediatriia no.* 10 (1989): 104. Topics discussed include: preschool children, Kyrgyzstan, heat, endurance.

1621. Cohen, P. "The influence on survival of season of onset of childhood acute lymphoblastic leukemia." *Chronobiol Int* 4, no. 2 (1987): 291-7. Topics discussed include: ethnic groups, Israel, epidemiology, mortality, cycles.

1622. Crum, F. S. "Geographical and Seasonal Variations in Infant Mortality in the United States 1917-1920." *Am J Pub Health* 10 (October 1920): 790. Topics discussed include: geography.

1623. Curtiss, J. et al. "Population characteristics and environmental factors that influence level and cause of mortality. A review." *J Environ Pathol Toxicol* 4, no. 2-3 (September 1980): 471-511. Topics discussed include: age effects, climatic effects, children, socioeconomics, urbanization, demography, pollution.

1624. Deacon, E. et al. "Some statistical and climatological aspects of the incidence of the sudden infant death syndrome." *Aust Paediatr J* 15, no. 4 (December 1979): 248-54. Topics discussed include: Australia, seasonality, epidemiology.

1625. ---. "The incidence of the sudden infant death syndrome in relation to climate." *Int J Biometeorol* 26, no. 3 (November 1982): 207-18. Topics discussed include: Australia, United Kingdom, cold climates, epidemiology, weather.

1626. Donelli, G. "Increase in infant mortality observed in the province of Naples in the autumn and winter 1978-1979. Introductory note." *Ann Ist Super Sanita* 17, no. 4 (1981): 699-703. Topics discussed include: Italy, respiratory diseases, epidemiology, seasonality.

1627. Douglas, A. et al. "Seasonality of disease in Kuwait." *Lancet* 337, no. 8754 (8 June 1991): 1393-7. Topics discussed include: cardiovascular, mortality, epidemiology, neoplasms, gender effects, socioeconomics, weather, infants.

1628. Faust, V. et al. "Weather and disease. 3. Weather dependent diseases conclusion." *Fortschr Med* 95, no. 6 (10 February 1977): 331-5. Topics discussed include: epidemiology, asthma, death sudden, epilepsy, Germany, infants, influenza, pregnancy, morbidity.

1629. Fyfe, T. et al. "Comparison of seasonal incidence and mortality of stroke and myocardial infarction." *Health Bull Edinb* 30, no. 2 (April 1972): 159-61. Topics discussed include: epidemiology, infants, myocardial infarction.

1630. Gonzalez, P. G. et al. "Seasonal variations in infant mortality in Cuba." *An Esp Pediatr* 28, no. 6 (June 1988): 521-5. Topics discussed include: Cuba.

1631. Gordon, J. et al. "Causes of death at different ages, by sex, and by season, in a rural population of the Punjab, 1957-1959: A field study." *Indian J Med Res* 53, no. 9 (September 1965): 906-16. Topics discussed include: India, infant mortality, rural areas, health.

1632. Greenberg, M. et al. "A study of the relationship between sudden infant death syndrome and environmental factors." *Am J Epidemiol* 98, no. 6 (December 1973): 412-22. Topics discussed include: Chicago, exposure, geography, humidity, population density, seasonality, socioeconomics, temperature, weather, infant mortality.

1633. Gudkov, V. et al. "An analysis of the seasonal rise in hepatitis A morbidity at a children's center and ways to improve the sanitary epidemiologic surveillance for this infection among organized preschoolers." *Zh Mikrobiol Epidemiol Immunobiol 1990* no. 3 (March 1990): 39-44. Topics discussed include: antigens, viral analysis, Byelarus, feces analysis, hepatitis, immunology, urban areas, day care centers.

1634. Guimaraes, N. D. C. "Seasonal variation in infant mortality rates in the City of Salvador, BA, Brazil." *Rev Saude Publica* 9, no. 3 (September 1975): 285-94. Topics discussed include: Brazil.

1635. Haberman, S. et al. "The seasonal variation in mortality from cerebrovascular disease." *J Neurol Sci* 52, no. 1 (October 1981): 25-36. Topics discussed include: age effects, bronchitis, emphysema, England, hypertension, infants, pneumonia, Wales.

1636. Hare, E. et al. "The changing seasonality of infant deaths in England and Wales 1912-78 and its relation to seasonal temperature." *J Epidemiol Community Health* 35, no. 2 (June 1981): 77-82. Topics discussed include: England, fetal death, epidemiology, history, medicine, pregnancy.

1637. Helmholtz, H. F. "Report on the Study of Temperature and Infant Health in Chicago." *Trans Am Assn Study and Prev Infant Mortality* 3 (1912): 282. Topics discussed include: health issues.

1638. Helweg, L. K. et al. "A statistical analysis of the seasonality in sudden infant death syndrome." *Int J Epidemiol* 14, no. 4 (December 1985): 566-74. Topics discussed include: age effects, communicable diseases, epidemiology, timing.

1639. Hereward, M. C. "An Investigation of the Seasonality of Sudden Infant Death Syndrome in the United States." *Dissertation Abstracts International, A: The Humanities and Social Sciences* 51, no. 5 (November 1990): 1785-A-1786-A. Topics discussed include: infant mortality, mortality rates, demography.

1640. Hesseling, P. "Onyalai at Rundu, Namibia 1981-1988: age, sex, morbidity, mortality and seasonal variation of 612 hospitalized patients." *Trans R Soc Trop Med Hyg* 84, no. 4 (July 1990): 605-7. Topics discussed include: hospitalization, length of stay, gender effects, epidemiology.

1641. Hoppenbrouwers, T. et al. "Seasonal relationship of sudden infant death syndrome and environmental pollutants." *Am J Epidemiol* 113, no. 6 (June 1981): 623-35. Topics discussed include: anoxia, California, risk, etiology, air pollution, poisoning, epidemiology.

1642. Hunt, V. et al. "Infant mortality and the environment of a lesser metropolitan county: a study based on births in one calendar year." *Environ Res* 9, no. 2 (April 1975): 135-51. Topics discussed include: Pennsylvania, pneumonia, pregnancy, prenatal care, respiratory diseases, seasonality, socioeconomics.

1643. Inokawa, T. et al. "Relationship between meteorological conditions and mortality. 14." *Nippon Ika Daigaku Zasshi* 37, no. 5 (1970): 368-75. Topics discussed include: age effects, air pollution, infants, gender effects, temperature, Tokyo, weather.

1644. Jakobovits, A. et al. "Seasonal incidence of conception, birth and perinatal death in multiple pregnancies." *Orv Hetil* 128, no. 18 (3 May 1987): 927-30. Topics discussed include: Hungary, infants, fertilization, fetal epidemiology, pregnancy.

1645. Janerich, D. et al. "Season of birth and neonatal mortality." *Am J Public Health* 61, no. 6 (June 1971): 1119-25. Topics discussed include: weight, infants, New York, pregnancy, delivery.

1646. Karrar, Z. et al. "Morbidity patterns among under five children in a rural community in Sudan." *Trop Geogr Med* 33, no. 1 (March 1981): 75-8. Topics discussed include: diarrhea, epidemiology, infants, measles, respiratory diseases, rural areas, population, seasonality, socioeconomics, Sudan.

1647. Kinmonth, A. "Review of the epidemiology of sudden infant death syndrome and its relationship to temperature regulation." *Br J Gen Pract* 40, no. 333 (April 1990): 161-3. Topics discussed include: England, pregnancy, risks, seasonality, socioeconomics, etiology, Wales, body regulation.

1648. Kraus, A. et al. "Sudden, unexpected death in infancy in Ontario. II. Findings regarding season, clustering of deaths on specific days and weather." *Can J Public Health* 58, no. 8 (August 1967): 364-71. Topics discussed include: newborns, Ontario, infants, diseases, epidemiology.

1649. Lave, L. et al. "Air pollution, climate, and home heating: their effects on U.S. mortality rates." *Am J Public Health* 62, no. 7 (July 1972): 909-16. Topics discussed include: economics, medicine, fetal death, fossil fuels, housing, infants, pregnancy, environment, health, heating.

1650. Lebedev, I. et al. "Method of analysis of seasonal fluctuations in infant mortality and their causes." *Sov Zdravookhr* no. 8 (1977): 44-9. Topics discussed include: respiratory infections, epidemiology, Ukraine.

1651. Lee, J. "Summer and death from neuroblastoma." *Br Med J* 2, no. 549 (13 May 1967): 404-7. Topics discussed include: aging, England, infants, neoplasms, mortality, gender, Wales, seasonality.

1652. Leiss, J. et al. "Age and season of birth in sudden infant death syndrome in North Carolina, 1982 1987: no interaction." *Am J Epidemiol* 137, no. 2 (15 January 1993): 207-12. Topics discussed include: age effects, weight, Caucasoid race, causes, epidemiology, maternal age, Negroid race, North Carolina, education.

1653. Lentzner, H. R. "Seasonal Patterns of Infant and Child Mortality in New York, Chicago, and New." *Dissertation Abstracts International, A: The Humanities and Social Sciences* 48, no. 3 (September 1987): 754-A. Topics discussed include: children, public health, New York, Chicago, Illinois, New Orleans, Louisiana.

1654. Lund, H. "Hyaline membrane disease in the newborn. Results from 10 year study. Seasonal variations in mortality folic acid deficiency?" *Ugeskr Laeger* 144, no. 23 (7 June 1982): 1671-4. Topics discussed include: Denmark, folic acid, etiology, infants, epidemiology.

1655. Madrigal, L. "Lack of birth seasonality in a nineteenth century agricultural population: Escazu, Costa Rica." *Hum Biol* 65, no. 2 (April 1993): 255-71. Topics discussed include: Catholicism, Costa Rica, epidemiology, history, medicine, infant mortality, rural areas.

1656. Masse, L. et al. "Seasonal variations of mortality in Dakar and implications in public health planning." *Br J Prev Soc Med* 27, no. 1 (February 1973): 67. Topics discussed include: infants, Senegal, public health.

1657. McGlashan, N. et al. "Sudden infant deaths and seasonality in Tasmania, 1970-1976." *Soc Sci Med* 17, no. 13 (1983): 885-8. Topics discussed include: Australia, temperature, epidemiology.

1658. Mitchell, E. et al. "Sudden infant death syndrome and weather temperature." *Paediatr Perinat Epidemiol* 6, no. 1 (January 1992): 19-28. Topics discussed include: New Zealand, epidemiology, seasonality, etiology.

1659. Momiyama, M. "Biometeorological study of the seasonal variation of mortality in Japan and other countries on the seasonal disease calendar." *Int J Biometeorol* 12, no. 4 (December 1968): 377-93. Topics discussed include: age effects, Denmark, Egypt, France, United Kingdom, infants, Japan, sociology, tuberculosis.

1660. Momiyama, M. et al. "Deseasonalization of mortality in the world." *Int J Biometeorol* 16, no. 4 (December 1972): 329-42. Topics discussed include: climatic effects, children, temperature, seasonality.

1661. Murphy, M. et al. "Sudden infant death syndrome and environmental temperature: an analysis using vital statistics." *J Epidemiol Community Health* 41, no. 1 (March 1987): 63-71. Topics discussed include: cold climates, England, temporal geography, epidemiology, Wales, etiology.

1662. Nelson, E. et al. "Climatic and social associations with postneonatal mortality rates within New Zealand." *N Z Med J* 101, no. 849 (13 July 1988): 443-6. Topics discussed include: ethnic groups, New Zealand, socioeconomics, temperature, sudden infant death.

1663. Oehmisch, W. et al. "Is there still a summer peak of infant mortality?" *Z Arztl Fortbild Jena* 67, no. 3 (1 February 1973): 149-53. Topics discussed include: Berlin, Germany, seasonality.

1664. Osmond, C. et al. "Seasonality in the sudden infant death syndrome." *Paediatr Perinat Epidemiol* 2, no. 4 (October 1988): 337-45. Topics discussed include: age effects, England, gender effects, Wales, epidemiology.

1665. Otto, W. "Sex distribution of live born, stillborn and mortality according to seasons." *Z Arztl Fortbild Jena* 59, no. 3 (1 February 1965): 151-9. Topics discussed include: infants, gender differences.

1666. Paulozzi, L. "The seasonality of mortality in Alaska." *Soc Sci Med D* 15, no. 3 (August 1981): 335-9. Topics discussed include: Alaska, demography, Eskimos, ethnic groups, infants, sociology, medicine.

1667. Peacock, J. et al. "Reported morbidity and the weather." *J R Coll Gen Pract* 25, no. 153 (April 1975): 247-51. Topics discussed include: England, infants, temperature, respiratory infections, epidemiology.

1668. Pearn, J. "Survival rates after serious immersion accidents in childhood." *Resuscitation* 6, no. 4 (1978): 271-8. Topics discussed include: age effects, Australia, cold climates, infants, seasonality, gender effects, drowning mortality.

1669. Peterson, D. et al. "Sudden infant death syndrome in epidemiologic perspective: etiologic implications of variation with season of the year." *Ann N Y Acad Sci* 533 (1988): 6-12. Topics discussed include: risks, epidemiology, Washington, etiology.

1670. Piraino, F. et al. "Echovirus 11 infections of newborns with mortality during the 1979 enterovirus season in Milwaukee, Wis." *Public Health Rep* 97, no. 4 (July 1982): 346-53. Topics discussed include: delivery, fetal death, epidemiology, infants, newborns, pregnancy, Wisconsin, epidemiology.

1671. Plotnikov, I. "Seasonal morbidity of acute leukemia." *Vrach Delo* no. 7 (July 1983): 53-4. Topics discussed include: acute disease, infants, Russia, leukemia, epidemiology.

1672. Ponsonby, A. et al. "Sudden infant death syndrome: seasonality and a biphasic model of pathogenesis." *J Epidemiol Community Health* 46, no. 1 (February 1992): 33-7. Topics discussed include: risks, gender effects, Tasmania, epidemiology.

1673. ---. "Climatic temperature and variation in the incidence of sudden infant death syndrome between the Australian states." *Med J Aust* 156, no. 4 (17 February 1992): 246-8, 251. Topics discussed include: Australia, epidemiology, seasonality.

1674. Raven, C. et al. "The sudden infant death syndrome: a possible hypersensitivity reaction determined by distribution of IgG in lungs." *J Forensic Sci* 23, no. 1 (January 1978): 116-28. Topics discussed include: seasonality, immunology, respiratory complications.

1675. Sadio, A. "Seasonal Hunger, Birth Timing, and Pregnancy Outcome in a Rural Region of the Sahel." *Dissertation Abstracts International, A: The Humanities and Social Sciences* 52, no. 3 (September 1991): 1089-A. Topics discussed include: infant mortality, sub-Saharan Africa, weight.

1676. Sakamoto, K. "Analysis of seasonal variation of perinatal death." *Nippon Eiseigaku Zasshi* 30, no. 1 (April 1975): 252. Topics discussed include: newborns, Japan, pregnancy, epidemiology, infant mortality.

1677. Salem, G. "Geography of mortality and birth in Pikine (Senegal): value and limitations of findings of the Civil State in African villages." *Soc Sci Med* 36, no. 10 (May 1993): 1297-311. Topics discussed include: demography, ecology, epidemiology, health, infants, seasonality, Senegal.

1678. Schereschowsky, J. W. "Heat and Infant Mortality." *U S Weekly Public Health Repts* 27 (5 December 1913): 2595. Topics discussed include: temperature.

1679. Schwartz, J. "What are people dying of on high air pollution days?" *Environ Res* 64, no. 1 (January 1994): 26-35. Topics discussed include: mortality, infants, London, lung diseases, Philadelphia, risks, smog.

1680. States, S. "Weather and death in Birmingham, Alabama." *Environ Res* 12, no. 3 (December 1976): 340-54. Topics discussed include: age effects, Alabama, cardiovascular diseases, humidity, infants, respiratory diseases, seasonality, temperature.

1681. Tabbane, C. et al. "Various aspects of child morbidity in the hospital environment: impact on medical education and the efficiency of peripheral medicine." *Tunis Med* 61, no. 6 (November 1983): 409-13. Topics discussed include: health care, education, medicine, infants, seasonality, Tunisia, hospitals.

1682. Taj, E. S. et al. "Heat illness in infants and small children in desert climates." *J Trop Med Hyg* 71, no. 4 (April 1968): 100-4. Topics discussed include: therapy, Iraq, water, desert climates, exhaustion, physiopathology.

1683. Teruel, J. et al. "Seasonal distribution of deaths caused by diarrhea and malnutrition in childhood." *Rev Inst Med Trop Sao Paulo* 15, no. 5 (September 1973): 289-97. Topics discussed include: Brazil, infants, child nutrition, diarrhea, infant mortality, nutrition.

1684. Thorburn, M. et al. "Causes of death in Jamaica, 1953 to 1964. An analysis of post mortem diagnosis at the University Hospital of the West Indies." *Trop Geogr Med* 20, no. 1 (March 1968): 35-49. Topics discussed include: cardiovascular diseases, infant mortality, Jamaica, nutrition, pregnancy, respiratory diseases, seasonality.

1685. Tolboom, J. et al. "Severe protein energy malnutrition in Lesotho, conditioning factors and death and survival in hospital." *Trop Geogr Med* 41, no. 1 (January 1989): 1-8. Topics discussed include: breast feeding, environment, infants, etiology, Lesotho, protein, malnutrition, risks, seasonality, socioeconomics.

1686. Tomkins, A. et al. "Nutritional status and risk of morbidity among young Gambian children allowing for social and environmental factors." *Trans R Soc Trop Med Hyg* 83, no. 2 (March 1989): 282-7. Topics discussed include: diarrhea, epidemiology, Gambia, infants, etiology, risks, seasonality, socioeconomics, urban areas, population, growth.

1687. Toyama, T. et al. "Seasonal variations of daily mortality in Tokyo City." *Keio J Med* 24, no. 3 (July 1975): 253-60. Topics discussed include: air pollution, climatic effects, coronary disease, infants, Tokyo.

1688. Trape, J. et al. "Malaria morbidity among children exposed to low seasonal transmission in Dakar, Senegal and its implications for malaria control in tropical Africa." *Am J Trop Med Hyg* 48, no. 6 (June 1993): 748-56. Topics discussed include: absenteeism, insects, malaria, Senegal.

1689. Underwood, J. "Seasonality of vital events in a Pacific Island population." *Soc Biol* 38, no. 1-2 (March 1991): 113-26. Topics discussed include: delivery, Guam, history, medicine, infant mortality, trends, marriage, religion, weather.

1690. Vakhitov, M. et al. "Seasonal variations of infant mortality." *Zdravookhr Ross Fed* 17, no. 2 (February 1973): 18-21. Topics discussed include: gastrointestinal diseases, pneumonia, diseases.

1691. Velcek, F. et al. "Traumatic death in urban children." *J Pediatr Surg* 12, no. 3 (June 1977): 375-84. Topics discussed include: accidents, traffic, ethnic groups, fires, New York City, seasonality, gender effects, socioeconomics, population, wounds, injuries, mortality.

1692. Victora, C. et al. "The seasonality of infant deaths due to diarrheal and respiratory diseases in southern Brazil, 1974 1978." *Bull Pan Am Health Organ* 19, no. 1 (1985): 29-39. Topics discussed include: Brazil, diarrhea, infant mortality, respiratory mortality.

1693. Williams, N. "Death in Its Season: Class, Environment and the Mortality of Infants in Nineteenth-Century Sheffield." *Social History of Medicine* 5, no. 1 (April 1992): 71-94. Topics discussed include: infants, England, nineteenth century, public health, urban areas, sociology, geography.

1694. Zamotin, B. et al. "Seasonality of Sonne dysentery among organized and unorganized groups of children in years with different morbidity levels." *Zh Mikrobiol Epidemiol Immunobiol no.* 11 (November 1986): 51-4. Topics discussed include: disease, propagation, dysentery, immunology, Russia, timing, urban areas, population, day care centers.

MIGRATION

1695. Aldasheva, A. et al. "Dependence of efficiency and fatigue on neuro psychological state in people working in long term high altitude migration." *Fiziol Cheloveka* 18, no. 5 (September 1992): 114-8. Topics discussed include: Commonwealth of Independent States, timing, physiology, fatigue, psychology, transients, migrants.

1696. Balakrishna, S. et al. "Seasonal Migration of Agricultural Labour: A Study in East Godavari District, Andhra Pradesh." *Behavioural Sciences and Rural Development* 4, no. 2 (July 1981): 231-245. Topics discussed include: rural areas, development, India, labor, migrants.

1697. Bompard, J. P. et al. "An Econometric Model of Seasonal Labour Migrations between Agriculture and Industry France 1860." *Annales d'Economie et de Statistique*, no. 19 (July 1990): 97-129. Topics discussed include: labor, mobility, demography, economics, history, Europe, factor prices, markets.

1698. ---. "Mobility and Unemployment in Mid-Nineteenth-Century France. Seasonal Migrations between Industry and Agriculture." *Annales* 45, no. 1 (January 1990): 55-76. Topics discussed include: labor, markets, competition, occupations, mobility, wages, labor supply, industrial lobor, competition, salaries.

1699. Boyd, R. T. et al. "Seasonal population movement along the lower Columbia River: the social and ecological context." *American Ethnologist* 14 (May 1987): 309-26. Topics discussed include: Indians, North America, migration, history, social networks.

1700. Breman, J. "Seasonal Migration and Co-Operative Capitalism: The Crushing of Cane and of Labour by the Sugar Factories of Bardoli, South Gujarat-Part 1." *Journal of Peasant Studies* 6, no. 1 (October 1978): 41-70. Topics discussed include: India, sugar factories, labor utilization, capitalism, development, production, agriculture, cooperatives.

1701. ---. "Seasonal Migration and Co-Operative Capitalism: The Crushing of Cane and of Labour by the Sugar Factories of Bardoli, South Gujarat-Part 2." *Journal of Peasant Studies* 6, no. 2 (January 1979): 168-209. Topics discussed include: India, sugar factories, labor, utilization, conditions, exploitation.

1702. Breman, J. C. "Seasonal Migration and Cooperative Capitalism: The Crushing of Cane and Labor by the Sugar Factories of Bardoli, South Gujarat." *Contractual Arrangements, Employment, and Wages in Rural Labor Markets in Asia Economic Growth Center, Yale University series*, 131-142. eds Hans P. Binswanger, and R. Rosenzweig. New Haven, London: Yale University Press, 1984. Topics discussed include: labor, agriculture.

1703. Cecil, R. G. et al. "The Caribbean Migrant Farm Worker Programme in Ontario: Seasonal Expansion of West Indian Economic Spaces." *International Migration, Migrations Internationales, Migraciones Internationales* 30, no. 1 (March 1992): 19-37. Topics discussed include: Caribbean, culture, agriculture, workers, employment, Canada.

1704. Collins, J. L. "Seasonal migration as a cultural response to energy scarcity at high altitude." *Current Anthropology* 24 (February 1983): 103-4. Topics discussed include: Peru, adaptation, culture.

1705. Cushing, B. J. "A Note on Specification of Climate Variables in Models of Population Migration." *Journal of Regional Science* 27, no. 4 (November 1987): 241-49. Topics discussed include: demography, economics, labor, mobility, migration.

1706. Daciuk, J. F. et al. "Health Concerns as a Deterrent to Seasonal Migration of Elderly Canadians." *Social Indicators Research* 22, no. 2 (March 1990): 181-197. Topics discussed include: Canada, health issues.

1707. Davila, A. "The seasonality of apprehensions of undocumented Mexican workers." *International Migration Review* 20 (December 1986): 986-91. Topics discussed include: alien labor, Mexicans.

1708. Dutt, J. et al. "Environment, migration and health in southern Peru." *Soc Sci Med* 12, no. 1B (January 1978): 29-38. Topics discussed include: age effects, altitude, culture, gender effects, health issues, transients.

1709. Emerson, R. D. "Trade in Products and International Migration in Seasonal Labor Markets." *American Journal of Agricultural Economics* 64, no. 2 (May 1982): 339-46. Topics discussed include: trade, labor, mobility, migration.

1710. Fournier, G. M. et al. "Elderly Migration: For Sun and Money." *Population Research and Policy Review* 7, no. 2 (1988): 189-99. Topics discussed include: demography, economics.

1711. Glover, T. "An Intertemporal Approach to Seasonal Agricultural Labor Markets: Discussion." *Seasonal Agricultural Labor Markets in the United States*, 249-53. ed Robert D. Emerson. Ames,: Iowa State University Press, 1984. Topics discussed include: labor, economics, farms, mobility, migration.

1712. Graves, P. E. "A Life-Cycle Empirical Analysis of Migration and Climate, by Race." *Journal of Urban Economics* 6, no. 2 (April 1979): 135-47. Topics discussed include: demography.

1713. ---. "Migration and Climate." *Journal of Regional Science* 20, no. 2 (May 1980): 227-37. Topics discussed include: demography, economics, urban areas, public policy.

1714. Griffith, D. C. "Social Organizational Obstacles to Capital Accumulation among Returning Migrants: The British West Indies Temporary Alien Labor Program." *Human Organization* 45, no. 1 (March 1986): 34-42. Topics discussed include: sociodemographics, labor, migration, public policy, Jamaica, economics, development, capital wealth.

1715. Guendelman, S. et al. "Migration Tradeoffs: Men's Experiences with Seasonal Lifestyles." *International Migration Review* 21, no. 3 (September 1987): 709-727. Topics discussed include: migrants, workers, family issues, California, Latin America, culture, Mexico, work, health issues.

1716. Happel, S. K. et al. "The Economic Impact of Elderly Winter Residents in the Phoenix Area." *Research on Aging* 10, no. 1 (March 1988): 119-133. Topics discussed include: migration, seasonality, Arizona, migration, costs, benefits, trailer parks.

1717. Haurin, D. R. "The Regional Distribution of Population, Migration, and Climate." *Quarterly Journal of Economics* 95, no. 2 (September 1980): 293-308. Topics discussed include: demography, regional economics, models, forecasts.

1718. Heaton, T. B. et al. "The Environment and Migration: Effects of Mild Climate, Bodies of Water, and Recreational Development." *Sociology and Social Research* 71, no. 1 (October 1986): 68-70. Topics discussed include: environment, weather, demography, patterns.

1719. Hertz, D. G. "Psychosomatic and psychosocial implications of environmental changes on migrants: A review of contemporary theories and findings." *Israel Journal of Psychiatry and Related Sciences* 19, no. 4 (1982): 329-338. Topics discussed include: environment, culture, overcrowding, migrant populations.

1720. Hogan, T. D. "Determinants of the seasonal migration of the elderly to sunbelt states." *Research on Aging* 9, no. 1 (March 1987): 115-133. Topics discussed include: elderly population.

1721. Hogan, T. D. et al. "Elderly Migration to the Sunbelt: Seasonal versus Permanent." *Journal of Applied Gerontology* 12, no. 2 (June 1993): 246-260. Topics discussed include: elderly population, seasonality, Arizona.

1722. Hori, S. et al. "Studies on heat tolerance of subtropical natives after migration to a temperate zone." *Int J Biometeorol* 22, no. 2 (June 1978): 82-93. Topics discussed include: body height, body weight, climate, Japan, acclimation, thermoregulation.

1723. Javiedes, L. "Reasons for Migration; Motivos para emigrar." *Revista de la Asociacion Latinoamericana de Psicologia Social* 1, no. 2 (July 1981): 289-310. Topics discussed include: migration, patterns, economics, motivations, Mexico.

1724. Kirchner, J. A. *Sugar & Seasonal Labor Migration: The Case of Tucuman, Argentina.* Chicago, IL: University of Chicago, Comittee on Geographical Studies, Research Papers No 192, 1980. Topics discussed include: government, political science, labor, industry, economy.

1725. Krout, J. A. "Seasonal migration of the elderly." *Gerontologist* 23, no. 3 (June 1983): 295-299. Topics discussed include: lifestyle.

1726. Lewin, H. "Brazilian Agriculture and Change in Labor Relations: The Growing Marginality of Seasonal and Migratory Wage Earners; Agricultura brasileira e mudancas nas relacoes de trabalho: a marginalidade crescente do assalariado temporario-o volante." *Desarrollo Rural en las Americas* 11, no. 2 (May 1979): 103-119. Topics discussed include: Brazil, employment, seasonality, migration.

1727. Longino, C. F., Jr. et al. "North American Research on Seasonal Migration." *Ageing and Society* 10, no. 2 (June 1990): 229-235. Topics discussed include: elderly, patterns, seasonality, Canada.

1728. Longino, C. F. et al. "On the nesting of snowbirds: A question about seasonal and permanent migrants." *Journal of Applied Gerontology* 10, no. 2 (June 1991): 157-168. Topics discussed include: migration, sociology, elderly, Canadia, decision making, social services, health care issues.

1729. Marshall, V. W. et al. "Canadian Seasonal Migrants to the Sunbelt: Boon or Burden?" *Journal of Applied Gerontology* 9, no. 4 (December 1990): 420-432. Topics discussed include: migration, Canada, Florida, economics, elderly, seasonality, sunbelt states.

1730. McHugh, K. "Seasonal migration as a substitute for, or precursor to, permanent migration." *Res Aging* 12, no. 2 (June 1990): 229-45. Topics discussed include: age effects, Arizona, epidemiology, Canada, health care programs.

1731. Mesic, M. "The European Migration Situation and Its Perspectives." *Migracijske teme* 4, no. 4 (December 1988): 371-394. Topics discussed include: Europe, economics, labor, trends.

1732. Mueller, W. et al. "A multinational Andean genetic and health program. VIII. Lung function changes with migration between altitudes." *Am J Phys Anthropol* 51, no. 2 (August 1979): 183-95. Topics discussed include: adaptation, Chile, ethnic groups, Indians, South America, development, occupations, gender effects, Spain, ethnology, histology, travel.

1733. Nelkin, D. *On the Season: Aspects of the Migrant Labor System*. Ithaca, NY: ILR Press, 1971. Topics discussed include: labor, industry, political science, migration.

1734. Ockerman, J. D. "Needs Assessment Report: Migratory and Seasonal Hispanic Farmworkers." (September 1990): 151 pp. Topics discussed include: Spanish speakers, migrants, welfare services, work, services, training.

1735. Oshiro, K. K. "Postwar seasonal migration from rural Japan." *The Geographical Review* 74 (April 1984): 145-56. Topics discussed include: Japan, agriculture, labor.

1736. "Oversight on issues affecting Hispanics and migrant and seasonal farmworkers : hearing before the Subcommittee on Employment, Poverty, and Migratory Labor of the Committee on Labor and Human Resources, United States Senate, Ninety-sixth Congress, second session, on examination on whether and to what extent existing federal programs affectively and efficiently meet the needs of Hispanics and migrant and seasonal farmworkers, Milwaukee, Wis., September 6, 1980." United States. Congress. Senate. Committee on Labor and Human Resources. Subcommittee on Employment, Poverty, and Migratory Labor. U.S. G.P.O., Washington, 1981. Topics discussed include: migrant agricultural laborers Wisconsin, hispanic America.

1737. Pampel, F. C. et al. "Retirement Migration Decision Making: The Integration of Geographic, Social, and Economic Preferences." *Research on Aging* 6, no. 2 (June 1984): 139-162. Topics discussed include: retirement, Iowa, planning.

1738. Pison, G. et al. "Seasonal migration: a risk factor for HIV infection in rural Senegal." *J Acquir Immune Defic Syndr* 6, no. 2 (February 1993): 196-200. Topics discussed include: age effects, blood transfusion, risks, Senegal, epidemiology, sexual behavior, gender effects, population, transients.

1739. Preucel, R. W. *Seasonal Circulation & Dual Residence in Pueblo Southwest: A Prehistoric Example from the Pajarito Plateau, New Mexico.* New York, NY: Garland Publishing, 1991. Topics discussed include: Indians, migration, antiquities.

1740. Regensteiner, J. et al. "Migration of the elderly from high altitudes in Colorado." *JAMA* 253, no. 21 (7 June 1985): 3124-8. Topics discussed include: Colorado, heart disease, epidemiology, lung disease, population trends, smoking, mobility.

1741. Reinders, J. E. "Tension Between Emmigration and Immigration Climate as a Central Element in Immigration Theory." *South African Journal of Sociology* 2 (April 1971): 32-40. Topics discussed include: migration, demography, immigration, climatic effectse.

1742. Rothermund, D. "A Survey of Rural Migration and Land Reclamation in India, 1885." *Journal of Peasant Studies* 4, no. 3 (April 1977): 230-242. Topics discussed include: rural areas, India, reclamation.

1743. Sakala, C. "Migrant and Seasonal Farmworkers in the United States: A Review of Health Hazards, Status, and Policy." *International Migration Review* 21, no. 3 (September 1987): 659-687. Topics discussed include: agriculture, workers, migrants, health isssues, services, hazards.

1744. Salih, T. et al. *Season of Migration to the North.* Colorado Springs, CO: Three Continents Press, 1991. Topics discussed include: migration.

1745. Santamaria, D. J. "Worker Migration and Interethnic Conflict. The Case of Temporary Indigenous Migrants to the Sugar Plantations of Salta and Jujuy; Migracion laboral y conflicto interetnico. El caso de los migrantes indigenas temporarios a los ingenios azucareros saltojujens." *Estudios Migratorios Latinoamericanos* 1, no. 3 (August 1986): 357-379. Topics discussed include: migrants, Argentina, traditional cultures, conflicts, Indians.

1746. Sautu, R. "Forms of Agrarian Organization, Seasonal Migrations and the Work of Women; Formas de organizacion agraria, migraciones estacionales y trabajo femenino." *Revista Paraguaya de Sociologia* 16, no. 46 (September 1979): 49-62. Topics discussed include: women, Argentina, production, organization, exploitation, capitalism.

1747. Schachter, J. et al. "Neighborhood Quality and Climate as Factors in U.S. Net Migration Patterns, 1974-76." *American Journal of Economics and Sociology* 41, no. 4 (October 1982): 387-400. Topics discussed include: neighborhoods, climatic effects, quality of life.

1748. Shabanova, M. A. "The Seasonal Migration of Construction Workers; Sezonnaya migratsiya stroitel'nykh rabochikh." *Sotsiologicheskie Issledovaniya* 16, no. 6 (1989): 79-84. Topics discussed include: labor, construction industry, seasonality, workers, USSR, Siberia.

1749. Spencer, B. et al. "Workload and Seasonal Variation in Birthrates: Some International Comparisons." *International Journal of Comparative Sociology* 17, no. 1-2 (March 1976): 48-58. Topics discussed include: work, seasonality, birth, birth rates, rural areas, urban areas, migration.

1750. Spreen, T. H. "An Intertemporal Approach to Seasonal Agricultural Labor Markets." *Seasonal Agricultural Labor Markets in the United States*, 230-49. ed Robert D. Emerson. Ames: Iowa State University Press, 1984. Topics discussed include: labor, economics, farms, mobility, migration.

1751. Steinnes, D. et al. "Take the money and sun: elderly migration as a consequence of gains in unaffordable housing markets." *J Gerontol* 47, no. 4 (July 1992): S197-203. Topics discussed include: climatic effects, costs, income, seasonality, economics, mobility.

1752. Sullivan, D. A. "The Ties that Bind: Differentials between Seasonal and Permanent Migrants to Retirement Communities." *Research on Aging* 7, no. 2 (June 1985): 235-250. Topics discussed include: migration, retirement, seasonality, permanent migration.

1753. Verhaeren, R. E. "The Role of Foreign Workers in the Seasonal Fluctuations of the French Economy." *International Migration Review* 20, no. 4 (December 1986): 856-874. Topics discussed include: migrants, workers, economics, seasonality, France.

1754. Visscher, B. et al. "Latitude, migration, and the prevalence of multiple sclerosis." *Am J Epidemiol* 106, no. 6 (December 1977): 470-5. Topics discussed include: California, China, ethnology, Europe, genetics, medicine, Israel, Japan, diagnosis, South Africa, Washington, demography, epidemiology, transients, migrants.

1755. Whiting, J. W. M. et al. "Winter Temperature as a Constraint to the Migration of Preindustrial Peoples." *American Anthropologist* 84, no. 2 (June 1982): 279-298. Topics discussed include: weather, preindustrial populations, winter.

1756. Wong, L. L. "Migrant Seasonal Agricultural Labour: Race and Ethnic Relations in the Okanagan Valley." *Dissertation Abstracts International, A: The Humanities and Social Sciences* 49, no. 12 (June 1989): 3882-A. Topics discussed include: workers, migration, race relations, ethnic groups, sociology, history, British Columbia, a literature review, Sydney.

1757. Wood, A. P. "The decline of seasonal labour migration to the coffee forests of South-west Ethiopia." *Geography* 68 (January 1983): 53-6. Topics discussed include: coffee, migrants, labor, Ethiopia.

1758. Yunus, H. S. *Seasonal migration in Jogjakarta*. 100 pages. Edited version ed. SEAPRAP research report; no. 86, Singapore: SEAPRAP, 1978. Topics discussed include: sociology, Indonesia, seasonality, labor issues, economics, rural areas, urban areas, demography.

MORTALITY, GENERAL

1759. Akslen, L. "Seasonal variation in melanoma deaths and the pattern of disease process. A preliminary report." *Chronobiologia* 16, no. 3 (July 1989): 257-63. Topics discussed include: brain neoplasms, epidemiology, melanoma, etiology, gender effects, skin neoplasms, sunlight.

1760. Anderson, T. "Seasonal variation in ischaemic heart disease mortality." *Lancet* 2, no. 683 (28 November 1970): 1140. Topics discussed include: age effects, climatic effects, temperature.

1761. Armstrong, B. "Environmental Factors and Cancer Incidence and Mortality in Different Countries with Special Reference to Dietary Practices." *Int J Cancer* 15 (1975): 617-631. Topics discussed include: environment, health issues, disease, death, geography.

1762. Aubenque, M. "Seasonal variations and chronological series in causes of death in France from 1900 to 1972." *Cah Sociol Demogr Med* 19, no. 1 (January 1979): 17-22. Topics discussed include: age effects, France, infants, mortality.

1763. Baker, B. A. "Winter weather and cardiovascular mortality in Minneapolis St. Paul." *Am J Public Health* 72, no. 3 (March 1982): 261-5. Topics discussed include: cardiovascular diseases, etiology, Minnesota, myocardial infarction, seasonality, snow, cold.

1764. Bako, G. "Seasonality of mortality from various diseases in Canada 1979-83." *Can J Public Health* 79, no. 5 (September 1988): 388-9. Topics discussed include: Canada, meteorology.

1765. Beard, C. "Daily and seasonal variation in sudden cardiac death, Rochester, Minnesota, 1950-1975." *Mayo Clin Proc* 57, no. 11 (November 1982): 704-6. Topics discussed include: coronary disease epidemiology, Minnesota, gender effects, mortality, epidemiology.

1766. Besson, M. L. "Rélations Entre les Eléments Météorologiques et la Mortalité." *Annales des Services Techniques d'Hygiene de la Ville de Paris* (1921): 2. Topics discussed include: weather, mortality, climate.

1767. Biersteker, K. "Ozone, temperature, and mortality in Rotterdam in the summers of 1974 and 1975." *Environ Res* 12, no. 2 (October 1976): 214-7. Topics discussed include: Netherlands, ozone, seasonality.

1768. Bolotovskii, V. "Effect of active immunization on the periodicity and seasonality of measles morbidity." *Zh Mikrobiol Epidemiol Immunobiol no.* 10 (October 1979): 29-33. Topics discussed include: disease propagation, epidemiology, immunization schedule, vaccines, USSR.

1769. Boyd, J. T. "Climate, Air Pollution, and Mortality." *British Journal of Preventive and Social Medicine* 14 (1960): 123-135. Topics discussed include: environment, health.

1770. ---. "Meteorological Conditions and Mortality." *Proc Roy Soc Med* 53 (1960): 107-108. Topics discussed include: weather, climatic effects, death.

1771. Brezowsky, H. "Morbidity and Weather." in *Medical Climatology*, 358-399. editor Sidney Licht. New Haven: Licht, Elisabeth, 1964. Topics discussed include: mortality, death, health.

1772. Bridger, C. A. "Mortality in St. Louis, Missouri, during Heat Waves in 1936, 1953, 1954, 1955 and 1966." *Environmental Research* 12 (1976): 38-48. Topics discussed include: temperature, environment, weather.

1773. Bridges, W. C. "Respiratory diseases in the Australian morbidity survey, 1969 1970." *Med J Aust* 2, no. 23 (4 December 1971): 1175-9. Topics discussed include: age effects, antibiotics therapeutic use, Australia, geography, respiratory tract treatment, seasonality, gender effects.

1774. ---. "The Australian general practice morbidity and prescribing survey, 1969 to 1974. 3. Workload and morbidity." *Med J Aust* 2, no. 1 Suppl (2 October 1976): 7-11. Topics discussed include: age effects, Australia, infants, seasonality, gender effects, urban areas, population.

1775. Buchan, A. "The Influence of Weather on Mortality from Different Diseases and at Different Ages." *J Scot Meteorol Soc Edinburgh and London N S* (1876): 5. Topics discussed include: climatic effects, health issues.

1776. Buechley, R. "Heat Island=Death Island?" *Environmental Research* 5 (1972): 85-92. Topics discussed include: weather, climatic effects, temperature, mortality.

1777. Bull, G. M. "Environment, Temperature and Death Rates." *Age and Ageing* 7 (1978): 210-224. Topics discussed include: environment, mortality.

1778. Bundesen, H. N. "Low Temperature, High Barometer and Sudden Death." *J Am Med Assoc* 87, no. 24 (1926): 1987-1989. Topics discussed include: weather, environment, mortality.

1779. Burton, A. "Correlation of cancer death rates with altitude and with the quality of water supply of the 100 largest cities in the United States." *J Toxicol Environ Health* 3, no. 3 (October 1977): 465-78. Topics discussed include: age effects, electric conductivity, water supply, neoplasms mortality.

1780. Cantrelle, P. "Seasonal mortality. Some data on West Africans." *Bull Soc Med Afr Noire Lang Fr* 11, no. 2 (1966): 239-42. Topics discussed include: Africa, climatic effects.

1781. Cassell, E. "Air pollution, weather, and illness in a New York population." *Arch Environ Health* 18, no. 4 (April 1969): 523-30. Topics discussed include: carbon monoxide, dust, exposure, hydrocarbons, New York City, sulfur dioxide, morbidity.

1782. Cech, I. "Meteorologic factors and temporal variations of cardiac mortality in an urban setting in a desert climatic zone." *Isr J Med Sci* 13, no. 5 (May 1977): 451-9. Topics discussed include: barometric pressure, coronary disease, humidity, Israel, seasonality, temperature, population, wind, desert climates, heart diseases, weather.

1783. ---. "Day to day and seasonal fluctuations of urban mortality in Houston, Texas." *Int J Biometeorol* 23, no. 2 (June 1979): 77-87. Topics discussed include: cycles, Texas, population, weather.

1784. Chen, C. "Analysis of the secular trend and seasonal variation of measles mortality rate in Taiwan." *Ann Acad Med Singapore* 13, no. 2 (April 1984): 136-41. Topics discussed include: measles, Taiwan, vaccination.

1785. Cliff, A. "The changing geographical coherence of measles morbidity in the United States, 1962 88." *Stat Med* 11, no. 11 (August 1992): 1409-24. Topics discussed include: cross-sectional analysis, measles transmission, seasonality, epidemiology.

1786. Coon, C. S. "Some Problems of Human Variability and Natural Selection in Climate and Culture." *Am Naturalist* 89 (1955): 257-280. Topics discussed include: culture, biology, mortality.

1787. Cordioli, E. "Influence of meteorological factors on the morbidity of acute myocardial infarct." *G Clin Med* 69, no. 5 (May 1988): 329-32. Topics discussed include: barometric pressure, Italy, temperature, myocardial infarction, epidemiology.

1788. Cucu, F. "Effects of a multifactorial prevention trial of coronary heart disease on the seasonal variation of the incidence of major cardiovascular events." *Med Interne* 30, no. 3 (July 1992): 175-85. Topics discussed include: cerebrovascular disorders, mortality, cardiac epidemiology, death, Romania.

1789. Daily, G. "An exploratory model of the impact of rapid climate change on the world food situation." *Proc R Soc Lond Biol* 241, no. 1302 (22 September 1990): 232-44. Topics discussed include: biometry, death, mathematics, food supply, population density.

1790. "Deadly cold: health hazards due to cold weather: hearing before the Subcommittee on Health and Long-Term Care of the Select Committee on Aging, House of Representatives, Ninety-eighth Congress, second session, February 2, 1984." U.S. Congress, House Select Committee on Aging. Subcommittee on Health and Long-Term Care. U.S. G.P.O., Washington, 1984. Topics discussed include: cold climates, physiology, hypothermia, age effects, mortality, diseases.

1791. DeLong, E. "Climate, socioeconomic status and Hodgkin's disease mortality in the United States." *J Chronic Dis* 37, no. 3 (1984): 209-13. Topics discussed include: education, social status, humidity, income, temperature, socioeconomics.

1792. Dembert, M. "Cold weather morbidity among United States Navy and Marine Corps Personnel." *Mil Med* 146, no. 11 (November 1981): 770-5. Topics discussed include: frostbite, epidemiology, naval medicine, cold climates, military medicine.

1793. Dines, W. H. "Weekly Death Rate and Temperature Curves." *Quart J Roy Met Soc London* (1901). Topics discussed include: mortality.

1794. Dockery, D. "Air pollution and daily mortality: associations with particulates and acid aerosols." *Environ Res* 59, no. 2 (December 1992): 362-73. Topics discussed include: environment, monitoring, Illinois, Missouri, rural areas, population, seasonality, temperature, Tennessee, urban areas.

1795. Dong, H. "The analysis of the affecting of season factor on mortality in Shanghai County (1975-1986)." *Chung Hua Yu Fang I Hsueh Tsa Chih* 23, no. 5 (September 1989): 263-5. Topics discussed include: age effects, China, epidemiology, rural areas, population.

1796. Douglas, A. "Seasonality of disease in Kuwait." *Lancet* 337, no. 8754 (8 June 1991): 1393-7. Topics discussed include: cardiovascular, mortality, epidemiology, neoplasms, gender effects, socioeconomics, weather, infants.

1797. ---. "Composition of seasonality of disease." *Scott Med J* 36, no. 3 (June 1991): 76-82. Topics discussed include: cerebrovascular disorders, mortality, physiopathology, coronary physiopathology, neoplasms, respiratory tract, epidemiology.

1798. Douglas, AS. "Seasonal, regional and secular variations of cardiovascular and cerebrovascular mortality in New Zealand." *Aust N Z J Med* 20, no. 5 (October 1990): 669-76. Topics discussed include: age effects, gender effects, temperature, coronary disease.

1799. Driscoll, D. "Base lines for measuring adverse effects of air pollution: some evidence for weather effects on mortality." *Environ Res* 4, no. 3 (August 1971): 233-42. Topics discussed include: Europe, geography, heat, humidity, morbidity, seasonality, temperature, urban areas, population, environment.

1800. ---. "The relationship between weather and mortality in ten major metropolitan areas in the United States, 1962-1965." *Int J Biometeorol* 15, no. 1 (March 1971): 23-39. Topics discussed include: urban areas, population.

1801. ---. "Weather influences on mortality and morbidity." *Rev Environ Health* 1, no. 4 (1974): 283-304. Topics discussed include: air movements, environment, exposure, Europe, humidity, meteorology, research methods, temperature, timing.

1802. Dudley, E. "Climate, water hardness and coronary heart disease." *J Chronic Dis* 22, no. 1 (June 1969): 25-48. Topics discussed include: Caucasoid race, environment, humidity, minerals, temperature, weather, mortality, water supply.

1803. Dunnigan, M. "Seasonal incidence and mortality of ischaemic heart disease." *Lancet* 2, no. 677 (17 October 1970): 793-7. Topics discussed include: age effects, hospitalization, Scotland, gender effects, temperature, coronary epidemiology.

1804. Ellis, F. P. "Mortality in the Elderly in Heat Wave in New York City, August, 1975." *Environmental Research* 15 (1978): 504-512. Topics discussed include: temperature, environment, biology, weather.

1805. Ellis, F. "Mortality from heat illness and heat aggravated illness in the United States." *Environ Res* 5, no. 1 (March 1972): 1-58. Topics discussed include: acclimation, air conditioning, air pollution, cerebrovascular disorders, climatic effects, coronary disease, seasonality, gender effects.

1806. Faust, V. "Weather and disease. 1. Weather dependent diseases." *Fortschr Med* 95, no. 4 (27 January 1977): 201-7. Topics discussed include: accidents, epidemiology, geography, Germany, labor complications, pregnancy, respiratory tract, skin, socioeconomics, morbidity.

1807. ---. "Weather and disease." *Fortschr Med* 95, no. 4 (27 January 1977): 208-11. Topics discussed include: air ionization, physiopathology, meteorology, nervous system, preventive medicine, skin, temperature, morbidity.

1808. ---. "Weather and disease. 3. Weather dependent diseases conclusion." *Fortschr Med* 95, no. 6 (10 February 1977): 331-5. Topics discussed include: epidemiology, asthma, death sudden, epilepsy, Germany, infants, influenza, pregnancy, morbidity.

1809. Fish, P. D. "Heatwave Morbidity and Mortality in Old Age." *Age Ageing* 14 (1985): 243-245. Topics discussed include: temperature, weather, health issues, illness.

1810. French, J. "Mortality from flash floods: a review of national weather service reports, 1969-81." *Public Health Rep* 98, no. 6 (November 1983): 584-8. Topics discussed include: epidemiology, drowning, disasters, rain.

1811. Frost, D. "Myocardial infarct death and temperature in Auckland, New Zealand." *Int J Biometeorol* 36, no. 1 (March 1992): 14-7. Topics discussed include: epidemiology, timing, cold climates.

1812. ---. "Myocardial infarct death, the population at risk, and temperature habituation." *Int J Biometeorol* 37, no. 1 (February 1993): 46-51. Topics discussed include: acclimation physiology, Australia, epidemiology, meteorology, Quebec, risk, seasonality.

1813. Fyfe, T. "Comparison of seasonal incidence and mortality of stroke and myocardial infarction." *Health Bull Edinb* 30, no. 2 (April 1972): 159-61. Topics discussed include: epidemiology, infants, myocardial infarction.

1814. ---. "Seasonal variation in serum lipids, and incidence and mortality of ischaemic heart disease." *J Atheroscler Res* 8, no. 3 (May 1968): 591-6. Topics discussed include: Scotland, cholesterol, blood, epidemiology, triglycerides.

1815. Galloway, P. R. "Annual Variations in Deaths by Age, Deaths by Causes, Prices, and Weather in London, 1670 to 1830." *Pop Stud* 39 (1985): 487-505. Topics discussed include: mortality, death, environment.

1816. Goldsmith, J. R. "Seasonal Variations in Mortality." in *Biometeorology*, 74. editors S. W. Tromp, and W. H. Weihe. Amsterdam: Swets and Zeitlinger N. V., 1967. Topics discussed include: biology, weather.

1817. Goldstein, I. "Time series analysis of morbidity data for assessment of acute environmental health effects." *Environ Res* 17, no. 2 (October 1978): 266-75. Topics discussed include: air pollution, asthma, mortality, New York, seasonality, timing, exposure, epidemiology.

1818. Golubchikova, B. "Seasonal variations of cancer mortality." *Vopr Onkol* 11, no. 10 (1965): 49-51. Topics discussed include: genital neoplasms, Ukraine.

1819. Gorokhovskii, B. "Environmental conditions and their effect on the incidence and mortality rate of myocardial infarct." *Sov Med* 34, no. 4 (April 1971): 105-10. Topics discussed include: Moscow, seasonality, magnetics, myocardial infarction, epidemiology, weather.

1820. Gover, M. "Mortality during Periods of Excessive Temperatures." *Public Health Reports* 53 (1938): 112-143. Topics discussed include: temperatures.

1821. Grubliak, V. "Meteorological characteristics of stroke morbidity in the city of Kamenets Podolski." *Vrach Delo* no. 2 (February 1981): 57-9. Topics discussed include: barometric pressure, Ukraine, cerebrovascular disorders, epidemiology, urban areas, population.

1822. Grushetskaia, E. "Effect of meteorological factors on the incidence of myocardial infarct and sudden death in acute coronary insufficiency." *Vrach Delo* no. 3 (March 1980): 51-5. Topics discussed include: acute disease, seasonality, Ukraine, coronary disease, mortality, epidemiology.

1823. Hajek, E. "Seasonality of mortality in human populations of chile as related to a climatic gradient." *Int J Biometeorol* 28, no. 1 (February 1984): 29-38. Topics discussed include: Chile, temperature.

1824. Hall, P. "The outdoor temperature and the mortality from cardiovascular diseases." *Lakartidningen* 67, no. 19 (6 May 1970): 2141-5. Topics discussed include: cold climates.

1825. Henschel, A. et al. "An Analysis of the Heat Deaths in St. Louis during July 1966." *Am J Public Health* 59 (1969): 2232-2242. Topics discussed include: temperature, mortality, death.

1826. Hewitt, D. "A Study of Temporal Variations in the Risk of Fetal Malformation and Death." *American Jornal of Public Health* 52 (October 1962): 1676-1688. Topics discussed include: seasonality, desease, health issues.

1827. Heyer, H. E. "The Increased Frequency of Acute Myocardial Infarction during Summer Months in Warm Climate." *Am Heart J* 45 (1953): 741-748. Topics discussed include: disease, health issues, mortality, death.

1828. Hippocrates. *Airs, Waters and Places*. 460 B.C. Topics discussed include: weather, climatic effects, mortality, health.

1829. Huntington, E. "The Interpretation of the Death Rate by Climographs." *Modern Medicine* 1 (1919): 13-22. Topics discussed include: mortality, climate.

1830. ---. "Temperature and Mortality in New York City." *Sta Bull Met Life Ins Co* 4 (February 1923): 6-23.

1831. ---. *Weather and health; a study of daily mortality in New York city*. 161 pages. National research council. Committee on the atmosphere and man. Bulletin of the National Research Council; no. 75, April 1930, Washington, DC: Published by the National Research Council of the National Academy of Sciences, 1930. Topics discussed include: physiology, medical climatology, New York.

1832. Imaizumi, Y. "Mortality rate of amyotrophic lateral sclerosis in Japan: effects of marital status and social class, and geographical variation." *Jinrui Idengaku Zasshi* 31, no. 2 (June 1986): 101-11. Topics discussed include: age effects, Japan, occupations, rural health issues, seasonality, gender effects, urban areas, health issues, marriage.

1833. Inokawa, T. "Relationship between meteorological conditions and mortality. 16." *Nippon Ika Daigaku Zasshi* 38, no. 2 (April 1971): 92-101. Topics discussed include: age effects, climatic effects, humidity, gender effects, temperature, weather.

1834. ---. "Relationship between meteorological conditions and mortality. 17." *Nippon Ika Daigaku Zasshi* 38, no. 4 (August 1971): 182-8. Topics discussed include: age effects, gender effects, Tokyo, weather, temperature.

1835. Isaacs, N. "Health and housing seasonality in New Zealand mortality." *Aust J Public Health* 17, no. 1 (March 1993): 68-70. Topics discussed include: New Zealand, climatic effects.

1836. Iuldashev, K. "Seasonal and circadian exacerbations of ischemic heart disease and their prognosis under the regional conditions of Tashkent." *Ter Arkh* 58, no. 1 (1986): 43-5. Topics discussed include: first aid, heart failure, congestive mortality, meteorology, epidemiology, Uzbekistan, circadian rhythms, urban areas, population.

1837. Jones, T. S. "Morbidity and Mortality Associated with the July 1980 Heatwave in St. Louis and Kansas City, Missouri." *JAMA* 247 (1982): 3327-3331. Topics discussed include: illness, temperature, weather, geography.

1838. Kalkstein, L. S. *The impacts of predicted climate changes upon human mortality*. 110 pages. Publications in climatology; v. 41, no. 1, Elmer, NJ: C.W. Thornthwaite Associates, Laboratory of Climatology, 1988. Topics discussed include: mathematical models.

1839. ---. "Weather and human mortality: an evaluation of demographic and interregional responses in the United States." *Annals of the Association of American Geographers* 79 (March 1989): 44-64.

1840. Kalkstein, L. "A new approach to evaluate the impact of climate on human mortality." *Environ Health Perspect* 96 (December 1991): 145-50. Topics discussed include: age effects, meteorology, epidemiology, seasonality, urban areas, population, air pollution.

1841. Kanellakis, A. P. "Seasonal Swing in Mortality in England and Wales." *British Journal of Preventive and Social Medicine* 12, no. 4 (1958):

1842. Katayama, K. "A Biometeorological Study of Mortality from Stroke and Heart Diseases: Its Geographical Differences in the United States." *Meteorology and Geohysics* 21 (1970): 127-139. Topics discussed include: weather, health issues, geography.

1843. Katsouyanni, K. "Evidence for interaction between air pollution and high temperature in the causation of excess mortality." *Arch Environ Health* 48, no. 4 (July 1993): 235-42. Topics discussed include: epidemiology, Greece, holidays, humidity, ozone, seasonality, smoke, air pollutants, environment, heat, urban areas, health issues, weather.

1844. Kavaler, L. *A matter of degree : heat, life, and death*. 226 pages. 1st ed ed. New York: Harper & Row, 1981. Topics discussed include: heat, physiology, adaptation, biology, sociology.

1845. Keatinge, W. "Medical problems of cold weather. The Oliver Sharpey lecture 1985." *J R Coll Physicians Lond* 20, no. 4 (October 1986): 283-7. Topics discussed include: blood cell count, body constitution, body temperature, diving, England, hypothermia mortality, physiopathology, immersion, memory.

1846. ---. "Seasonal mortality among elderly people with unrestricted home heating." *Br Med J Clin Res Ed* 293, no. 6549 (20 September 1986): 732-3. Topics discussed include: seasonality, heating.

1847. ---. "Changes in seasonal mortalities with improvement in home heating in England and Wales from 1964 to 1984." *Int J Biometeorol* 33, no. 2 (June 1989): 71-6. Topics discussed include: automobiles, cold climates, fats, England, prescriptions, smoking, timing, Wales, heating, mortality.

1848. ---. "Increases in platelet and red cell counts, blood viscosity, and arterial pressure during mild surface cooling: factors in mortality from coronary and cerebral thrombosis in winter." *Br Med J Clin Res Ed* 289, no. 6456 (24 November 1984): 1405-8. Topics discussed include: body temperature, cerebral embolism, thrombosis, cholesterol, blood, oxygen consumption, seasonality, blood pressure, blood viscosity, skin temperature.

1849. Keller, C. "Seasonal patterns in perinatal mortality and preterm delivery." *Am J Epidemiol* 118, no. 5 (November 1983): 689-98. Topics discussed include: epidemiologic methods, gestational age, Minnesota, pregnancy, probability, fetal death, epidemiology, labor.

1850. Kersten, E. "Effect of different climates on the work capacity and morbidity of sailors." *Gig Sanit* 36, no. 4 (April 1971): 79-82. Topics discussed include: blood pressure, thermoregulation, naval medicine.

1851. Kevan, S. M. "Season of Life-Season of Death." *Social Science and Medicine* 13D, no. 4 (December 1979): 227-232. Topics discussed include: mortality, dying, birth, Canada.

1852. Khaavel', A. "Nature of the relation of acute appendicitis morbidity to meteorological and heliogeophysical factors." *Vestn Khir* 120, no. 4 (April 1978): 67-70. Topics discussed include: acute disease, climatic effects, Estonia, physics, seasonality, appendicitis epidemiology, sunlight.

1853. Khilkov, V. "The effect of the time of year on skin disease morbidity in northern climates." *Vestn Dermatol Venerol* 45, no. 3 (1971): 73-6. Topics discussed include: Russia, cold climates, seasonality, skin epidemiology.

1854. Kim, Y. "Air pollution, climate, socioeconomic status and total mortality in the United States." *Sci Total Environ* 42, no. 3 (April 1985): 245-56. Topics discussed include: age effects, timing, socioeconomics.

1855. Knox, E. "Meteorological associations of cerebrovascular disease mortality in England and Wales." *J Epidemiol Community Health* 35, no. 3 (September 1981): 220-3. Topics discussed include: air pollution, England, seasonality, Wales, cerebrovascular disorders.

1856. Korochkin, I. "Adaptation and morbidity among Soviet engineers in tropical countries." *Sov Med* no. 8 (1990): 56-9. Topics discussed include: Nigeria, work-related diseases, etiology, seasonality, USSR, ethnology, acclimation, physiology, epidemiology, transients, migrants.

1857. Kunst, A. "The decline in winter excess mortality in The Netherlands." *Int J Epidemiol* 20, no. 4 (December 1991): 971-7. Topics discussed include: cardiovascular diseases, influenza, Netherlands, epidemiology, seasonality.

1858. ---. "Outdoor air temperature and mortality in The Netherlands: a time series analysis." *Am J Epidemiol* 137, no. 3 (1 February 1993): 331-41. Topics discussed include: air pollution, cause of death, humidity, influenza, seasonality, timing, wind, environment, exposure, population.

1859. Kupferschmidt, H. "The morbidity of middle Europeans in the subtropical climate of the Near East." *Dtsch Gesundheitsw* 26, no. 15 (8 April 1971): 704-7. Topics discussed include: conjunctivitis, epidemiology, Egypt, Europe, gastroenteritis, respiratory tract infections, skin diseases, tropical medicine.

1860. Kutschenreuter, C. "A Study of the Effect of Weather on Mortality." *Transactions of the New York Academy of Sciences* 22, no. 2 (1959): 126-138. Topics discussed include: death, climate.

1861. Kutschenreuter, P. H. "Weather does Affect Mortality." *ASHRAE J* (1963): 39-43.

1862. Landers, J. "Mortality, weather and prices in London 1675-1825: a study of short-term fluctuations." *Journal of Historical Geography* 12 (October 1986): 347-64. Topics discussed include: London, climatic effects, United Kingdom, history, prices.

1863. ---. "Burial Seasonality and Causes of Death in London 1670-1819." *Population Studies* 42, no. 1 (March 1988): 59-83. Topics discussed include: economics, history, Europe, factor prices, markets, demography.

1864. Lantukh, V. "Geographical factors and eye morbidity." *Vestn Oftalmol no.* 2 (March 1982): 56-63. Topics discussed include: adaptation, altitude, climatic effects, eye physiology, geography, meteorology, prognosis, seasonality, USSR, eye diseases, epidemiology.

1865. Larsen, U. "The effects of monthly temperature fluctuations on mortality in the United States from 1921 to 1985." *Int J Biometeorol* 34, no. 3 (December 1990): 136-45. Topics discussed include: seasonality, epidemiology, weather.

1866. ---. "Short-Term Fluctuations in Death by Cause, Temperature, and Income in the United States, 1930-1985." *Social Biology* 37, no. 3-4 (September 1990): 172-187. Topics discussed include: weather, heart diseases, mortality rates, seasonality, cardiovascular disease, income.

1867. Lebowitz, M. "A comparative analysis of the stimulus response relationship between mortality and air pollution weather." *Environ Res* 6, no. 1 (March 1973): 106-18. Topics discussed include: humidity, seasonality, temperature, urban areas, population, environment, health.

1868. ---. "A comparison of the relationships of mortality and morbidity with air pollution weather and the implications for further research." *Sci Total Environ* 2, no. 2 (November 1973): 191-5. Topics discussed include: air movements, chronic disease, common cold, lung diseases, obstructive epidemiology, New York City, population, seasonality, temperature.

1869. ---. "The relationship between air pollution and weather as stimuli and daily mortality as responses in Tokyo, Japan, with comparisons with other cities." *Environ Res* 6, no. 3 (September 1973): 327-33. Topics discussed include: New York City, sulfur dioxide poisoning, temperature, Tokyo.

1870. Leonida, D. "Morbidity patterns reflected in a school health program during an influenza epidemic season." *Ill Med J* 137, no. 3 (March 1970): 262-4. Topics discussed include: absenteeism, disease propagation, Illinois, immunization, influenza epidemiology, health services.

1871. Liedermann, A. "Studies on mortality, air pollution and climate in Stuttgart." *Offentl Gesundheitswes* 40, no. 9 (September 1978): 596-602. Topics discussed include: environment, health issues, Germany, seasonality, weather.

1872. Lipa, B. J. "Search for Correlation between Geomagnetic Disturbances and Mortality." *Nature* 259 (1976): 302-304. Topics discussed include: environment, biology.

1873. Loewenstein, J. "Effect of atmospheric pollution (SO2 dust) and meteorological conditions on mortality in Paris between 1969 and 1976." *Rev Epidemiol Sante Publique* 31, no. 2 (1983): 143-61. Topics discussed include: barometric pressure, seasonality, temperature, urban areas, population, weather, air pollution, dust, sulfur dioxide.

1874. Lye, M. "Effects of Heat Wave on Mortality Rates in Elderly Patients." *Lancet* 1 (1977): 529-531. Topics discussed include: temperature, biology, weather.

1875. Lyster, W. R. "The Altered Seasons of Death in America." *Journal of Biosocial Science* 4, no. 2 (April 1972): 145-151. Topics discussed include: mortality, demography, public health issues, seasonality.

1876. Macdonald, E. "Demographic variation in cancer in relation to industrial and environmental influence." *Environ Health Perspect* 17 (October 1976): 153-66. Topics discussed include: age effects, air analysis, pollution, climatic effects, ethnic groups, neoplasms mortality, gender effects, Texas, exposure, industry.

1877. Macey, S. M. "Deaths from excessive heat and excessive cold among the elderly." *Gerontologist* (August 1993): 497-500. Topics discussed include: death rates, death causes, tolerance, acclimation, gender differences, racial differences, minority groups.

1878. Macfarlane, A. "Daily mortality and environment in English conurbations. II. Deaths during summer hot spells in greater London." *Environ Res* 15, no. 3 (June 1978): 332-41. Topics discussed include: age effects, London, wind, heat, seasonality.

1879. ---. "Short-Term Increases in Mortality during Heat Waves." *Nature* 264 (1976): 434-436. Topics discussed include: seasonality, temperature.

1880. Mackenbach, J. "Seasonal variation in mortality in The Netherlands." *J Epidemiol Community Health* 46, no. 3 (June 1992): 261-5. Topics discussed include: cause of death, Netherlands epidemiology.

1881. ---. "Air pollution, lagged effects of temperature, and mortality: The Netherlands 1979 87." *J Epidemiol Community Health* 47, no. 2 (April 1993): 121-6. Topics discussed include: cardiovascular diseases, neoplasms, Netherlands, epidemiology, respiratory diseases, weather, sulfur dioxide.

1882. Madai, L. "The seasonal character of mortality in Europe." *Demografia* 15, no. 3-4 (1972): 434-448. Topics discussed include: Europe, death, Hungary, seasonality.

1883. Madzhidov, N. "The effect of meteorological factors on the mortality from strokes on the plains of Central Asia." *Zh Nevropatol Psikhiatr* 91, no. 11 (1991): 48-9. Topics discussed include: age effects, barometric pressure, humidity, gender effects, urban areas, population, Uzbekistan, cerebrovascular disorders.

1884. Malinskii, D. "Correlation between meteorological factors and the respiratory morbidity in organized collectives under the conditions of the polar regions." *Gig Sanit no.* 11 (November 1974): 83-5. Topics discussed include: acute disease, arctic regions, USSR, respiratory tract diseases epidemiology, weather.

1885. Malysheva, A. "Importance of industrial microclimate in human seasonal adaptation." *Gig Sanit* 35, no. 11 (November 1970): 24-9. Topics discussed include: environment, exposure, morbidity, work-related diseases, temperature, USSR, ventilation, adaptation, immunity, work-related medicine.

1886. Mannino, J. "Environmental.temperature and mortality from acute myocardial infarction." *Int J Biometeorol* 33, no. 1 (January 1989): 32-5. Topics discussed include: epidemiologic methods, seasonality, Wisconsin, myocardial infarction.

1887. Mao, Y. "Seasonality in epidemics of asthma mortality and hospital admission rates, Ontario, 1979 86." *Can J Public Health* 81, no. 3 (May 1990): 226-8. Topics discussed include: asthma, epidemiology, Ontario.

1888. Marcuzzi, G. "Seasonality of death in the period 1889 1988 in the Val di Scalve (Bergamo Pre Alps, Lombardia, Italy)." *Hum Biol* 64, no. 2 (April 1992): 215-22. Topics discussed include: Italy, epidemiology, mortality, trends, population characteristics.

1889. Marmor, M. "Heat Wave Mortality in New York City, 1949 to 1970." *Archives of Environmental Health* 30 (1975): 131-136. Topics discussed include: temperature, weather, environment.

1890. Marshall, R. "An analysis of the seasonal variation of coronary heart disease and respiratory disease mortality in New Zealand." *Int J Epidemiol* 17, no. 2 (June 1988): 325-31. Topics discussed include: age effects, ethnology, etiology, epidemiology, New Zealand, gender effects.

1891. Mason, B. "Association between season of first detection of breast cancer and disease progression." *Breast Cancer Res Treat* 9, no. 3 (1987): 227-32. Topics discussed include: mortality, light, breast neoplasms, epidemiology.

1892. Matousek, J. "The summer 1976 period of extra high air temperatures and morbidity with some diseases." *Cas Lek Cesk* 118, no. 43-44 (2 November 1979): 1338-42. Topics discussed include: cardiovascular epidemiology, coronary epidemiology, Czechoslovakia, hospitalization, seasonality, temperature.

1893. McCarroll, J. "Morbidity associated with cold weather training." *Mil Med* 144, no. 10 (October 1979): 680-4. Topics discussed include: diagnosis, cold climates, military medicine.

1894. Miasnikov, I. "Tularemia morbidity during winter spring farm labor and its prevention." *Zh Mikrobiol Epidemiol Immunobiol* 46, no. 10 (October 1969): 90-6. Topics discussed include: agriculture, workers, Russia, seasonality.

1895. Mills, C. A. "Coronary Occlusion, Heart Failure, and Environmental Temperatures." *Amer Heart J* 16, no. 6 (1938): 701-713. Topics discussed include: temperature, environment, disease, health issues, mortality, death.

1896. Missenard, A. "Influence des Conditions Thermiques Ambiantes sur la Capacité de Travail, la Morbidité et la Mortalité des ouvriers et la Construction des Lieux de Travail." *Publ Circul Comm d'Organis du Bâtiment et des Travaux* 8, (5 February 1944). Topics discussed include: temperature, labor, economy, mortality, morbidity, disease, health issues, industry.

1897. Momiyama, M. "Seasonal effects on human mortality." *Prog Biometeorol* 1 Pt 1B, no. 709 (1974): 521-548. Topics discussed include: age effects, climatic effects, Egypt, Europe, heart diseases, Japan, neoplasms, New York City, pneumonia, temperature, tuberculosis.

1898. ---. *Seasonality in human mortality : a medico-geographical study*. 181 pages. Tokyo: University of Tokyo Press, 1977. Topics discussed include: diseases, seasonality, medicine.

1899. Momiyama, S. M. "A Medical-Climatological Study in the Seasonal Variations of Mortality in the United States of America." *Papers in Meteorology and Geophysics* 17, 18 (1966): 276-286, 209-232. Topics discussed include: health issues.

1900. Mulloev, A. "Organization of ambulatory aid for workers in the shoe industry in the case of diseases with a temporary loss of work capacity." *Sov Zdravookhr* 0, no. 7 (1974): 38-42. Topics discussed include: absenteeism, morbidity, Russia, seasonality, ambulatory care, disability evaluation, work-related treatment, shoes.

1901. Murray, M. A. "The Geography of Death in England and Wales." *Annals of the Association of American Geographers* 52 (1962): 130-149. Topics discussed include: geography, mortality.

1902. ---. "The Geography of Death in the United States and the United Kingdom." *Annals of the Association of American Geographers* 57 (1966): 301-314. Topics discussed include: geography, mortality.

1903. Nagata, H. "Meteorological conditions, air pollution and daily morbidity in summer." *Nippon Eiseigaku Zasshi* 33, no. 6 (February 1979): 772-7. Topics discussed include: acute disease, Japan, seasonality, gastrointestinal diseases, epidemiology, respiratory tract diseases.

1904. Naumenko, V. "Effect of environmental factors on sudden death in the Far North." *Sud Med Ekspert* 28, no. 1 (January 1985): 9-12. Topics discussed include: arctic regions, coronary disease mortality, seasonality, USSR, cold climates, coronary disease etiology, meteorology.

1905. Nicolau, G. "Circadian, weekly, and seasonal variations in cardiac mortality, blood pressure, and catecholamine excretion." *Chronobiol Int* 8, no. 2 (1991): 149-59. Topics discussed include: blood pressure, catecholamines urine, circadian rhythms physiology, heart diseases, cycles.

1906. Novikova, K. F. "The Effect of Solar Activity on the Development of Myoricardial Infarction and Mortality Resulting Therefrom." *Cardiology* 4 (1968): 109-112. Topics discussed include: environment, health issues, biology, disease.

1907. Nurgaleeva, R. "Effect of weather conditions in an epidemic season on the nature of the morbidity of hemorrhagic fever with renal syndrome." *Zh Mikrobiol Epidemiol Immunobiol no.* 3 (March 1988): 40-4. Topics discussed include: age effects, disease reservoirs, disease vectors, epidemic transmission, rural areas, population, gender effects, disease outbreak.

1908. Oechsli, F. W. "Excess Mortality Associated with Three Los Angeles September Hot Spells." *Environmental Research* 3 (1970): 277-284. Topics discussed include: temperature, weather, seasonality, biology.

1909. Oehmisch, W. "Seasonal rhythm of mortality. Mortality according to age groups and seasons." *Dtsch Gesundheitsw* 24, no. 22 (29 May 1969): 1046-52. Topics discussed include: accidents, arteriosclerosis, diabetes mellitus, Germany, lung diseases, neoplasms, cycles, urologic diseases.

1910. Otto, W. "On the singularity of seasonal mortality." *Z Arztl Fortbild Jena* 62, no. 21 (1 November 1968): 1198-204. Topics discussed include: cardiovascular diseases, central nervous system diseases, respiratory tract diseases, weather.

1911. Parnell, R. W. "Notes on Climate and Morbidity." *J Hyg* 46 (1948): 311-314. Topics discussed include: health issues, illness, disease.

1912. Pearl, R. "Seasonal Fluctuations of the Vital Index of Population." *Proc Natl Acad Sci* 8 (1922): 77-78. Topics discussed include: demography, mortality.

1913. Peretjagina, N. "The upper tolerance limits of nonepidemic daily morbidity for influenza and other acute respiratory diseases in the epidemic season." *Bull World Health Organ* 55, no. 6 (1977): 761-3. Topics discussed include: probability, influenza, respiratory tract infections, epidemiology.

1914. Piergentili, P. "Seasonal variations and trends in cardiovascular disease mortality in Naples, 1974-1978." *G Ital Cardiol* 13, no. 12 (December 1983): 393-8. Topics discussed include: Italy.

1915. Pope, C. D. "Daily mortality and PM10 pollution in Utah Valley." *Arch Environ Health* 47, no. 3 (May 1992): 211-7. Topics discussed include: environment, cardiovascular diseases, causality, respiratory tract diseases, seasonality, weather.

1916. Porksheian, O. "Sudden death in ischemic heart disease, hypertension and arteriosclerosis and its dependence on meteorological factors in different regions of the USSR." *Sud Med Ekspert* 25, no. 2 (April 1982): 23-6. Topics discussed include: seasonality, urban areas, population, USSR, arteriosclerosis mortality, coronary mortality, hypertension mortality.

1917. Posse, P. "Effect of the weather on morbidity dynamics in a large city. An investigation on the population of Munich." *MMW Munch Med Wochenschr* 117, no. 11 (14 March 1975): 425-30. Topics discussed include: circadian rhythms, emergency medical services, Germany, prognosis.

1918. Post, J. D. *Food Shortage, Climatic Variability, & Epidemic Disease in Preindustrial Europe: The Mortality Peak in the Early 1740s*. Ithaca, NY: Cornell University Press, 1985. Topics discussed include: climates, nutrition, malnutrition, Europe.

1919. Putzke, H. "The incidence of fatal lung emboli with special reference to the underlying disease and the effect of weather." *Z Gesamte Inn Med* 44, no. 4 (15 February 1989): 106-10. Topics discussed include: Germany, neoplasms, pulmonary embolisms, etiology, risk, causes of death, mortality, seasonality.

1920. Ramlow, J. "Effects of the summer heat wave of 1988 on daily mortality in Allegheny County, PA." *Public Health Rep* 105, no. 3 (May 1990): 283-9. Topics discussed include: age effects, Pennsylvania, seasonality, gender effects, death.

1921. Rao, M. N. "On 'Weather and human mortality' by Kalkstein and Davis." *Annals of the Association of American Geographers* 80 (June 1990): 295-7.

1922. Reisz, O. "Seasonal incidence and mortality due to myocardial infarct." *Fysiatr Revmatol Vestn* 50, no. 5 (October 1972): 265-72. Topics discussed include: barometric pressure, Czechoslovakia, myocardial infarction, temperature, weather, epidemiology.

1923. Retailliau, H. "Excess mortality and influenza surveillance in Taiwan." *Int J Epidemiol* 7, no. 3 (September 1978): 223-9. Topics discussed include: absenteeism, disease propagation, orthomyxoviridae isolation, seasonality, Taiwan, influenza, microbiology.

1924. Rogot, E. "Associations between coronary mortality and the weather, Chicago, 1967." *Public Health Rep* 89, no. 4 (July 1974): 330-8. Topics discussed include: age effects, Chicago, humidity, racial differences, rain, respiratory tract diseases, gender effects, snow, sunlight, temperature, wind, coronary disease.

1925. ---. "Associations of cardiovascular mortality with weather in Memphis, Tennessee." *Public Health Rep* 85, no. 1 (January 1970): 25-39. Topics discussed include: cerebrovascular disorders, coronary disease, neoplasms, respiratory tract diseases, temperature, Tennessee.

1926. ---. "Associations of coronary and stroke mortality with temperature and snowfall in selected areas of the United States, 1962 1966." *Am J Epidemiol* 103, no. 6 (June 1976): 565-75. Topics discussed include: age effects, gender effects, cerebrovascular disorders, coronary disease, snow, weather.

1927. Rose, G. "Cold weather and ischaemic heart disease." *Br J Prev Soc Med* 20, no. 2 (April 1966): 97-100. Topics discussed include: coronary mortality, epidemiology, myocardial infarction, seasonality.

1928. Rosenwaike, I. "Seasonal Variation of Deaths in the United States 1951-1960." *Journal of the American Statistical Association* 61 (September 1966): 706-719. Topics discussed include: demography, death.

1929. Sakamoto-Momiyama, M. *Seasonability in Human Mortality: A Medico Geographical Study*. Tokyo: University of Tokyo Press, 1977. Topics discussed include: health issues, geography.

1930. ---. "Statistical Analysis of Seasonal Variation in Mortality." *Journal of the Meteorological Society of Japan* 49 (1971): 494-509. Topics discussed include: statistics.

1931. ---. "Changes in the Seasonality of Human Mortality: A Medico-Geographical Study." *Social Science and Medicine* 12, no. 1D (March 1978): 29-42. Topics discussed include: death, disease, civilization, biometeorology, Europe, North America, Japan.

1932. Salem, G. "Geographie de la mortalite et de la natalite a Pikine (Senegal): interets et limites des donnees d'Etat Civil dans les villes africaines." *Social Science & Medicine* 36, no. 10: 1297-311

(May 1993). Topics discussed include: Senegal, spatial mmodels, medical geography, seasonality, mortality, urban areas, health.

1933. Salvage, A. "Old people in hot weather." *Age Ageing* 20, no. 4 (July 1991): 233-5. Topics discussed include: thermoregulation, mortality, heat, meteorology.

1934. Sällström, T. "Seasonal Variations in Morbidity in Cases of Gastric Ulcer and Duodenal Ulcer." *Acta Med Scand* 121, no. 4 (1945): 369-381. Topics discussed include: disease, health issues, ulcer.

1935. Schastnyi, E. "Multiyear and seasonal hepatitis A morbidity in different population age groups in certain areas." *Zh Mikrobiol Epidemiol Immunobiol no.* 6 (June 1988): 44-8. Topics discussed include: age effects, cycles, Uzbekistan, disease propagation, hepatitis, epidemiology.

1936. Schickele, E. "Environment and Fatal Heat Stroke: An Analysis of 157 Cases Occuring in the Army in the U. S. during World War II." *Military Surgeon* 100 (1947): 235-256. Topics discussed include: environment, temperature, mortality, health issues, death, weather.

1937. Schieche, M. "Effect of seasonal vacation on worker's morbidity." *Z Gesamte Hyg* 18, no. 7 (July 1972): 527-8. Topics discussed include: Germany, work-related medicine, recreation, holidays.

1938. Schnabel, K. "Significance of air pressure decrease for the fetal oxygen supply in mothers not adapted to high altitudes in high mountain regions." *Verh Dtsch Ges Inn Med* 80, (1974): 916-20. Topics discussed include: adaptation, blood pressure, pregnancy, temperature, anoxia etiology, fetal death etiology, mountaineering, oxygen consumption.

1939. Schumann, S. "Patterns of Urban Heat-wave Deaths and Implications for Prevention: Data from New York and St. Louis During July, 1966." *Environmental Research* 5 (1972): 59-75. Topics discussed include: climates, weather, temperature, mortality, death.

1940. Scragg, R. "Seasonality of cardiovascular disease mortality and the possible protective effect of ultra violet radiation." *Int J Epidemiol* 10, no. 4 (December 1981): 337-41. Topics discussed include: altitude, calcium metabolism, cardiovascular, temperature, vitaminphysiology, etiology, ultraviolet rays.

1941. Serowy, C. "On the seasonal influence on the morbidity and effect and results of climatotherapy on the Baltic Sea in endogenous eczema." *Dermatol Wochenschr* 153, no. 45 (11 November 1967): 1233-44. Topics discussed include: eczema, epidemiology, Germany, treatment, thalassotherapy.

1942. ---. "Seasonal rhythmic behavior of morbidity, treatment effectiveness and treatment results in endogenous eczema." *Dermatol Monatsschr* 160, no. 9 (September 1974): 735-41. Topics discussed include: age effects, climatic effects, dermatitis, atopic therapy.

1943. Shumway, R. H. "Statistical analysis of daily London mortality and associated weather and pollution effects." University of California, Davis. Statistical Laboratory, and California. Biological Effects and Air Standards Branch. Statistical Laboratory, Division of Statistics, University of California, Davis, CA, 1983. Topics discussed include: air pollution, London, physiology, climate.

1944. Shumway, R. H. "Modeling mortality fluctuations in Los Angeles as functions of pollution and weather effects." *Environ Res* 45, no. 2 (April 1988): 224-41. Topics discussed include: air pollutants, pharmacology, California.

1945. Sinel'Shchikov, V. "Sudden death from ischemic heart disease under the climatic conditions of regions of the Far North, western Siberia and Central Asia." *Sud Med Ekspert* 24, no. 3 (July 1981): 26-8. Topics discussed include: arctic regions, Asia, meteorology, Siberia, sunlight, coronary mortality.

1946. Slepushkin, A. "Anti influenza immunity and the prediction of influenza morbidity in the upcoming epidemic season." *Vopr Virusol no.* 3 (May 1983): 290-2. Topics discussed include: antibodies, antibodies, viral analysis, antibody formation, influenza epidemiology, Moscow, immunology, prognosis.

1947. Sozinsky, T. S. "Relation of Age to the Seasonal Death Rate." *Med and Surg Reporter* 41 (1879): 92-113. Topics discussed include: mortality.

1948. Spann, W. "The Weather and Mortality Rate." *Germ Med Monthly* 2, no. 4 (1957): 116-118. Topics discussed include: climatic effects, death.

1949. Spivak, M. "Effect of meteorological and heliogeophysical factors on the acute pneumonia morbidity of the adult population of Arkhangelsk in 1981 1983." *Ter Arkh* 58, no. 1 (1986): 91-3. Topics discussed include: acute disease, cold climates, Russia, pneumonia, epidemiology, sunlight.

1950. States, S. J. "Weather and Deaths in Pittsburgh, Pennsylvania: A Comparison with Birmingham, Alabama." *International Journal of Biometeorology* 21 (1977): 7-15. Topics discussed include: mortality, geography, climate.

1951. "Statistical analysis of daily London mortality and associated weather and pollution effects : executive summary." University of California, Davis. Statistical Laboratory, and California. Biological Effects and Air Standards Branch. The Laboratory, Davis, CA, 1983. Topics discussed include: air pollution, London, physiology, climate.

1952. Stocks, P. "High Barometer and Sudden Deaths." *Brit Med J* 2 (1925): 1188. Topics discussed include: weather, environment, mortality, death.

1953. Stojkov, N. *Influence of seasonal and other factors on mortality of population of Yugoslavia 1950-1969*. 75 pages. Yugoslavia. Savezni zavod za statistiku, Studije analize i prikazi; 59, Beograd: Savezni zavod za statistiku, 1972. Topics discussed include: weather, physiological effects, Yugoslavia.

1954. Svishchev, A. "Effect of age, sex and season factors on sudden death in ischemic heart disease." *Arkh Patol* 40, no. 10 (1978): 14-8. Topics discussed include: age effects, coronary pathology, Moscow, gender effects, urban areas, health issues, coronary mortality.

1955. Sygit, M. "Temporary work disability (work absenteeism) for reasons of sickness among workers in the shipbuilding industry." *Sante Publique Bucur* 30, no. 1 (January 1987): 35-44. Topics discussed include: morbidity, seasonality, absenteeism, ships.

1956. Tanabe, N. "Seasonal variation of sudden death based on analysis of death certificate information." *Nippon Koshu Eisei Zasshi* 40, no. 3 (March 1993): 196-204. Topics discussed include: age effects, Japan.

1957. Teng, H. C. "The Relationship between Sudden Changes in Weather and the Occurence of Acute Myocardial Infarction." *American Heart Journal* 49 (1955): 9-20. Topics discussed include: health issues, disease, death.

1958. Thurston, G. "Reexamination of London, England, mortality in relation to exposure to acidic aerosols during 1963-1972 winters." *Environ Health Perspect* 79 (February 1989): 73-82. Topics discussed include: aerosols, London, risk, smog, acid rain, air pollutants, respiratory tract diseases, seasonality.

1959. Tomari, T. "Seasonal variation of mortality from cerebro cardiovascular diseases effect of ambient temperature on death." *Nippon Koshu Eisei Zasshi* 38, no. 5 (May 1991): 315-23. Topics discussed include: Japan, epidemiology, cerebrovascular disorders, heart problems.

1960. Tout, D. "The Discomfort Index, mortality and the London summers of 1976 and 1978." *Int J Biometeorol* 24, no. 4 (December 1980): 323-8. Topics discussed include: cerebrovascular disorders, humidity, respiratory insufficiency, cardiovascular diseases, heat, seasonality, weather.

1961. Traute, V. "Correletion between Geomagnetic Storms and Death Rate." *Deutsche Medizinische Wochenschrift* (January 1935): 95. Topics discussed include: geography, environment, mortality, statistics.

1962. Tsuji, T. "Epidemiologic studies on seasonal variation in mortality and morbidity of stroke." *Nippon Eiseigaku Zasshi* 30, no. 1 (April 1975): 185. Topics discussed include: cerebrovascular disorders, Japan, epidemiology.

1963. Vacheva, D. "Seasonal characteristics and prediction of morbidity with temporary disability in work shifts." *Sov Zdravookhr* 28, no. 8 (1969): 53-9. Topics discussed include: prognosis, temperature, textile industry, timing, USSR, disability evaluation, work-related medicine.

1964. Vashchenkova, A. "Effect of seasonal factors on the dynamics of psoriasis morbidity." *Vestn Dermatol Venerol no.* 5 (May 1976): 71-4. Topics discussed include: etiology, Siberia, psoriasis epidemiology.

1965. VonAllmen, S. "Summer mortality surveillance from selected city and county medical examiners." *MMWR Surveill Summ* 32, no. 1 (February 1983): 1SS-6SS. Topics discussed include: heat, seasonality.

1966. Weiss, K. "Seasonal trends in US asthma hospitalizations and mortality." *JAMA* 263, no. 17 (2 May 1990): 2323-8. Topics discussed include: data collection, morbidity, patient discharge statistics, epidemiology, asthma, hospitalization statistics, cycles.

1967. Weiss, M. H. "Seasonal Incidence of Myocardial Infarction in Kentuckiana Area." *J Kentucky Med Assoc* 51 (1953): 14-15. Topics discussed include: disease, health issues, mortality, death.

1968. Welch, C. E. "The timing of death in Athens, Georgia, 1974 1980: A research note." *Omega Journal of Death and Dying* 13, no. 4 (1982): 389-394. Topics discussed include: dying, age differences, biological rhythms, seasonality, frequency distribution, gender differences, time.

1969. West, R. "Mortality from ischaemic heart disease association with weather." *Br J Prev Soc Med* 27, no. 1 (February 1973): 36-40. Topics discussed include: calcium analysis, England, socioeconomics, temperature, Wales, water analysis, water supply.

1970. ---. "Mortality from ischaemic heart disease inter town variation and its association with climate in England and Wales." *Int J Epidemiol* 5, no. 2 (June 1976): 195-201. Topics discussed include: age effects, England, rain, gender effects, socioeconomics, temperature, Wales.

1971. Williams, E. "Mortality rates from multiple sclerosis: geographical and temporal variations revisited." *J Neurol Neurosurg Psychiatry* 54, no. 2 (February 1991): 104-9. Topics discussed include: cross-sectional analysis, United Kingdom, epidemiology, risk, seasonality, climatic effects, multiple sclerosis.

1972. Wong, M. "Sudden unexplained death syndrome. A review and update." *Trop Geogr Med* 44, no. 4 (October 1992): S1-19. Topics discussed include: Asia, autopsy, causality, culture, pathology, ethnic groups, risk, seasonality, gender effects, Singapore, socioeconomics.

1973. Wood, F. C. "The Seasonal Incidence of Acute Coronary Occlusion in Philadelphia." *Med Clinics North America* 19, no. 1 (1935): 151-157. Topics discussed include: disease, health issues, mortality, death.

1974. Woodhouse PR. "Seasonal variations of plasma fibrinogen and factor VII activity in the elderly: winter infections and death from cardiovascular disease." *Lancet* 343, no. 8895 (19 February 1994): 435-9. Topics discussed include: mortality, cycles, blood, United Kingdom, neutrophils, respiratory tract infections immunology, physiopathology, risk, seasonality.

1975. Wyndham, C. "Are mortality rates for respiratory diseases in the RSA affected by climate?" *S Afr Med J* 69, no. 4 (15 February 1986): 223-6. Topics discussed include: age effects, seasonality, South Africa, temperature, tropical pneumonia.

1976. ---. "Climate and disease." *S Afr Med J* 53, no. 26 (24 June 1978): 1051-61. Topics discussed include: age effects, altitude, Caucasoid race, United Kingdom, Negroid race, seasonality, South Africa, temperature, coronary mortality.

1977. Zavgorud'Ko, V. "Morbidity with temporary loss of work capacity in relation to the natural and climatic conditions in the region of the construction of the eastern portion of the Baikal Amur mainline." *Zdravookhr Ross Fed* no. 10 (1978): 12-4. Topics discussed include: geography, seasonality, Siberia, absenteeism, railroads.

1978. Zherikova, A. "The epidemiology of typhoid in the Mongolian People's Republic (the dynamics of morbidity, its territorial and seasonal distribution)." *Zh Mikrobiol Epidemiol Immunobiol* 49, no. 8 (August 1972): 42-6. Topics discussed include: Mongolia.

1979. Zhou, Y. "Statistical analysis of the effect of meteorological factors on the mortality of urban populations." *Acta Acad Med Wuhan* 3, no. 3 (1983): 156-60. Topics discussed include: China.

SUICIDE

1980. Abel, E. "Temporal variation in violent death in Erie County, New York, 1973-1983." *Am J Forensic Med Pathol* 8, no. 2 (June 1987): 107-11. Topics discussed include: accidents, traffic, alcohol drinking, homicide, suicide, epidemiology, timing, seasonality.

1981. Allen, J. "Depressive symptoms and family history in seasonal and nonseasonal mood disorders." *Am J Psychiatry* 150, no. 3 (March 1993): 443-8. Topics discussed include: alcoholism, body weight, morbidity, eating disorders diagnosis, sleep disorders diagnosis, suicide, psychology.

1982. Araki, S. "Seasonal variation in suicide rates by cause and sex." *J Biosoc Sci* 18, no. 4 (October 1986): 471-8. Topics discussed include: Japan, gender effects, epidemiology.

1983. ---. "Suicide in Japan: Socioeconomic Effects on Its Secular and Seasonal Trends." *Suicide and Life Threatening Behavior* 17, no. 1 (March 1987): 64-71. Topics discussed include: Japan, socioeconomics.

1984. Bagley, C. "Ecological Variation of Three Types of Suicide." *Psychological Medicine* 6 (1976): 423-427. Topics discussed include: environment, mental disorders.

1985. ---. "Social Structure and the Ecological Distribution of Mental Illness, Suicide and Delinquency." *Psychological Medicine* 3 (1973): 177-187. Topics discussed include: environment, mental disorders, crime, society.

1986. Baker, G. "Seasonal births, academic achievement and psychopathology (suicide and homicide): A regional analysis." *Psychological Reports* 59, no. 2, Pt 2 (October 1986): 742. Topics discussed include: academic performance, high school, epidemiology, adolescence.

1987. Barraclough, B. M. "Monthly Variation of Suicidal, Accidental and Undetermined Poisoning Deaths." *British Journal of Psychiatry* 132 (1978): 279-282. Topics discussed include: seasonality, death.

1988. Bazas, T. "Incidence and seasonal variation of suicide mortality in Greece." *Comprehensive Psychiatry* 20, no. 1 (January 1979): 15-20. Topics discussed include: age, gender differences, urban areas, rural areas, environment.

1989. Beck, A. T. "Attempted Suicide and Month of Birth." *Psychol Rep* 33 (1973): 506.

1990. Besancon, F. "Weather conditions in myocardial infarction, cerebrovascular strokes and suicide attempts." *Sem Hop Paris* 57, no. 39-40 (18-25 October 1981): 1609-11. Topics discussed include: humidity, temperature, etiology.

1991. Bollen, K. A. "Temporal variations in mortality: a comparison of U.S. suicides and motor vehicle fatalities, 1972-1976." *Demography* 20 (February 1983): 45-59. Topics discussed include: traffic accidents, seasonality, mental illness.

1992. Bouma, J. "Daily, Monthly, and Yearly Fluctuations in Total Number of Suicides and Suicide Attempts in the Western Part of the Netherlands." *J Interdis Cycle Res* 3 (1972): 269. Topics discussed include: seasonality.

1993. Bray, J. J. "An Investigation of Demographic Factors in Suicide Notes of Milwaukee County and Suicide Note Intent: Implications for Prevention and Education." Ph.D. diss., The University of Wisconsin, 1991. Topics discussed include: psychology, sociology, demography, health, seasonality; available from UMI.

1994. Breuer, H. "Social, toxicological and meteorological data on suicide attempts." *Eur Arch Psychiatry Neurol Sci* 235, no. 6 (1986): 367-70. Topics discussed include: alcohol, psychology, poisoning, Germany, sociology, weather.

1995. ---. "Suicide and weather." *Dtsch Med Wochenschr* 109, no. 45 (9 November 1984): 1716-20. Topics discussed include: Germany, seasonality, psychology.

1996. Brewerton, T. D. "Fluoxetine induced suicidality, serotonin, and seasonality." *Biological Psychiatry* 30, no. 2 (July 1991): 190-196. Topics discussed include: depression, fluoxetine, therapy, suicidal ideation.

1997. Cerbus, G. "Seasonal variation in some mental health statistics: suicides, homicides, psychiatric admissions, and institutional placement of the retarded." *J Clin Psychol* 26, no. 1 (January 1970): 61-3. Topics discussed include: mental retardation, hospitalization, hospitals, mental disorders epidemiology.

1998. Chiu, L. P. "Do weather, day of the week, and address affect the rate of attempted suicide in Hong Kong?" *Social Psychiatry and Psychiatric Epidemiology* 23, no. 4 (October 1988): 229-235. Topics discussed include: atmospheric effects, timing.

1999. Danneel, R. "The influence of geophysical factors on the frequency of suicides." *Arch Psychiatr Nervenkr* 219, no. 2 (1974): 153-7. Topics discussed include: astronomy, geography, Germany, magnetics, seasonality, epidemiology.

2000. ---. "Seasonal differences in the frequency of suicide in youth and old age." *Archiv fur Psychiatrie und Nervenkrankheiten* 221, no. 1 (1975): 11-13. Topics discussed include: Germany, environment, age differences, young adults.

2001. Dexter, E. G. "Suicide and the Weather." *Pop Sci Monogr* 58 (1901): 604. Topics discussed include: climate.

2002. Digon, E. "Suicide and Climatology." *Archives of Environmental Health* 12 (1966): 279-286. Topics discussed include: death, climate.

2003. Dixon, K. "A statistical investigation into the relationship between meteorological parameters and suicide." *Int J Biometeorol* 27, no. 2 (July 1983): 93-105. Topics discussed include: New York City, epidemiology.

2004. Dreyer, K. "Comparative Suicide Statistics: II. Death Rates from Suicide in Denmark since 1921, and Seasonal Variations since 1835." *Dan Med Bull* 6 (1959): 75. Topics discussed include: weather, climate.

2005. Durkheim, E. *Suicide: A Study in Sociology*. New York: Free Press, 1961. Topics discussed include: climate, death.

2006. Eastwood, M. R. "Seasonal patterns of suicide, depression and electroconvulsive therapy." *British Journal of Psychiatry* 129 (November 1976): 472-475. Topics discussed include: hospital admissions, therapy, Canada.

2007. Farrington, M. K. "The Seasonality of Pathology: A Sociological Analysis." Ph.D. diss., University of New Hampshire, 1980. Topics discussed include: sociology, social welfare, suicide, mortality, disease, behavior, culture, economy, climatic effects; available from UMI.

2008. Faust, V. "Suicide and weather." *Fortschr Med* 91, no. 4 (8 February 1973): 154-8. Topics discussed include: alcoholism, climatic effects, depression, neurotic disorders, schizophrenia, substance dependence, Switzerland.

2009. ---. "Suicide and the weather. Risk of suicide in mental disorders and possibilities for prevention." *Fortschr Med* 94, no. 7 (4 March 1976): 370-8. Topics discussed include: alcoholism, depression, Germany, neurotic disorders, personality, risk, schizophrenia, substance dependence.

2010. ---. "Weather and disease. 2. Weather dependent diseases." *Fortschr Med* 95, no. 5 (3 February 1977): 268-72. Topics discussed include: accidents, epidemiology, Germany, infection, psychophysiologic disorders, reaction time, suicide, morbidity.

2011. Fiallo, S. A. "Suicidal attempt and some climatological parameters." *Revista del Hospital Psiquiatrico de La Habana* 27, no. 1 (January 1986): 75-79. Topics discussed include: atmospheric effects, humidity, Cuba, temperature.

2012. Flosz, E. "Epidemiology and meteoropathology of the suicide occurrence in the district of Wernigerode." *Z Gesamte Hyg* 16, no. 2 (February 1970): 124-33. Topics discussed include: age effects, atmospheric effects, Germany, humidity, seasonality, temperature, weather.

2013. Frank, M. L. "Geophysical variables and behavior: LI. Temporal variation of suicide in teens and young adults." *Perceptual and Motor Skills* 67, no. 2 (October 1988): 586. Topics discussed include: seasonality, holidays, age differences.

2014. Fuchs, H. "Seasonal Rhythm of Suicides - an International Study." *Statist Nachricht* (1959): 14. Topics discussed include: weather.

2015. Gabennesch, H. "When promises fail: A theory of temporal fluctuations in suicide." *Social Forces* 67, no. 1 (September 1988): 129-145. Topics discussed include: seasonality, holiday fluctuations, timing, alternative theories.

2016. Garcia, P. A. "Epidemiologic aspects of suicide in Asturias in relation to the climate, the seasons and other factors." *Actas Luso Esp Neurol Psiquiatr Cienc Afines* 19, no. 4 (July 1991): 185-90. Topics discussed include: age effects, lunar effects, gender effects, Spain, epidemiology, wind, meteorology.

2017. Gostynski, M. "Meteorotropic activity of the weather conditions and their role in suicide epidemiology. I. Communication." *Zentralbl Bakteriol Orig B* 161, no. 2 (October 1975): 158-64. Topics discussed include: atmospheric pressure, Poland, seasonality.

2018. ---. "Weather stress in the epidemiology of selected biological phenomena." *Pol Tyg Lek* 30, no. 45 (10 November 1975): 1869-71. Topics discussed include: Poland, stress, accidents, suicide.

2019. Gowitt, G. "Suicide in Fulton County, Georgia (1975-1984)." *J Forensic Sci* 31, no. 3 (July 1986): 1029-38. Topics discussed include: age effects, Georgia, seasonality, gender effects, epidemiology.

2020. Hanhart, E. "Contribution of genetic factors in suicide based on temporary increase in the rate of familial suicides in Swiss isolates." *Arch Julius Klaus Stift Vererbungsforsch Sozialanthropol Rassenhyg* 43-44, (1968): 305-76. Topics discussed include: depression, pregnancy, religion, psychology, schizophrenia, seasonality, twins.

2021. Imre, O. "Studies on the meterological correlations of suicide. II. Correlations between suicides and meteorological changes." *Morphol Igazsagugyi Orv Sz* 12, no. 2 (April 1972): 132-6. Topics discussed include: Hungary, epidemiology, weather.

2022. Kevan, S. "Perspectives on season of suicide: a review." *Soc Sci Med Med Geogr* 14, no. 4 (December 1980): 369-78. Topics discussed include: age effects, gender effects, urban areas, epidemiology.

2023. ---. "Weather climate criminality." *ZFA Stuttgart* 52, no. 5 (20 February 1976): 257-64. Topics discussed include: accidents, suicide, crime.

2024. Kok, L. "Season, climate and suicide in Singapore." *Med Sci Law* 33, no. 3 (July 1993): 247-52. Topics discussed include: Singapore, epidemiology, psychology, cross-cultural issues.

2025. Kubacki, A. "Suicide weather? Special Issue: Canadian Academy of Child Psychiatry: A Canadian perspective." *Canadian Journal of Psychiatry* 31, no. 6 (August 1986): 602-604. Topics discussed include: atmospheric effects, gender differences, seasonal variations.

2026. Kunz, P. "Relationship Between Suicide and Month of Birth." *Psychological Reports* 42 (1978): 794. Topics discussed include: seasonality.

2027. Larcan, A. "Suicidal behaviour and meteorological conditions." *Annales Medico Psychologiques* 1, no. 1 (January 1976): 51-61. Topics discussed include: weather, attempt timing, atmospheric effects, environment.

2028. Laszlo, T. "Studies on the meteorological correlations of suicides. I. The rate of suicides at different meteorological situations." *Morphol Igazsagugyi Orv Sz* 12, no. 2 (April 1972): 121-31. Topics discussed include: Hungary, epidemiology, weather.

2029. Lester, D. "Geophysical variables and behavior: XLI. Regional variation of suicide and homicide rates by different methods." *Percept Mot Skills* 64, no. 3 Pt 2 (June 1987): 1074. Topics discussed include: cross-sectional analysis, homicide, epidemiology.

2030. ---. "Geophysical variables and behavior: XLVIII. Climate and personal violence (suicide and homicide): a cross cultural study." *Percept Mot Skills* 66, no. 2 (April 1988): 602. Topics discussed include: cross-cultural issues, homicide, epidemiology.

2031. ---. "Geophysical variables and behavior: LVI. The spatial autocorrelation of states suicide and homicide rates." *Percept Mot Skills* 68, no. 1 (February 1989): 218. Topics discussed include: cross-sectional analysis, geography, homicide, epidemiology.

2032. ---. "Seasonal variation in suicidal deaths." *Br J Psychiatry* 118, no. 547 (June 1971): 627-8. Topics discussed include: age effects, New York, gender effects, epidemiology.

2033. ---. "Seasonal variation in suicide rates." *Lancet* 1, no. 803 (17 March 1973): 611-2. Topics discussed include: temperature, epidemiology.

2034. ---. "The seasonal variation of suicide in Zimbabwe." *Percept Mot Skills* 73, no. 1 (August 1991): 18. Topics discussed include: epidemiology, Zimbabwe, cross-cultural issues.

2035. ---. "Seasonality in suicide and economic growth." *Percept Mot Skills* 77, no. 1 (August 1993): 10. Topics discussed include: Japan, epidemiology, economics.

2036. ---. "Temporal Variations in Suicide and Homicide." *Amer J Epidemiol* 109 (1979): 517-520. Topics discussed include: seasonality, homicide.

2037. ---. "Variation in suicide and homicide by latitude and longitude." *Percept Mot Skills* 51, no. 3 Pt 2 (December 1980): 1346. Topics discussed include: epidemiology.

2038. ---. "The seasonal variation in suicide in urban and rural areas: comment on Micciolo, Zimmerman Tansella, Williams and Tansella." *J Affect Disord* 19, no. 3 (July 1990): 171. Topics discussed include: cross-sectional analysis, Italy, epidemiology, cross-cultural issues.

2039. ---. "The Climate of Urban Areas in the United States and Their Rates of Personal Violence (Suicide and Homicide)." *Death Studies* 15, no. 6 (November 1991): 611-616. Topics discussed include: environment, weather.

2040. ---. "Geophysical variables and behavior: XXXVI. Regional variation in suicide and homicide: Controls for weather and southern culture." *Perceptual and Motor Skills* 64, no. 2 (April 1987): 430. Topics discussed include: geography, epidemiology, atmospheric effects.

2041. ---. "Seasonal variation in suicidal deaths by each method." *Psychological Reports* 56, no. 2 (April 1985): 650. Topics discussed include: methods, epidemiology.

2042. ---. "Suicide and homicide rates: Their relationship to latitude and longitude and to the weather." *Suicide and Life Threatening Behavior* 16, no. 3 (September 1986): 356-359. Topics discussed include: rain, temperature, geography, mortality, atmospheric effects, urban areas.

2043. ---. "Temporal variation in suicide and homicide." *American Journal of Epidemiology* 109, no. 5 (1979): 517-520. Topics discussed include: violence.

2044. ---. "Geophysical variables and behavior: LIX: The seasonal variation in suicide rates: Three studies." *Perceptual and Motor Skills* 70, no. 1 (February 1990): 57-58. Topics discussed include: geography, marriage, economics.

2045. ---. "Seasonal variation in suicide rates in the United States." *Journal of Clinical Psychiatry* 49, no. 9 (September 1988): 371. Topics discussed include: age differences, gender differences, epidemiology.

2046. ---. "Sex differences in the seasonal distribution of suicides." *British Journal of Psychiatry* 153 (July 1988): 115-117. Topics discussed include: gender differences.

2047. Linkowski, P. "Effect of some climatic factors on violent and non violent suicides in Belgium." *J Affect Disord* 25, no. 3 (July 1992): 161-6. Topics discussed include: accidents, Belgium, homicide, risk, seasonality, psychology, weather.

2048. Lyster, W. "Seasonal variations in suicide rates." *Lancet* 1, no. 805 (31 March 1973): 725. Topics discussed include: United Kingdom, epidemiology.

2049. MacMahon, K. "Short term temporal cycles in the frequency of suicide. United States, 1972-1978." *Am J Epidemiol* 117, no. 6 (June 1983): 744-50. Topics discussed include: seasons, timing, epidemiology.

2050. Maes, M. "Seasonality in violent suicide but not in nonviolent suicide or homicide." *Am J Psychiatry* 150, no. 9 (September 1993): 1380-5. Topics discussed include: age effects, Belgium, epidemiology, cycles, gender effects, violence.

2051. Maldonado, G. "Variation in suicide occurrence by time of day, day of the week, month, and lunar phase." *Suicide Life Threat Behav* 21, no. 2 (June 1991): 174-87. Topics discussed include: California, epidemiology, lunar effects, risk, seasonality, psychology.

2052. Marneros, A. "Long term course of schizoaffective disorders. Part III: Onset, type of episodes and syndrome shift, precipitating factors, suicidality, seasonality, inactivity of illness, and outcome." *Eur Arch Psychiatry Neurol Sci* 237, no. 5 (1988): 283-90. Topics discussed include: psychology.

2053. Meares, R. "A sex difference in the seasonal variation of suicide rate: A single cycle for men, two cycles for women." *British Journal of Psychiatry* 138 (April 1981): 321-325. Topics discussed include: gender differences, United Kingdom, gender differences, epidemiology.

2054. Metnieks, A. "Suicide and meteorologic and cosmic factors. Statistical findings." *Minerva Medicoleg* 85, no. 3 (May 1965): 97-104. Topics discussed include: Italy, weather.

2055. Micciolo, R. "Geographical and urban rural variation in the seasonality of suicide: some further evidence." *J Affect Disord* 21, no. 1 (January 1991): 39-43. Topics discussed include: Italy, epidemiology, gender effects.

2056. ---. "Geographical variation in the seasonality of suicide." *J Affect Disord* 15, no. 2 (September 1988): 163-8. Topics discussed include: climate, Italy, demography, epidemiology.

2057. ---. "Seasonal variation in suicide: is there a sex difference?" *Psychol Med* 19, no. 1 (February 1989): 199-203. Topics discussed include: Italy, gender effects, epidemiology.

2058. Mills, C. A. "Suicide and Homicides in Their Relationship to Weather Changes." *American Journal of Psychiatry* 91 (1934): 669-677. Topics discussed include: climate, homicide.

2059. Miner, J.R. "Suicide and its relation to climatic and other factors." Baltimore, MD, *The American Journal of Hygiene*, 1922.

2060. Miner, J. R. "Suicide and Its Relations to Climatic and Other Factors: A Statistical Study." Ph.D. diss., The John Hopkins University, 1922. Topics discussed include: psychology, climate.

2061. Nayha, S. "The bi-seasonal incidence of some suicides: Experience from Finland by marital status, 1961-1976." *Acta Psychiatrica Scandinavica* 67, no. 1 (January 1983): 32-42. Topics discussed include: seasonality, marital status, Finland, epidemiology, adolescents.

2062. ---. "Short and Medium-Term Variations in Mortality in Finland: A Study on Cyclic Variations, Annual and Weekly Periods and Certain Irregular Changes in Mortality in Finland During the Period 1868-1972." Ph.D. diss., Oulun Yliopisto (Finland), 1980. Topics discussed include: health, seasonality, suicide.

2063. Ogle, W. "Suicides in England and Wales in Relation to Age, Sex, Season and Occupation." *J R Statist Soc* 49 (1886): 101. Topics discussed include: behavior, biology.

2064. Parker, G. "Seasonal variation in depressive disorders and suicidal deaths in New South Wales." *British Journal of Psychiatry* 140 (June 1982): 626-632. Topics discussed include: affective disorders, Australia, self inflicted wounds.

2065. Phillips, D. P. "The Frequency of Suicides around Major Public Holidays: Some Surprising Findings." *Suicide and Life Threatening Behavior* 10, no. 1 (March 1980): 41-50. Topics discussed include: suicide, holidays, correlations.

2066. Poikolainen, K. "Heavy alcohol use and seasonal variation of suicide rates." *Psychiatria Fennica* suppl (1983): 93-96. Topics discussed include: Finland, epidemiology, alcoholism.

2067. Pokorny, A. D. "Suicide and Weather." *Arcives of Environmental Health* 13 (1966): 255-256. Topics discussed include: climate.

2068. ---. "Sunspots, Suicide and Homicide." *Diseases of the Nervous System* 27 (1966): 347-348. Topics discussed include: homicide, weather, climate.

2069. Pokorny, A. "Suicide and weather. Effect of barometric pressure and humidity." *Arch Environ Health* 13, no. 2 (August 1966): 255-6. Topics discussed include: atmospheric effects, humidity.

2070. Pokorny, A. D. "Suicide, Suicide Attempts, and Weather." *American Journal of Psychiatry* 120 (1963): 377-381. Topics discussed include: climatic effects.

2071. Porubsky, V. "Seasonal incidence of committed suicides." *Bratisl Lek Listy* 67, no. 4 (April 1977): 416-26. Topics discussed include: Czechoslovakia, gender effects, timing, epidemiology.

2072. Prokupek, J. "Seasonal occurrence of suicides." *Cesk Psychiatr* 64, no. 1 (January 1968): 13-20. Topics discussed include: Czechoslovakia, epidemiology.

2073. Reid, P. "Seasonal variation in Irish suicidal deaths." *Psychological Reports* 46, no. 1 (February 1980): 306. Topics discussed include: death, Ireland, gender differences, mortality.

2074. Roberts, A. R. W. "Social Indicators of and Seasonal Variations in Sex and Age Specific Suicide Rates in Georgia." Ph.D. diss., Emory University, 1980. Topics discussed include: sociology, death; available from UMI.

2075. Rotton, J. "Determinism redux: Climate and cultural correlates of violence." *Environment and Behavior* 18, no. 3 (May 1986): 346-368. Topics discussed include: sociology, suicide, homicide, atmospheric effects.

2076. Ruiz, L. J. "Do environmental factors influence suicide attempts?" *Rev Enferm* 13, no. 140 (April 1990): 69-72. Topics discussed include: age effects, emergencies, meteorological effects, gender effects, Spain.

2077. Sanborn, D. D. "Suicide: seasonal patterns and related variables." *Dis Nerv Syst* 31, no. 10 (October 1970): 702-4. Topics discussed include: atmospheric effects, New Hampshire, epidemiology.

2078. Sanborn, D. E. "Sex, season, and suicide." *Psychological Reports* 42, no. 3, Pt 2 (June 1978): 1332. Topics discussed include: seasonality, gender differences.

2079. Souetre, E. "Seasonality of suicides: environmental, sociological and biological covariations." *J Affect Disord* 13, no. 3 (November 1987): 215-25. Topics discussed include: cross-sectional analysis, France, socioeconomics, hormones, blood.

2080. ---. "Influence of environmental factors on suicidal behavior." *Psychiatry Res* 32, no. 3 (June 1990): 253-63. Topics discussed include: cross-sectional analysis, France, epidemiology, light, seasonality, socioeconomics, weather.

2081. Spiers, P. "Seasonal variation in suicide rates." *Lancet* 2, no. 774 (26 August 1972): 428-9. Topics discussed include: temperature, epidemiology.

2082. Stefanini, P. "Analysis of temporal distribution of suicides. Methodological notes on comparison of dates of death and dates of suicidal act." *Epidemiol Prev* 12, no. 42 (March 1990): 23-9. Topics discussed include: Italy, seasonality.

2083. Takahashi, E. "Seasonal Variation of Conception and Suicide." *Journal of Experimental Medicine* 84 (1964): 829-839. Topics discussed include: fertility, climate.

2084. Thorson, J. A. "Sunshine and suicide: Possible influences of climate on behavior. Special Issue: Suicide practical, developmental, and speculative issues." *Death Education* 8, no. Suppl (1984): 125-136. Topics discussed include: regional variation, epidemiology, atmospheric effects, environment, illumination, Europe.

2085. Tromp, S. W. "Study of the Possible Relationship Between Atmospheric Environment, Suicide and Suicide Attempts in the Western Part of the Netherlands (Period 1954-1969)." *Monograph Ser Biometeorol Res Centre* 22 (1973): 24. Topics discussed include: environment, weather.

2086. Trucco, M. "Suicides in Santiago. II. Seasonal variation." *Rev Med Chil* 105, no. 1 (January 1977): 47-9. Topics discussed include: Chile, epidemiology.

2087. Waehrens, J. "Connection between the incidence of suicide and atmospheric pressure." *Ugeskr Laeger* 143, no. 45 (2 November 1981): 2993-6. Topics discussed include: Denmark, seasonality, atmospheric effects, epidemiology.

2088. Warren, C. W. "Seasonal Variation in Suicide and Homicide: A Question of Consistency." *Journal of Biosocial Science* 15, no. 3 (July 1983): 349-356.

2089. Wenz, F. V. "Effects of seasons and sociological variables on suicidal behavior." *Public Health Reports* 92, no. 3 (May 1977): 233-239. Topics discussed include: demography.

2090. ---. "Seasonal Suicide Attempts and Forms of Loneliness." *Psychological Reports* 40, no. 3 (June 1977): 807-810. Topics discussed include: loneliness, seasonality.

2091. Wilmotte, J. "Suicidal behavior in the prison environment." *Acta Psychiatr Belg* 73, no. 3 (May 1973): 411-26. Topics discussed include: age effects, family factors, homicide, marriage, seasonality, gender effects, timing.

2092. Wood, J. B. "The Effect of Birthday on Deathday of Suicides and Other Deaths: Its Magnitude and Variations." American Statistical Association 1979 Proceedings of Social Statistics Section, Washington, D. C., 1979. Topics discussed include: death, mortality.

2093. Yoshida, K. "Clustering of suicides under age 20 seasonal trends and the influence of newspaper reports." *Nippon Koshu Eisei Zasshi* 38, no. 5 (May 1991): 324-32. Topics discussed include: Japan, newspapers.

2094. Zimmermann, T. C. "Seasonal variation in suicide in Italy." *Epidemiology and the prevention of mental disorders*, pp. 185-194. Routledge, London, England; 367 pages. eds Tomas Helgason Brian Cooper. 1989. Topics discussed include: Italy, gender differences, social environments.

2095. Zung, W. W. "Seasonal variation of suicide and depression." *Archives of General Psychiatry* 30, no. 1 (January 1974): 89-91. Topics discussed include: disorders, emotion.

6

Economic Effects: A Select Bibliography

CONSUMER & INDUSTRIAL BEHAVIOR

Agriculture

2096. Abee, C. et al. *A First Report on the Relations Between Climates & Crops: U S Department of Agriculture, Weather Bureau Bulletin, No 36.* Salem, NH: Ayer Company Publishers, 1978.

2097. Adams, R. M. et al. "Implications of Global Climate Change for Western Agriculture." *Western Journal of Agricultural Economics* 13, no. 2 (December 1988): 348-56. Topics discussed include: conservation, pollution.

2098. "Agro-Climatic Atlas of Europe." editors P. Thran, and S. Brockhuizen. 294 Agro-Ecological Atlas Ser., Elsevier Science Publishing Company, 1965. Topics discussed include: agriculture, climate.

2099. Alden, D.M. "Climate-change and agricultural production - an extensive averaged spatial-ecologic approach." *Environment & Planning A* ; 26, no. 1 (January 1994): 121-136.

2100. Anderson, J. R. "Impacts of Climatic Variability in Australian Agriculture: A Review." *Review of Marketing and Agricultural Economics* 47, no. 3 (December 1979): 147-77. Topics discussed include: supply, rural areas.

2101. ---. "Sparse Data, Climatic Variability, and Yield Uncertainty in Response Analysis." *American Journal of Agricultural Economics* 55, no. 1 (February 1973): 77-82. Topics discussed include: supply, demand.

2102. ---. "Sparse Data, Climatic Variability, and Yield Uncertainty in Response Analysis - Reply." *American Journal of Agricultural Economics* 56, no. 3 (August 1974): 647. Topics discussed include: supply, demand, economic geography.

2103. Anderton, D. et al. "Demographic seasonality and development: the effects of agricultural colonialism in Taiwan, 1906-1942." *Demography* 27, no. 3 (August 1990): 397-411. Topics discussed include: developing countries, history, medicine, demography, politics.

2104. Arthur, L. M. "Potential Adjustments to Climate Change." *Canadian Journal of Agricultural Economics* 38, no. 4 (December 1990): 711-16. Topics discussed include: conservation, pollution.

2105. Arthur, L. M. et al. "Potential Effects of Climate Change on Agriculture in the Prairie Region of Canada." *Western Journal of Agricultural Economics* 13, no. 2 (December 1988): 216-24. Topics discussed include: long run effects.

2106. Asaduzzaman, M. "The Seasonal Variation in Price of Rice: Bangladesh 1950-72." *Bangladesh Economic Review* 1, no. 2 (April 1973): 213-20. Topics discussed include: agriculture, supply.

2107. Babcock, B. A. "The Value of Weather Information in Market Equilibrium." *American Journal of Agricultural Economics* 72, no. 1 (February 1990): 63-72. Topics discussed include: supply, demand.

2108. Bailey, R. et al. "The ecology of birth seasonality among agriculturalists in central Africa." *J Biosoc Sci* 24, no. 3 (July 1992): 393-412. Topics discussed include: newborn infants, pregnancy, tropical climates, Zaire, developing countries.

2109. Baker, D. G. et al. "Agriculture and the recent 'benign climate' in Minnesota." *Bulletin of the American Meteorological Society* 74 (January 1993): 1035-40. Topics discussed include: Minnesota, history.

2110. Bandyopadhyaya, J. "Climate as an Obstacle to Development in the Tropics." *International Social Science Journal* 30, no. 2 (1978): 339-352. Topics discussed include: agriculture, industry, trade, technology.

2111. Beighley, H. P. et al. "The Effect of a Marketing Order on Winter Carrot Prices: Comment." *American Journal of Agricultural Economics* 51, no. 4 (November 1969): 929-33. Topics discussed include: supply, demand.

2112. Beveridge, W. "Weather and Harvest Cycles." *The Economic Journal* (December 1921): 421-453. Topics discussed include: economy, agriculture.

2113. Bivings, E. L. "Price Seasonality and Trade Liberalization: A Dynamic Spatial of the Mexican Feedgrains Sector." Ph.D. diss., Stanford University, 1992. Topics discussed include: agriculture, supply, demand.

2114. Black, J. R. et al. "Some Evidence on Weather-Crop-Yield Interaction." *American Journal of Agricultural Economics* 60, no. 3 (August 1978): 540-43. Topics discussed include: supply, demand.

2115. Black, R. "Weather Variation as a Cost-Price Uncertainty Factor as It Affects Corn and Soybean Production." *American Journal of Agricultural Economics* 57, no. 5 (December 1975): 940-44. Topics discussed include: supply, demand.

2116. Blumenstock, D. I. et al. "Climate and the World Pattern." in *Climate and Man*, 98-127. U. S. Department of Agriculture. Yearbook of Agriculture for 1941, Washington, DC: U.S. Government Printing Office, 1941.

2117. Brannon, R. H. et al. "Irrigated Dry Season Crop Production in Northeast Thailand: A Case Study." *Journal of Developing Areas* 14, no. 2 (January 1980): 191-200. Topics discussed include: agriculture, production, productivity, farming.

2118. Brown, B. G. et al. "On the economic value of seasonal-precipitation forecasts: the fallowing, planting problem." *Bulletin of the American Meteorological Society* 67 (July 1986): 833-41. Topics discussed include: agriculture, rain, forecasting, weather.

2119. Bryson, R. A. et al. *Climatic Change and Agricultural Responses: A Statement on Research and Technological Priorities Between Now and the Year 2000 A Report to the National Science Foundation*, Institute for Environmental Studies, University of Wisconsin-Madison, 1974. Topics discussed include: economy.

2120. Buller, O. et al. "Measuring the Effect of Weather on Crop Production." *Canadian Journal of Agricultural Economics* 17, no. 1 (February 1969): 91-98. Topics discussed include: supply, natural resources.

2121. Burgos, J. J. *Renewable Resources and Agriculture in Latin America in Relation to the Stability of Climate*. 1979. Topics discussed include: climate.

2122. Burt, O. R. et al. "Economic Analysis of Irrigation in Subhumid Climate." *American Journal of Agricultural Economics* 53, no. 1 (February 1971): 33-46. Topics discussed include: agribusiness management, supply, demand.

2123. Caine, R. J. H. et al. "Adjusting the Seasonality of Milk Shipments in Canada: Problems, Economic Impacts and Potential Policies." *Canadian Journal of Agricultural Economics* 31, no. 3 (November 1983): 331-50. Topics discussed include: policy, markets, marketing, cooperatives, management issues.

2124. Carlson, J. D. "The importance of agricultural weather information: a Michigan survey." *Bulletin of the American Meteorological Society* 70 (April 1989): 366-80. Topics discussed include: economic issues.

2125. Chang, J. H. *Climate and Agriculture*. Chicago: Aldine, 1968.

2126. Chowdhury, N. "Seasonality of Foodgrain Price and Procurement Programme in Bangladesh since Liberation - An Exploratory Study." *Bangladesh Development Studies* 15, no. 1 (March 1987): 105-28. Topics discussed include: agriculture, supply, demand.

2127. Cloud, D. S. "House to environmentalists: no more." *Congressional Quarterly Weekly Report* 48 (29 September 1990): 3112. Topics discussed include: agriculture, ecology, policy, environment.

2128. Cumming, G. et al. "Seasonal variation of cardiorespiratory fitness of grain farmers." *J Occup Med* 16, no. 2 (February 1974): 91-3. Topics discussed include: body weight, Canada, heart rate, respiration, spirometry, sports, agriculture, cereals, heart physiology, physical fitness, respiratory system physiology.

2129. Currie, R. "Lunar tides and the wealth of nations." *New Scientist* 120, no. 5 (1988): 52-5. Topics discussed include: tides, climatic changes, meteorology, agriculture.

2130. Danok, A. B. et al. "Machinery Selection Modeling: Incorporation of Weather Variability." *American Journal of Agricultural Economics* 62, no. 4 (November 1980): 700-708. Topics discussed include: agribusiness management, development, irrigation policy.

2131. Davis, B. et al. "Structure of the U.S. and West Texas Early Summer Onion Markets." *Southern Journal of Agricultural Economics* 11, no. 1 (July 1979): 107-12. Topics discussed include: supply, demand.

2132. Davis, J. et al. "Economic Aspects of Changing the Seasonality of Milk Production and Pricing." *Irish Journal of Agricultural Economics and Rural Sociology* 10, no. 2 (1984): 97-108. Topics discussed include: supply, markets, marketing, cooperatives.

2133. Desai, P. "Rise in Variability of Weather-Adjusted Grain Yields under Brezhnev." *Desai, Padma The Soviet economy Problems and prospects*, 230-245. Oxford: Blackwell, 1987. Topics discussed include: agriculture, supply, demand.

2134. Dev, S. M. et al. "Estimation of Performance Probabilities for Aggregate Yields: An Analysis of Agro-climatic Regions in India." *Journal of Quantitative Economics* 6, no. 1 (January 1990): 105-26. Topics discussed include: agriculture, supply, demand.

2135. Doll, J. P. "Obtaining Preliminary Bayesian Estimates of the Value of a Weather Forecast." *American Journal of Agricultural Economics* 53, no. 4 (November 1971): 651-55. Topics discussed include: farm management.

2136. Doran, H. E. et al. "Harmonic Analysis of Seasonal Data: Some Important Properties." *American Journal of Agricultural Economics* 54, no. 4 (November 1972): 646-51. Topics discussed include: times series analysis, spectral analysis models.

2137. Dorfman, J. H. et al. "State-Space Modeling of Cyclical Supply, Seasonal Demand, and Agricultural Inventories." *American Journal of Agricultural Economics* 73, no. 3 (August 1991): 829-40. Topics discussed include: supply, prices, markets, marketing, cooperatives, agribusiness.

2138. Drynan, R. G. et al. "Sparse Data, Climatic Variability, and Yield Uncertainty in Response Analysis - Comment." *American Journal of Agricultural Economics* 56, no. 3 (August 1974): 646. Topics discussed include: supply, economic geography.

2139. Dudley, N. J. et al. "Irrigation Planning 2: Choosing Optimal Acreages within a Season." *Water Resources Research* 7, no. 5 (October 1971): 1051-63. Topics discussed include: agriculture, management issues, development, irrigation policy.

2140. Easterling, W. E. et al. "Adapting Future Agriculture to Changes in Climate." *Greenhouse warming: Abatement and adaptation Proceedings of a workshop held in Washington, D.C., June 14-15, 1988*, 91-104. eds Norman J. Rosenberg et al. Washington, DC: Resources for the Future, 1989. Topics discussed include: conservation, pollution, supply, demand.

2141. "Effects of additional precipitation on agricultural production, the environment, the economy, and human society in South Dakota : final report : prepared for the Division of Atmospheric Water Resources Management, Bureau of Reclamation, United States Department of Interior." South Dakota Agricultural Experiment Station, and United States. Division of Atmospheric Water Resources Management. The Division, Denver, 1973. Topics discussed include: crops, climatic effects, South Dakota, rain, meteorology, bioclimatology.

2142. Ethridge, D. E. et al. "Cattle Ranching Production and Marketing Strategies under Combined Price and Weather Risks." *Western Journal of Agricultural Economics* 15, no. 2 (December 1990): 175-85. Topics discussed include: agribusiness management, supply, demand.

2143. Foote, R. J. et al. "Seasonal variation : methods of measurement and tests of significance." Washington, DC, U.S. Dept. of Agriculture, Bureau of Agricultural Economics, 1952. Topics discussed include: economics, statistics.

2144. Fosse, E. R. et al. "Potential impacts of shifts in climate on the crop insurance industry." *Bulletin of the American Meteorological Society* 74 (September 1993): 1703-8. Topics discussed include: insurance, agriculture, climatic changes.

2145. Fuller, S. W. et al. "Structure of the Fresh Onion Market in the Spring Season: A Focus on Texas and Its Competition." *Western Journal of Agricultural Economics* 16, no. 2 (December 1991): 404-16. Topics discussed include: markets, marketing, cooperatives, agribusiness, microeconomics, family farms, households, management.

2146. Gao, X. M. et al. "A Bio-economic Dynamic Programming Analysis of the Seasonal Supply Response by Florida Dairy Producers." *Southern Journal of Agricultural Economics* 24, no. 2 (December 1992): 211-20. Topics discussed include: corporate farming, family farms, factor inputs, management issues.

2147. Gill, G. J. *Seasonality and agriculture in the developing world: a problem of the poor and powerless*, Cambridge, England; New York: Cambridge University Press, 1991. Topics discussed include: economics, developing countries, crops, climatic effects, rural areas, poverty.

2148. ---. "Seasonality and agriculture in the developing world." *The Quarterly Review of Biology* 67, no. 343 (December 1992): 549-50.

2149. Giri, A. K. et al. "Impact of Variations in Rainfall and Weather Conditions on Production of Crops in West Bengal." *Economic Affairs* 24, no. 1-4 (January 1979): 57-62. Topics discussed include: agriculture, supply, demand.

2150. Griffin, R. C. et al. "Seasonality in Community Water Demand." *Western Journal of Agricultural Economics* 16, no. 2 (December 1991): 207-17. Topics discussed include: renewable resources, conservation, environment management, water, air, conservation, pollution, natural resources.

2151. Grivetti, L. E. "Geographical location, climate and weather, and magic: Aspects of agricultural success in the Eastern Kalahari, Botswana." *Social Science Information* 20, no. 3 (1981): 509-536. Topics discussed include: magic ritual, success, Botswana, Africa, ethnic groups, environment effects, geography, rites, workers.

2152. Hansen, L. "Farmer Response to Changes in Climate: The Case of Corn Production." *Journal of Agricultural Economics Research* 43, no. 4 (September 1991): 18-27. Topics discussed include: corporate farming, family farms, factor inputs, management issues.

2153. Hassan, R. M. et al. "Unequal Wealth Accumulation and Income Inequality in a Unimodal Agriculture: Sudan's Radad Irrigation Scheme." *Journal of Development Studies* 26, no. 1 (October 1989): 120-130. Topics discussed include: family farms, production, income, savings, wealth, tenant farmers, irrigation.

2154. Hassan, Z. A. et al. "Quarterly Demands for Meat in Canada with Alternative Seasonality Hypotheses." *Canadian Journal of Agricultural Economics* 31, no. 1 (March 1983): 77-94. Topics discussed include: consumer behavior, expenditures, seasonal patterns, consumption.

2155. Hensley, M. et al. "Seasonal variation in non specific bronchial reactivity: a study of wheat workers with a history of wheat associated asthma." *Thorax* 43, no. 2 (February 1988): 103-7. Topics discussed include: asthma, etiology, physiopatholoy, agriculture, workers' diseases, physiopathology, wheat.

2156. Hirshleifer, D. "Seasonal Patterns of Futures Hedging and the Resolution of Output Uncertainty." *Journal of Economic Theory* 53, no. 2 (April 1991): 304-27. Topics discussed include: pricing, futures markets, uncertainty, agriculture, marketing.

2157. Huss, A. R. "Seasonal patterns of birth and conception in rural highland Lesotho." *Hum Biol* 60, no. 3 (June 1988): 493-506. Topics discussed include: agriculture, coitus, environment, rural areas, population, socioeconomics, fertilization.

2158. Iakimets, V. "Adjustment of Regional Agriculture to Expected Climatic Changes." *Agriculture in a turbulent world economy Proceedings of the nineteenth international conference of agricultural economists, held at Malaga, Spain, 26 August 4 September 1985*, 160-67. eds Allen Renborg. Ulf, Maunder. Aldershot, U.K., and Brookfield, Vt.: Gower, 1986. Topics discussed include: supply, demand.

2159. Ikeda, M. et al. "Seasonal variation in hemoglobin concentration in non agricultural populations under various climatic conditions." *Hum Biol* 58, no. 2 (April 1986): 189-96. Topics discussed include: humidity, India, Japan, Korea, temperature, Thailand.

2160. International, I. F. A. S. A. et al. "The Impact of climatic variations on agriculture." *New Scientist* 123 2v (2 September 1989): 63-4.

2161. Jackson, I. J. *Climate, Water and Agriculture in the Tropics.* London: Longmans, 1977. Topics discussed include: environment, economy.

2162. James, A. et al. "Symptoms and longitudinal changes in lung function in young seasonal grain handlers." *Br J Ind Med* 43, no. 9 (September 1986): 587-91. Topics discussed include: Australia, physiopathology, agriculture, workers' diseases, cereals.

2163. Janssen, M. et al. "Seasonal and Regional Differences in Agricultural Supply Response in Benin." *European Review of Agricultural Economics* 17, no. 4 (1990): 407-20. Topics discussed include: supply, demand.

2164. Johnson, S. R. et al. "Agricultural Land Price Differentials and Their Relationship to Potentially Modifiable Aspects of the Climate." *Review of Economics and Statistics* 52, no. 2 (May 1970): 173-80. Topics discussed include: natural resources, regional issues, economics, forecasts.

2165. Jose, H. D. et al. "Consideration of Weather Risk in Forage Machinery Selection." *Canadian Journal of Agricultural Economics* 19, no. 1 (July 1971): 98-109. Topics discussed include: farm management.

2166. Kaiser, H. et al. *Agricultural Dimensions of Global Climate Change*. Delray Beach, FL: Saint Lucie Press, 1993.

2167. Kaylen, M. S. et al. "Stochastic Trend, Weather and U.S. Corn Yield Variability." *Applied Economics* 24, no. 5 (May 1992): 513-18. Topics discussed include: prices, agriculture, supply, demand.

2168. Keane, M. "Product Flexibility and Seasonality in Dairying." *Irish Journal of Agricultural Economics and Rural Sociology* 8, no. 1 (1980): 51-72. Topics discussed include: supply, demand.

2169. ---. "Seasonality and Optimum Plant Capacity and Stockholding." *Irish Journal of Agricultural Economics and Rural Sociology* 6, no. 2 (1976): 129-45. Topics discussed include: farm management.

2170. Kinnucan, H. et al. "Seasonality in the Consumer Response to Milk Advertising with Implications for Milk Promotion Policy." *American Journal of Agricultural Economics* 68, no. 3 (August 1986): 562-71. Topics discussed include: markets, marketing, cooperatives, consumer behavior, expenditures, seasonality, consumption.

2171. Kueh, Y. Y. "Weather Cycles and Agricultural Instability in China." *Journal of Agricultural Economics* 37, no. 1 (January 1986): 101-04.

2172. Lanier, S. *Florida: its scenery, climate, and history*, 336 pages. Philadelphia: J. B. Lippincott & co., 1876. Topics discussed include: agriculture, Florida, fruit culture.

2173. Lebow, W. M. et al. "Forecasting Applications of GMDH in Agricultural and Meteorological Time Series." *Self organizing Methods in Modeling: GMDH Type Algorithms Statistics: Textbooks and Monographs series*, 121-47. ed Stanley J. Farlow. New York and Basel: Dekker, 1984. Topics discussed include: supply, demand, forecasting, models.

2174. Lings, S. "Pesticide lung: a pilot investigation of fruit growers and farmers during the spraying season." *Br J Ind Med* 39, no. 4 (November 1982): 370-6. Topics discussed include: agriculture, workers, diseases, physiopathology, lung diseases, pneumonia.

2175. Malick, W. M. et al. "Stock Effects and Seasonality in the FCOJ Futures Basis." *Journal of Futures Markets* 7, no. 2 (April 1987): 157-67. Topics discussed include: agriculture markets, marketing, cooperatives, capital markets, regulation.

2176. Maunder, W. J. "Climatic Variations in Agricultural Production in New Zealand." *New Zealand Geographer* 22 (1966): 55-69. Topics discussed include: climate, weather.

2177. ---. "Climatic Constraints to Agricultural Production: Prediction and Planning in the New Zealand Setting." *New Zealand Agricultural Science* 11 (1977): 110-119. Topics discussed include: climate, economy.

2178. ---. "Weather and Climate as Factors in Forecasting National Dairy Production." in Proceedings of the Symposium on the Management of Dynamic Systems in New Zealand Agriculture, 101-126. Lower Hutt: D. S. I. R., 1977. Topics discussed include: economy.

2179. McNew, K. P. et al. "Sources and uses of weather information for agricultural decision makers." *Bulletin of the American Meteorological Society* 72 (April 1991): 491-8. Topics discussed include: economic issues.

2180. Melichar, E. "Seasonal Discount Assistance to Rural Banks: Evaluation of a Federal Reserve Proposal." *Agricultural Finance Review* 30 (July 1969): 44-57. Topics discussed include: finance, commercial banking.

2181. Michalson, E. L. "Impact of Weather and Technology on Net Return Estimates." *Agricultural Economics Research* 21, no. 1 (January 1969): 19-22. Topics discussed include: farm management.

2182. Mjelde, J. W. et al. "Obtaining Lower and Upper Bounds on the Value of Seasonal Climate Forecasts as a Function of Risk Preferences." *Western Journal of Agricultural Economics* 13, no. 2 (December 1988): 285-93. Topics discussed include: agribusiness management, forecasting, models.

2183. Mnatsakanian, A. "Effect of seasonal factors on indices of physical development of newborn infants in the agricultural regions of Armenia (Ekhegnadzor, Sisian, Goris)." *Pediatriia* 50, no. 8 (August 1971): 86-7. Topics discussed include: Armenia, climatic effects, rural areas, population, birth weight, growth.

2184. Mooney, S. et al. "The Impacts of Climate Change on Agriculture in Manitoba." *Canadian Journal of Agricultural Economics* 38, no. 4 (December 1990): 685-94. Topics discussed include: supply, demand, conservation, pollution.

2185. Morales, R. "Drinking Patterns among Seasonal Agricultural Workers, 1982." *Dissertation Abstracts International, A: The Humanities and Social Sciences* 46, no. 11 (May 1986): 3500-A.

2186. Motha, R. P. et al. "The Joint Agricultural Weather Facility's operational assessment program." *Bulletin of the American Meteorological Society* 67 (September 1986): 1114-22. Topics discussed include: economic issues.

2187. Murphy, J. A. "The Seasonality of Risk and Return on Agricultural Futures Positions." *American Journal of Agricultural Economics* 69, no. 3 (August 1987): 639-46. Topics discussed include: capital markets, regulation, marketing, cooperatives.

2188. Murshid, K. A. S. "Weather, New Technology and Instability in Foodgrain Production in Bangladesh." *Bangladesh Development Studies* 15, no. 1 (March 1987): 31-56. Topics discussed include: agriculture, supply, demand.

2189. Nemec, J. "Climate and Development: Agriculture Practices and Water Resources." *Climate and development: Climatic change and variability and the resulting social, economic and technological implications*, 397-408. eds H. J. Karpe, D. Otten, and S. C. Trinidade. New York; Berlin; London and Tokyo: Springer in cooperation with the United Nations, 1990. Topics discussed include: conservation, pollution, natural resources.

2190. Ottichilo, W. "Weathering the storm : climate change and investment in Kenya." 90 pages. African Centre for Technology Studies (ACTS) environmental policy series; no. 3, Nairobi, Kenya Stockholm, Sweden: ACTS Press, African Centre for Technology Studies ; Stockholm Environment Institute, 1991. Topics discussed include: economic issues, Kenya, agriculture, climatic effects.

2191. Oury, B. "Weather and Economic Development." *Finance and Development* 6, no. 2: 24-29 (June 1969) Topics discussed include: agriculture, development models, theories.

2192. Park, W. I. "The Effects of Data Aggregation on Econometric Estimates of Climate Change Impact on Corn and Soybean Production in the Midwest." Ph.D. diss., University of Illinois, 1992. Topics discussed include: agriculture, supply, demand, prices.

2193. Parry, M. L. *Climatic Change and Agricultural Settlement*. Folkstone, England: Dawson, and Hamden, Conn., Archon Books, 1978. Topics discussed include: climate, economy.

2194. Parry, M. L. et al. *The Impact of Climatic Variations on Agriculture, Vol 1 : Assessments in Cool Temperate & Cold Regions*. Norwell, MA: Kluwer Academic Publishers, 1988. Topics discussed include: climate, economy.

2195. Parry, M. "Climate change and world agriculture." *London, Earthscan in association with the International Institute for Applied Systems Analysis and the United Nations Environment Programme*, 157 pages. 1990. Topics discussed include: conservation, pollution.

2196. Parry, M. L. et al. "Some Strategies of Response in Agriculture to Changes of Climate." *Climate and development: Climatic change and variability and the resulting social, economic and technological implications*, 152-72. eds H. J. Karpe, D. Otten, and S. C. Trinidade. New York; Berlin; London and

Tokyo: Springer in cooperation with the United Nations, 1990. Topics discussed include: policy, conservation, pollution.

2197. Paul, A. H. "Weather and the Daily Use of Outdoor Recreation Areas in Canada." in *Weather Forecasting for Agriculture and Industry*, editor J. A. Taylor. Newton Abbot, UK: David and Charles, 1972. Topics discussed include: culture, society, behavior.

2198. Paxson, C. H. "Using Weather Variability to Estimate the Response of Savings to Transitory Income in Thailand." *American Economic Review* 82, no. 1 (March 1992): 15-33. Topics discussed include: life cycle effects, saving behavior, family farms, agriculture, management, living standards, consumer behavior, economics.

2199. Peppler, R. A. et al. "Tropospheric static stability and central North American growing season rainfall." *Monthly Weather Review* 117 (June 1989): 1156-80. Topics discussed include: meteorology, agriculture, rain, North America.

2200. Pesaran, M. H. et al. "Persistence, Seasonality and Trend in the UK Egg Production." *Applied Economics* 23, no. 3 (March 1991): 479-84. Topics discussed include: agriculture, supply, demand, prices.

2201. Petr, J. *Weather & Yield*. New York, NY: Elsevier Science Publishing Company, 1991. Topics discussed include: climate, crops, agriculture.

2202. Prevatt, J. W. et al. "Measuring the Effect of Capital Structure and Seasonality on Expected Returns and Risk - The Fresh Market Vegetable Case." *Southern Journal of Agricultural Economics* 24, no. 1 (July 1992): 171-78. Topics discussed include: corporate farming, family farms, factor inputs, management issues.

2203. Provansal, D. et al. "Production and Reproduction in Eastern Andalusia; Production et reproduction en Andalousie orientale." *Social Science Information, Information sur les Sciences Sociales* 28, no. 3 (September 1989): 483-519. Topics discussed include: agriculture development, Spain, twentieth century, agriculture transformation, Mediterranean Sea.

2204. Rahman, S. H. "An Analysis of Seasonal Jute Price Behaviour." *Bangladesh Development Studies* 15, no. 3 (September 1987): 43-61. Topics discussed include: agriculture, supply, demand.

2205. Reid, G. K. R. et al. "Pastoral Production, Stocking Rate and Seasonal Conditions." *Quarterly Review of Agricultural Economics* 26, no. 4 (October 1973): 217-27. Topics discussed include: regional issues, economics, supply, demand, management issues.

2206. Reilly, J. M. et al. *Economic Issues in Global Climate Change: Agriculture, Forestry, & Natural Ressources*. Boulder, CO: Westview Press, 1992. Topics discussed include: economy, forestry.

2207. Reilly, J. et al. "Climate change and agriculture: the role of international trade." *The American Economic Review* 83 (May 1993): 306-12. Topics discussed include: global models, crops, temperature changes.

2208. Reynolds, A. "Analyzing Fresh Vegetable Consumption from Household Survey Data." *Southern Journal of Agricultural Economics* 22, no. 2 (December 1990): 31-38. Topics discussed include: consumer behavior, standards of living, supply, demand.

2209. Roll, R. "Orange Juice and Weather." *American Economic Review* 74, no. 5 (December 1984): 861-80. Topics discussed include: prices, capital markets, regulation, corporate farming.

2210. Romero, R. F. "An Evaluation of the Economic Potential of Using Warm-Season Grasses for Livestock Feeding in West Virginia Utilizing the Linear Programming Quasi-optimal Solution Technique." Ph.D. diss., West Virginia University, 1986. Topics discussed include: agriculture, farm management.

2211. Rosenzweig, M. R. et al. "Wealth, Weather Risk and the Composition and Profitability of Agricultural Investments." *Economic Journal* 103, no. 416 (January 1993): 56-78. Topics discussed include: development, natural resources, corporate farming, family farms, factor inputs, development models, management issues.

2212. Samarasinghe, V. et al. "Income and Wealth Disparities in a Land Settlement of the Sri Lanka Dry Zone." *Understanding Green Revolutions Agrarian Change and Development Planning in South Asia: Essays in Honour of B H Farmer*, 173-93. eds Tim P. Bayliss Smith, and Sudhir Wanmali. New York; Sydney: Cambridge University Press, 1984. Topics discussed include: agriculture, agribusiness management, wages, fringe benefits.

2213. "Seasonality in the Australian beef industry." Australia. Bureau of Agricultural Economics. Canberra, 1970. Topics discussed include: cattle trade, seasonality, beef, Australia.

2214. Sevak, R. D. et al. "An Evaluation of 'Drought Prone Area Programme' in Rajasthan." *Artha Vikas* 14, no. 2 (July 1978): 42-57. Topics discussed include: economics, policy, agriculture, rural areas, regional issues.

2215. Shafer, C. E. "The Effect of a Marketing Order on Winter Carrot Prices: Reply." *American Journal of Agricultural Economics* 51, no. 4 (November 1969): 933-34. Topics discussed include: supply, demand.

2216. Shonkwiler, J. S. et al. "Imports and the Supply of Winter Tomatoes: An Application of Rational Expectations." *American Journal of Agricultural Economics* 64, no. 4 (November 1982): 634-41. Topics discussed include: trade relations, supply, demand, management issues.

2217. Singh, J. et al. "An Analysis of Farm Resource Use Pattern in Different Agro-climate Regions of the Punjab State." *Economic Affairs* 23, no. 11 (November 1978): 409-15. Topics discussed include: agriculture, agribusiness management, development, irrigation policy.

2218. Siriski, J. et al. "Butyrylcholinesterase activity in agricultural workers in seasons when pesticides are not used." *Arh Hig Rada Toksikol* 40, no. 2 (June 1989): 215-9. Topics discussed include: blood, environment, exposure, pesticides.

2219. Snyder, R. W. "Using Local Comprehensive Planning to Control Lakewater Pollution in Seasonal Home Communities." *American Journal of Agricultural Economics* 51, no. 5 (December 1969): 1583-87. Topics discussed include: conservation, pollution, development, irrigation policy.

2220. Stevens, S. C. "Evidence for a Weather Persistence Effect on the Corn, Wheat, and Soybean Growing Season Price Dynamics." *Journal of Futures Markets* 11, no. 1 (February 1991): 81-88. Topics discussed include: pricing, options markets, agriculture, supply, demand, capital markets, regulation, portfolio composition.

2221. Tang, A. M. "Trend, Policy Cycle, and Weather Disturbance in Chinese Agriculture, 1952-1978." *American Journal of Agricultural Economics* 62, no. 2 (May 1980): 339-48. Topics discussed include: supply, demand, policy, international issues.

2222. Taylor, J. *Weather & Agriculture*. Elkins Park, PA: Franklin Book Company, 1967. Topics discussed include: meteorology.

2223. ---. *Weather Economics: Papers & Disc Symposium Welsh Plant Breeding Station*. Elkins Park, PA: Franklin Book Company, 1970. Topics discussed include: meteorology, agriculture.

2224. Tobey, J. et al. "Economic Implications of Global Climate Change for World Agriculture." *Journal of Agricultural and Resource Economics* 17, no. 1 (July 1992): 195-204. Topics discussed include: renewable resources, environment, management, water, air, conservation, pollution, natural resources.

2225. Van Kooten, G. C. "Climate Change Impacts on Forestry: Economic Issues." *Canadian Journal of Agricultural Economics* 38, no. 4 (December 1990): 701-10. Topics discussed include: natural resources, conservation, pollution.

2226. Vandeveer, L. R. et al. "Irrigation and Potential Diversification Benefits in Humid Climates." *Southern Journal of Agricultural Economics* 21, no. 2 (December 1989): 167-74. Topics discussed include: agribusiness management, development, irrigation policy, regional issues, economics, forecasts.

2227. Venkateswaran, M. "Impact of Weather and Crop Price Information on Grain Marketing Decisions under Risk and Uncertainty." Ph.D. diss., Purdue University, 1989. Topics discussed include: agriculture, markets, marketing, cooperatives.

2228. Visher, S. "Weather Influences on Crop Yields." *Econ Geogr* 16 (1940): 437-443. Topics discussed include: climatic effects, agriculture.

2229. Weiss, A. *Climate & agriculture : systems approaches to decision making : proceedings.* 304 pages. American Meteorological Society. 1989. Topics discussed include: crops, climatic effects, meteorology, climatology, economic issues, bioclimatology.

2230. White, G. M. et al. "The Effect of Economic and Seasonal Conditions on Grazing Industry Producer Cash Incomes to 1977-78." *Quarterly Review of Agricultural Economics* 30, no. 4 (October 1977): 310-31. Topics discussed include: supply, demand, management issues.

2231. Wilks, D. S. et al. "The value of seasonal precipitation forecasts in a haying, pasturing problem in western Oregon." *Monthly Weather Review* 113 (October 1985): 1738-45. Topics discussed include: meteorology, agriculture, rain forecasting.

2232. Williams, G. D. V. "Weather and Prairie Wheat Production." *Canadian Journal of Agricultural Economics* 17, no. 1 (February 1969): 99-109. Topics discussed include: supply, demand.

2233. Wong, R. K. W. "Towards a strategy for adapting to climate change in Alberta." 79 pages. Alberta. Alberta Agriculture. Edmonton, Alberta, Canada: Resources Technologies Dept., Alberta Research Council, 1989. Topics discussed include: sociology, Alberta.

2234. Wood, J. E. "Analysis of Potential Hedging Criteria for Live Hogs Using Seasonal Indices." *American Journal of Agricultural Economics* 54, no. 5 (December 1972): 972-75. Topics discussed include: supply, demand.

2235. Woodbury, R. "Climatic Changes and Prehistoric Agriculture in the South-Western United States." *N Y A S* 95, no. art I (5 October 1961): 705-709. Topics discussed include: climate, history, civilization.

2236. Yaron, D. et al. "Optimal Allocation of Farm Irrigation Water during Peak Seasons." *American Journal of Agricultural Economics* 64, no. 4 (November 1982): 681-89. Topics discussed include: agribusiness management, development, irrigation policy.

2237. Yoshino, M. M. *Climate & Agricultural Land Use in Monsoon Asia.* New York, NY: Columbia University Press, 1984. Topics discussed include: Asia.

2238. Zagaitov, I. "Fluctuations in Weather Conditions and the Planning of Agricultural Production." *Problems of Economics* 25, no. 8 (December 1982): 55-73. Topics discussed include: supply, demand.

Alcohol & Other Beverages

2239. Abel, E. L. et al. "Seasonal, monthly, and day of week trends in homicide as affected by alcohol and race." *Alcoholism Clinical and Experimental Research* 9, no. 3 (May 1985): 281-283. Topics discussed include: consumption, victims, ethnic differences, drinking patterns.

2240. Bagirov, B. "A new beverage consisting of a decoction of Iandak and green tea for use in hot climates." *Gig Sanit* 32, no. 9 (September 1967): 114-6. Topics discussed include: body weight, perspiration, water electrolyte balance, thermoregulation, grasses, heat, work-related medicine.

2241. Basu, J. E. "Looking at the climate--California's wine country." *Weatherwise* 38 (April 1985): 87-94. Topics discussed include: agriculture, grapes, meteorology, California.

2242. Bowman, E. H. "Strategy and the Weather." *Sloan Management Review* 17, no. 2 (December 1976): 49-62. Topics discussed include: manufacturing, food processing, tobacco, beverages.

2243. Casselman, T. P. "Body temperature and the breathalyzer boobytrap." *Michigan Bar Journal* (September 1982): 721-722. Topics discussed include: alcohol, testing, criminals, drunk driving.

2244. Cenci, P. et al. "Biologically active substances in nonalcholic beverages. I. The survey and statistical study of consumption in a group of young people." *Ann Ig* 4, no. 4 (July 1992): 211-8. Topics discussed include: carbonated drinks, citrus fruits, coffee, consumer satisfaction, Italy, milk, seasonality, tea, behavior.

2245. Collins, W. "Performance effects of alcohol intoxication and hangover at ground level and at simulated altitude." *Aviat Space Environ Med* 51, no. 4 (April 1980): 327-35. Topics discussed include: anxiety psychology, emotions drug effects, reaction time, drug effects, gender effects, aerospace medicine, motor skills.

2246. De, C. J. "Seasonal rhythms of human nutrient intake and meal pattern." *Physiol Behav* 50, no. 1 (July 1991): 243-8. Topics discussed include: alcohol, drinking, psychology, nutrition, caloric intake, eating behavior, food preferences, hunger.

2247. Dossi, E. et al. "Alcoholic psychoses in relation to the geological and climatic environment of the Alto Adige region." *Riv Sper Freniatr* 95, no. 3 (30 June 1971): 613-20. Topics discussed include: alcoholism, etiology, Italy, epidemiology.

2248. Earll, J. M. et al. "The influence of temperature on alcohol absorption rates in humans." *Military Medicine* 150, no. 11 (November 1985): 612-613. Topics discussed include: hot drinks, cold drinks, absorption, beverages, physiology.

2249. Faust, V. et al. "Weather and mental disease. A 13 year long retrospective coincidental study on 17,268 hospitalized patients using the weather condition classification system." *Dtsch Med Wochenschr* 98, no. 18 (4 May 1973): 924-9. Topics discussed include: adjustment disorders, alcoholism, anxiety disorders, depression, Germany, neurotic disorders, schizophrenia, temperature.

2250. ---. "Season and psychiatric disease." *Med Klin* 70, no. 11 (14 March 1975): 467-73. Topics discussed include: alcoholism, epidemiology, personality, anxiety, depression, Germany, meteorology, schizophrenia, gender effects, weather.

2251. Fitzgerald, J. L. et al. "Distribution of alcohol consumption and problem drinking: Comparison of sales records and survey data." *Journal of Studies on Alcohol* 39, no. 5 (May 1978): 879-893. Topics discussed include: consumption, drinking variations, beverages, economics, seasonality, social influences.

2252. ---. "Seasonal changes in alcohol consumption and related problems in Iowa, 1979-1980." *Journal of Studies on Alcohol* 45, no. 4 (July 1984): 363-368. Topics discussed include: seasonal sales, problem drinking, Iowa.

2253. Fitzgerald, J. et al. "Drinking frequency changes in selected contexts by season in Iowa, 1979 1980." *J Stud Alcohol* 47, no. 4 (July 1986): 311-5. Topics discussed include: psychology, Iowa, social behavior, social environment, drinking, epidemiology.

2254. Gentilini, M. et al. "Urbanism and public health in the tropics." *Bull Soc Pathol Exot Filiales* 76, no. 3 (May 1983): 276-84. Topics discussed include: Africa, alcoholism, developing countries, food supply, industry, smoking, substance dependence, water supply.

2255. Gersovitz, M. et al. "Seasonality in Regression: An Application of Smoothness Priors." *Journal of the American Statistical Association* 73, no. 362 (June 1978): 264-73. Topics discussed include: single equation systems, time series analysis, spectral analysis models, tobacco, beverages.

2256. Graham, T. "Alcohol ingestion and man's ability to adapt to exercise in a cold environment." *Can J Appl Sport Sci* 6, no. 1 (March 1981): 27-31. Topics discussed include: alcohol, ethyl blood, body temperature, carbon dioxide metabolism, oxygen consumption, timing, adaptation, physiology, physical work.

2257. Graham, T. et al. "Effect of alcohol ingestion on man's thermoregulatory responses during cold water immersion." *Aviat Space Environ Med* 51, no. 2 (February 1980): 155-9. Topics discussed include: blood, glucose, metabolism, shivering, alcohol, ethyl pharmacology, thermoregulation.

2258. Juselius, K. "Seasonality in dynamic regression models : an application to the aggregate demand for beverages." Ph.D. diss., Swedish School of Economics and Business Administration, 1983. Topics discussed include: industry, Finland, mathematical models, economics.

2259. Keane, M. "An Analysis of Seasonal Pricing Schemes for Manufacturing Milk." *Irish Journal of Agricultural Economics and Rural Sociology* 8, no. 2 (1981): 213-33. Topics discussed include: manufacturing, food processing, tobacco, beverages.

2260. Knight, P. "Weather and beer: a holiday sampler." *Weatherwise* 41 (December 1988): 328-9. Topics discussed include: beer production.

2261. Kodama, H. et al. "Biorhythm of core temperature in depressive and nondepressive alcoholics." *Drug and Alcohol Dependence* 21, no. 1 (February 1988): 1-6. Topics discussed include: circadian rhythms, Japan, drug withdrawal, biological rhythms, major depression, alcoholism.

2262. Kupari, M. et al. "Seasonal variation in occurrence of acute atrial fibrillation and relation to air temperature and sale of alcohol." *Am J Cardiol* 66, no. 20 (15 December 1990): 1519-20. Topics discussed include: acute diseases, drinking, emergencies, Finland, epidemiology, risks, beverages, cold climates.

2263. "Labour and social problems arising out of seasonal fluctuations of the food products and drink industries." Tripartite Technical Meeting for the Food Products and Drink Industries, 2d, Geneva, 1978. International Labour Office, Geneva, 1978. Topics discussed include: food industry, trade, seasonality, beverage industry, labor supply, wages.

2264. Lantis, M. "Environmental stresses on human behavior. Summary and suggestions." *Arch Environ Health* 17, no. 4 (October 1968): 578-85. Topics discussed include: acclimation, acculturation, alcoholism, epidemiology, arctic regions, cold climates, communication, culture, disease propagation, Eskimos, personality.

2265. Lemmens, P. et al. "Seasonal variation in survey and sales estimates of alcohol consumption." *J Stud Alcohol* 54, no. 2 (March 1993): 157-63. Topics discussed include: drinking, psychology, motivation, Netherlands, epidemiology, beverages.

2266. Lifton, L. R. "Rational Firm Behavior in Response to Environmental Uncertainty: Collective Marketing Activities in the New York State Wine Grape Industry." *Dissertation Abstracts International, A: The Humanities and Social Sciences* 50, no. 11 (May 1990): 3762-A. Topics discussed include: agriculture, economics, weather, seasonality, uncertainty, rational behavior, grape growing.

2267. London, W. P. et al. "Alcohol consumption and latitude in the United States." *The American Journal of Psychiatry* 142 (May 1985): 657. Topics discussed include: seasonality, drinking behavior.

2268. London, W. "Alcoholism: theoretical consideration of season of birth and geographic latitude." *Alcohol* 4, no. 2 (March 1987): 127-9. Topics discussed include: genetics, order, laterality, etiology, geography.

2269. McGrath, R. et al. "Preliminary data on seasonally related alcohol dependence." *J Clin Psychiatry* 54, no. 7 (July 1993): 260-2. Topics discussed include: threatment, hospitalization, phototherapy, psychiatry.

2270. Owen, M. et al. "Effects of ingesting carbohydrate beverages during exercise in the heat." *Med Sci Sports Exerc* 18, no. 5 (October 1986): 568-75. Topics discussed include: dehydration, carbohydrates, pharmacology, running.

2271. Pawlowska, H. "Days with increased incidence of accidents: effect of alcohol and season on the frequency of maxillary injuries as related to meterological conditions." *Czas Stomatol* 24, no. 11 (November 1971): 1315-24. Topics discussed include: traffic, intoxication, Poland, drinking, injuries, epidemiology, weather.

2272. Poikolainen, K. "Seasonality of alcohol related hospital admissions has implications for prevention." *Drug and Alcohol Dependence* 10, no. 1 (September 1982): 65-69. Topics discussed include: seasonality, hospitals, Finland, alcoholism.

2273. Riley, R. "A dietary survey of downward and Indian migrants and long term coastal residents living in Southern Coastal Peru." *Arch Latinoam Nutr* 29, no. 1 (March 1979): 69-102. Topics discussed include: alcohol drinking, altitude, caloric intake, diet standards, food supply, geography, nutrition, Peru, seasonality, socioeconomics.

2274. Risbo, A. et al. "Human body temperature and controlled cold exposure during moderate and severe experimental alcohol intoxication." *Acta Anaesthesiol Scand* 25, no. 3 (June 1981): 215-8. Topics discussed include: oxygen consumption, shivering, alcohol, ethyl pharmacology, intoxication physiopathology, cold climates.

2275. Rolls, B. J. et al. "Effects of temperature and mode of presentation of juice on hunger, thirst and food intake in humans." *Appetite* 15, no. 3 (December 1990): 199-208. Topics discussed include: hunger, thirst, food intake, beverages.

2276. Roy, A. et al. "Cerebrospinal fluid variables among alcoholics lack seasonal variation." *Acta Psychiatr Scand* 84, no. 6 (December 1991): 579-82. Topics discussed include: alcoholism, treatment, serotonin, physiology.

2277. Szefel, A. "Season related admissions to a mental hospital for acute alcoholic psychoses." *Psychiatr Pol* 17, no. 3 (May 1983): 209-14. Topics discussed include: acute diseases, hospitalization, Poland, urban areas, population, psychoses, alcoholics, epidemiology.

2278. Winkler, G. et al. "Seasonal variations in nutrition behavior: results of the dietary survey of 1984, 85 of the MONICA project in Augsburg." *Z Ernahrungswiss* 31, no. 1 (March 1992): 19-24. Topics discussed include: alcohol drinking, diet, Germany, food preferences.

2279. Wirths, W. "Surveys on the nutritional habits in agricultural households in the German Federal Republic from the nutritional physiological aspect. IV. Classification based on manpower." *Z Ernahrungswiss* 9, no. 4 (July 1969): 249-70. Topics discussed include: alcoholic beverages, drinking behavior, food, preferences, preservation, supply, refrigeration, seasonality, students, timing, vegetables, food habits.

2280. Yen, P. "Summer beverages. Quenching and nutritious?" *Geriatr Nurs New York* 9, no. 4 (July 1988): 245. Topics discussed include: carbonation, fruit.

2281. Zellner, D. A. et al. "Effect of temperature and expectations on liking for beverages." *Physiology and Behavior* 44, no. 1 (1988): 61-68. Topics discussed include: preferences, nonalcoholic beverages, taste perception, expectations.

Architecture & Housing

2282. Abdel-Latif, M. A. et al. "Settling the desert with advanced industrial technology: two recent Saudi Arabian settlement projects." *Ekistics* 52 (March 1985): 162-75. Topics discussed include: technology, Saudi Arabia, architecture, climate.

2283. Armstrong, G. H. "Designing Self-Sufficient Housing for Temperate Climate." *Impact of Science on Society* 29, no. 3 (1979): 253-260. Topics discussed include: energy, England, United Kingdom.

2284. Aronin, J. *Climate and Architecture*. New York: Van Nostrand Reinhold, 1984. Topics discussed include: meteorology, science.

2285. Binford, L. R. "Mobility, Housing and Environment: A Comparative Study." *Journal of Anthropological Research* 46, no. 2 (June 1990): 119-152. Topics discussed include: hunting, gathering cultures, sociology, mobility, seasonality, cross-cultural effects.

2286. Building Research Advisory Board. "Weather and Building Industry." Proceedings of B R A B Conference, Washington, 1950. Washington DC: U. S. National Academy of Sciences, 1950. Topics discussed include: climate, economy, society, culture.

2287. Cammarota, M. T. "The Impact of Unseasonable Weather on Housing Starts." *American Real Estate and Urban Economics Association Journal* 17, no. 3 (September 1989): 300-313. Topics discussed include: construction.

2288. Colwell, P. F. et al. "Seasonality and Size Effects - The Case of Real-Estate-Related Investment." *Journal of Real Estate Finance and Economics* 3, no. 3 (September 1990): 251-59. Topics discussed include: capital markets, regulatory policy, mortgage market, urban housing, nonurban housing, services, real estate.

2289. "Construction industry seasonality : man-hour report." Chicago Construction Coordinating Committee, U.S. Dept. of Labor, Chicago, IL, 1975. Topics discussed include: construction industry, seasonality, time.

2290. Cooper, W. B. et al. *Warm air heating for climate control.* 3rd ed., William B. Cooper et al. ed. 564 pages. New Jersey: Englewood Cliffs, NJ: Prentice Hall, 1994. Topics discussed include: heating, heat pumps, dwellings, ventilation.

2291. Dewees, D. N. et al. "Cold Houses and Warm Climates Revisited: On Keeping Warm in Chicago, or Paradox Lost: Comment." *Journal of Political Economy* 98, no. 3 (June 1990): 656-63. Topics discussed include: microeconomics, theory, household demand, consumer demand, housing economics.

2292. Evans, R. D. "Residential Construction Volatility: A Seasonal and Cyclical Inventory Adjustment Analysis." *American Real Estate and Urban Economics Association Journal* 9, no. 1 (March 1981): 74-82. Topics discussed include: economic cycles, housing.

2293. Foster, H. G. "Labor-Force Adjustments to Seasonal Fluctuations in Construction." *Industrial and Labor Relations Review* 23, no. 4 (July 1970): 528-40. Topics discussed include: construction, unemployment levels.

2294. Fournol, A. *Climats et Habitation*, Paris, 1956. Topics discussed include: climate, civilization, building.

2295. Friedman, D. "Cold Houses in Warm Climates and Vice Versa: A Paradox of Rational Heating." *Journal of Political Economy* 95, no. 5 (October 1987): 1089-97. Topics discussed include: microeconomics, household comsumption, consumer demand, housing economics.

2296. Givoni, B. *Man, Climate & Architecture*. New York: Elsevier Science Publishing Company, 1976. Topics discussed include: architecture, arts, sociology, anthropology, archaelogy, climate.

2297. Goodman, J. L., Jr. "Housing and the Weather." *American Real Estate and Urban Economics Association Journal* 15, no. 1 (March 1987): 638-63. Topics discussed include: construction.

2298. Hess, C. et al. "Field and laboratory tests of etched track detectors for 222Rn: summer vs winter variations and tightness effects in Maine houses." *Health Phys* 49, no. 1 (July 1985): 65-79. Topics discussed include: air pollutants analysis, radioactive analysis, housing, radon analysis, seasonality.

2299. Hinze, J. et al. "Weather in construction contracts." *Journal of Construction Engineering and Management* 115 (June 1989): 270-83. Topics discussed include: economic issues, government contracts, laws, regulations.

2300. Housing and Home Finance Agency. *Application of Climatic Data to Home Design.* Washington, DC: U.S. Government Printing Office, 1954. Topics discussed include: climate, weather, society, culture.

2301. Howenstine, E. J. "Programs for Providing Winter Jobs in Construction." *Monthly Labor Review* 94, no. 2 (February 1971): 24-32. Topics discussed include: construction, public policy, government.

2302. Koehn, E. et al. "Climatic effects on construction." *Journal of Construction Engineering and Management* 111 (June 1985): 129-37. Topics discussed include: labor productivity, construction workers, temperature, physiological effects.

2303. Lee, M. "Economical cooling of hot weather concrete." *Concrete Construction* 34 (September 1989): 791+. Topics discussed include: concrete cooling, industry.

2304. Loftness, V. "Climate and Architecture." *Weatherwise* 31 (December 1978): 212-217. Topics discussed include: industry, culture, housing.

2305. Lozanov, L. "Changes in the physical environmental factors in modern buildings related to the use of polymeric construction materials." *Probl Khig* 10 (1985): 117-25. Topics discussed include: electricity, heating, hygiene, temperature, climate, construction materials, housing, microclimates.

2306. MacDonald, D. M. et al. "Flood Hazard Pricing and Insurance Premium Differentials: Evidence from the Housing Market." *Journal of Risk and Insurance* 57, no. 4 (December 1990): 654-63. Topics discussed include: economics, housing, insurance.

2307. Mamedov, K. "Responses of adolescent construction workers to physical work under hot climate conditions." *Gig Sanit no.* 11 (November 1976): 57-9. Topics discussed include: Uzbekistan, exertion, heat, work-related medicine, work.

2308. Maunder, W. J. et al. "The Effect of Weather on Road Construction: Applications of a Simulation Model." *Monthly Weather Review* 99 (1971): 946-953. Topics discussed include: industry, construction.

2309. Moe, R. J. et al. "Residential Energy Use and Improved Building Thermal Efficiency: An Econometric Analysis." *Analytic Techniques for Energy Planning: Proceedings of the First Symposium, Organized by the Operations Research Society of America Special Interest Group on Energy, Natural Resources and the Environment*, 135-49. ed Benjamin Lev et al. Amsterdam; New York and Oxford, North Holland: Elsevier Science, 1984. Topics discussed include: conservation, pollution, consumer behavior, expenditures, cycles, consumption.

2310. Musgrave, J. C. "Measuring the Influences of Weather on Housing Starts." *Construction Review* 14 (1968): 4-7. Topics discussed include: industry, construction.

2311. Pasha, H. A. "Housing for Temporary Residents: A Study of Pilgrims to Makkah." *Urban Studies* 24, no. 4 (August 1987): 312-323. Topics discussed include: crowding, seasonality, Saudi Arabia.

2312. ---. "Seasonal Housing Markets: The Pilgrimage to Mecca." *International Regional Science Review* 10, no. 3 (December 1986): 193-210. Topics discussed include: urban economics.

2313. Power, F. B. et al. "A Status Report on the National Flood Insurance Program-Mid 1978." *Journal of Risk and Insurance* 46, no. 2 (June 1979): 61-76. Topics discussed include: insurance, housing, economics.

2314. Ramamohan Rao, T. V. S. "Seasonal Unemployment, Secondary Labor and the Phillips Curve." *American Economist* 15, no. 2 (September 1971): 25-31. Topics discussed include: labor markets, wage levels, employment levels, construction, labor economics.

2315. Rhodes, S. L. "Rethinking Desertification: What Do We Know and What Have We Learned?" *World Development* 19, no. 9 (September 1991): 1137-43. Topics discussed include: renewable resources, conservation, environmental management, housing economics, pollution, natural resources.

2316. Rosen, K. T. *Seasonal Cycles in the Housing Market: Patterns, Costs and Politics.* Cambridge (Mass.): The MIT Press, 1979. Topics discussed include: economy.

2317. Russell, J. L. et al. "Seasonality in Construction: A Continuing Problem." *Monthly Labor Review* 92, no. 12 (December 1969): 3-8. Topics discussed include: construction, unemployment levels, vacancy levels, retirement.

2318. Salikhodzhaev, S. et al. "Physiological justification of a rational work and rest regimen for construction workers in the summer under the climatic conditions of Central Asia." *Gig Tr Prof Zabol no.* 10 (October 1974): 43-5. Topics discussed include: seasonality, Siberia, work-related medicine, rest, work.

2319. Shilling, J. D. et al. "Flood Insurance, Wealth Redistribution, and Urban Property Values." *Journal of Urban Economics* 26, no. 1 (July 1989): 43-53. Topics discussed include: housing.

2320. Smith, A. "Enclosures keep workers warm and more productive in cold weather." *Aberdeen's Concrete Construction* 35 (October 1990): 857-8+. Topics discussed include: concrete construction, winter, workers, health, hygiene.

2321. Speyrer, J. F. et al. "Housing Prices and Flood Risk: An Examination Using Spline Regression." *Journal of Real Estate Finance and Economics* 4, no. 4 (December 1991): 395-407. Topics discussed include: supply, markets, insurance.

2322. Terjung, W. "Urban climatology (with reference to the interrelationship between external weather and the microclimate in houses and buildings)." *Prog Biometeorol* 1, no. 1A (1974): 168-80, 624-31. Topics discussed include: architecture, environment design, humidity, radiation monitoring, rain, snow, temperature, wind, environment, housing.

2323. Tschetter, J. et al. "Employment Changes in Construction: Secular, Cyclical, and Seasonal." *Monthly Labor Review* 106, no. 3 (March 1983): 11-17. Topics discussed include: construction, unemployment levels, vacancy levels, retirement patterns.

2324. Turan, M. *Vernacular Architecture: Paradigms of Environmental Response.* Broekfield, VT: Ashgate Publishing Company, 1990. Topics discussed include: architecture, climate.

2325. US Housing & Home Finance Agency. *Application of Climatic Data to House Design.* New York, NY: AMS Press, 1977. Topics discussed include: climate, architecture, dwellings.

2326. Vakalo, E. et al. "Orientation and the thermal behavior of housing in hot, dry climates." *Ekistics* 57 (September-October-November-December '90): 337-41. Topics discussed include: power resources, energy, development, architecture, solar radiation.

2327. Wittrock, J. *Reducing seasonal unemployment in the construction industry: methods of stabilising construction activity and employee income.* 248 pages. Developing job opportunities, 4, Paris: Organisation for Economic Cooperation and Development, 1967. Topics discussed include: construction, workers, employment, seasonality, unemployment.

2328. Woodhouse, W. "Housing in the tropics: some design problems." *R Soc Health J* 86, no. 6 (November 1966): 341-6. Topics discussed include: ventilation, weather, tropical climate.

2329. Wylson, A. *Aquatecture: Architecture & Water.* Stoneham, MA: Butterworth-Heinemann, 1986. Topics discussed include: architecture, climate.

2330. Young, O. R. "International cooperation: Building regimes for natural resources and the environment." *Cornell Studies in Political Economy series,* 248 pages. Ithaca and London: Cornell University Press, 1989. Topics discussed include: sociology, conservation, pollution.

Clothing

2331. Adam, J. "Climate and clothing." *Practitioner* 198, no. 187 (May 1967): 645-50. Topics discussed include: acclimation.

2332. Afanas'Eva, R. "Hygienic prerequisites for creating women's clothing for protection from the cold." *Gig Sanit no.* 5 (May 1977): 17-22. Topics discussed include: body temperature regulation, humidity, gender effects, skin temperature.

2333. Afanas'Eva, R. et al. "Summer working clothes for oilmen working under conditions of hot climate." *Gig Tr Prof Zabol no.* 10 (October 1975): 57-8. Topics discussed include: fuel oils, Kazakhstan, seasonality, work-related medicine.

2334. ---. "Physiological and hygienic requirements for women's clothing for cold climate regions." *Gig Sanit no.* 7 (July 1980): 20-2. Topics discussed include: body temperature regulation, microseasons, skin temperature, standards.

2335. ---. "The heat conserving properties of winter clothing when exposed to various wind velocities and temperature of the surrounding air." *Gig Sanit* 33, no. 11 (November 1968): 93-5. Topics discussed include: cold climates, body regulation, weather.

2336. ---. "Hygienic requirements of winter clothing for girls living in the Far North." *Gig Sanit no.* 11 (November 1980): 35-7. Topics discussed include: arctic regions, hygiene, seasonality, Siberia, cold climate.

2337. Agadzhanov, V. et al. "Changes of functional state of the human body and the dynamics of its recovery in hard physical work in protective devices in various air temperatures." *Gig Tr Prof Zabol no.* 1 (1992): 18-21. Topics discussed include: acid base equilibrium, blood circulation, central nervo system physiology, work capacity evaluation, adaptation, thermoregulation physiology, clothing.

2338. Antunano, M. et al. "Heat stress in protective clothing: validation of a computer model and the heat humidity index (HHI)." *Aviat Space Environ Med* 63, no. 12 (December 1992): 1087-92. Topics discussed include: energy metabolism physiology, exercise test, oxygen physiology, exhation physiopathology.

2339. Arad, M. et al. "Effects of heat exercise stress, NBC clothing, and pyridostigmine treatment on psychomotor and subjective measures of performance." *Military Medicine* 157, no. 4 (April 1992): 210-214. Topics discussed include: protective, subjective experience, healthy acclimatized males, safety devices, cholinomimetic drugs, thermal acclimation, war.

2340. Auliciems, A. et al. "Cold stress in Canada. A human climatic classification." *Int J Biometeorol* 20, no. 4 (December 1976): 287-94. Topics discussed include: Canada, clothing, thermoregulation, cold climate.

2341. Bakhramova, L. "Experience with the physiologic classification of weather for evaluation of the buildings of Baku and the winter clothing of the populace." *Gig Sanit no.* 2 (February 1987): 75-6. Topics discussed include: Azerbaijan, health, residence characteristics, urban health.

2342. Barannikov, V. "Health evaluation of the heat protecting properties of special clothing worn by potassium miners." *Gig Tr Prof Zabol* 16, no. 2 (February 1972): 39-42. Topics discussed include: cold climates, mining, potassium, USSR.

2343. Barker, R. L. et al. "Thermal insulation properties of fabrics in high-intensity heat exposures." *Tappi Journal* 68 (December 1985): 62-7. Topics discussed include: textile fabrics heat transmission, clothing, protective, textile fibers, synthetic.

2344. Bavro, G. et al. "On the possibilities of expanding the climatic limits of use of special clothing made of synthetic materials." *Gig Sanit* 32, no. 9 (September 1967): 16-21. Topics discussed include: body temperature, cotton, humidity, skin physiology, temperature, wool, polymers.

2345. Blum, H. "The Solar Heat Load: Its Relationship to the Ideal Heat Load and Its Relative Importance to the Design of Clothing." *J Clin Invest* 24 (1945): 712-721. Topics discussed include: temperature, climatic effects, weather.

2346. Brockmann, H. "Development of revised simplex fabric for summer flying gloves. NADC MR 6607." *NADC MR Rep* 19 (1966): 1-3. Topics discussed include: aviation, clothing, heat.

2347. Budd, G. "Skin temperature, thermal comfort, sweating, clothing and activity of men sledging in Antarctica." *J Physiol Lond* 186, no. 1 (September 1966): 201-15. Topics discussed include: Antarctic regions, acclimation, body regulation, cold climates, skin physiology, perspiration.

2348. Catenacci, G. et al. "Relation between thermal comfort indices and various values of thermal impedence for clothes." *Boll Soc Ital Biol Sper* 59, no. 12 (30 December 1983): 1966-72. Topics discussed include: energy metabolism, seasonality, temperature, climatic effects, microclimates, satisfaction.

2349. Cena, K. et al. "Transfer of heat through animal coats and clothing." *Int Rev Physiol* 20 (1979): 1-42. Topics discussed include: climates, metabolism, pressure, species specificity, thermodynamics, water, thermoregulation, hair.

2350. ---. *Bioengineering, thermal physiology, and comfort.* 289 pages. Politechnika Wroclawska. Studies in environmental science; 10, Amsterdam; New York: Elsevier Scientific Pub. Co.; distribution for the U.S.A. and Canada, Elsevier, North-Holland, 1981. Topics discussed include: heat, physiology, insulation, bioengineering, protective clothing, adaptation, thermoregulation.

2351. Clothier, J. "Work in hostile climates." *Ann Occup Hyg* 14, no. 2 (June 1971): 197-203. Topics discussed include: acclimation, cold climates, environment, health, housing, tropical climates, work-related medicine.

2352. Coghlan, A. "Clothes that change colour in the heat of the moment." *New Scientist* 130 (11 May 1991): 25. Topics discussed include: clothing, dress.

2353. Consolazio, C. et al. "Energy requirement and metabolism during exposure to extreme environments." *World Rev Nutr Diet* 18 (1973): 177-94. Topics discussed include: age effects, altitude, temperature, body weight, climatic effects, clothing, cold climates, heart rate, heat, humidity, respiration, gender effects, acclimation.

2354. Dawson B. et al. "Artificially induced heat acclimation of team game players with sweat clothing . 1. Responses to wearing sweat clothing during exercise in cool conditions." *Journal of Human Movement Studies* 15, no. 4 (1988): 171-183.

2355. ---. " Artificially induced heat acclimation of team game players with sweat clothing. 2 Training in sweat clothing in cool conditions to improve heat tolerance." *Journal of Human Movement Studies* 19, no. 3 (1990): 101-117. Topics discussed include: perspiration, tolerance, acclimation, thermoregulation, exercise.

2356. Dawson, B. et al. "Improvements in heat tolerance induced by interval running training in the heat and in sweat clothing in cool conditions." *J Sports Sci* 7, no. 3 (December 1989): 189-203. Topics discussed include: blood volume, body temperature, cold climates, heart rate, perspiration, acclimation physiology, physical education, training methods, running.

2357. Day, M. et al. "Durability of firefighters' protective clothing to heat and light." *Textile Research Journal* 58 (March 1988): 141-7. Topics discussed include: textile fibers, light, fire resistant materials.

2358. De, F. C. "Human climates of northern China." *Atmos Environ* 13, no. 1 (1979): 71-7. Topics discussed include: China, theory, seasonality, thermoregulation, clothing, cold climate.

2359. Farnworth, B. "Mechanisms of heat flow through clothing insulation." *Textile Research Journal* 53 (December 1983): 717-25. Topics discussed include: feathers, blended fabrics, textile fabrics, heat transmission.

2360. ---. "A numerical model of the combined diffusion of heat and water vapor through clothing." *Textile Research Journal* 56 (November 1986): 653-65. Topics discussed include: textile fabrics, heat transmission, moisture absorption, winter.

2361. Gagge, A. P. et al. "Influence of Clothing on the Physiological Reactions of the Human Body to Varying Environmental Temperatures." *Am Journ Physiol* 124 (1938): 30-51. Topics discussed include: biology, climatic effects, environment, behavior.

2362. Gavhed, D. et al. "Thermoregulatory and subjective responses of clothed men in the cold during continuous and intermittent exercise." *Eur J Appl Physiol* 63, no. 1 (1991): 29-35. Topics discussed include: clothing, energy metabolism, oxygen consumption, sensation physiology, skin temperature, thermoregulation, exercise physiology.

2363. Gisolfi, C. et al. "Work heat tolerance of distance runners." *Ann N Y Acad Sci* 301 (1977): 139-50. Topics discussed include: body temperature, thermoregulation, clothing, heart rate, oxygen consumption, physical education, training, perspiration, weather, acclimation.

2364. Givoni, B. et al. "Predicting rectal temperature response to work, environment, and clothing." *J Appl Physiol* 32, no. 6 (June 1972): 812-22. Topics discussed include: body regulation, heat, humidity, metabolism, stress physiopathology, weather, work.

2365. Gonzalez, R. "Biophysical and physiological integration of proper clothing for exercise." *Exerc Sport Sci Rev* 15 (1987): 261-95. Topics discussed include: biophysics, thermoregulation, skin physiology, perspiration, thermal conductivity, physical work.

2366. Greig, G. B. *Seasonal fluctuations in employment in the women's clothing industry in New York.* 164 pages. Studies in history, economics and public law, no. 554, New York: AMS Press, 1968. Topics discussed include: workers, New York.

2367. Haisman, M. et al. "Physiological evaluations of armoured vests in hot wet and hot dry climates." *Ergonomics* 17, no. 1 (January 1974): 1-12. Topics discussed include: air movements, perspiration, body temperature, desert climates, heart rate, protective clothing, tropical climate.

2368. Haller, K. "On wearing tests for clothing." *Arch Hyg Bakteriol* 151, no. 1 (May 1967): 53-7. Topics discussed include: climates, cotton, environment, seasonality.

2369. Holmer, I. "Protective clothing against cold performance standards as method for preventive measures." *Arctic Med Res* 51 Supp, no. 7 (1992): 94-8. Topics discussed include: cold climates, standards.

2370. ---. "Recent trends in clothing physiology." *Scand J Work Environ Health* 15 Supp, no. 1 (1989): 58-65. Topics discussed include: forecasting, manikins, ventilation, thermoregulation.

2371. Holmer, I. et al. "Resultant clothing insulation during exercise in the cold." *Arctic Med Res* 50 Supp, no. 6 (1991): 94-8. Topics discussed include: skin temperature, cold climates, exercise.

2372. ---. "Quantification of heat balance during work in three types of asbestos protective clothing." *Int Arch Occup Environ Health* 64, no. 4 (1992): 243-9. Topics discussed include: asbestosis physiopathology, exercise test, heart rate, physiology, oxygen, thermoregulation, work-related exposure.

2373. Ikuta, N. "Hygienic studies of foundation garments. 1. Effects of subjective observations, clothing climate, and clothing pressure." *Nippon Eiseigaku Zasshi* 25, no. 4 (October 1970): 344-9. Topics discussed include: fatigue, humidity, posture, temperature, pressure.

2374. Jenkins, G. "Safely warm this winter. 2." *Geriatr Nur Home Care* 7, no. 12 (December 1987): 22-3. Topics discussed include: body temperature, clothing.

2375. Jeong, W. et al. "Effects of wearing two different types of clothing on body temperatures during and after exercise." *Int J Biometeorol* 33, no. 2 (June 1989): 77-81. Topics discussed include: exertion, skin temperature, timing, thermoregulation.

2376. Jette, M. et al. "Effect of inspiratory resistance to prolonged exercise in a hot environment wearing protective clothing." *Int J Biometeorol* 36, no. 3 (August 1992): 130-5. Topics discussed include: exercise physiology, tropical climates, airway resistance physiology, heat.

2377. Joy, R. et al. "A method of relating physiology and military performance. A study of some effects of vapor barrier clothing in a hot climate." *Mil Med* 133, no. 6 (June 1968): 458-70. Topics discussed include: exertion, heat, military medicine.

2378. Kaechele, E. "The effect of heat retaining textile segments on workers in cold conditions." *Schriftenr Zentralbl Arbeitsmed Arbeitsschutz Prophyl Ergonomie* 6, (1981): 221-3. Topics discussed include: pain treatment, clothing, cold climates, work-related medicine, textiles.

2379. Kamon, E. et al. "Physiological and biophysical limits to work in the heat for clothed men and women." *J Appl Physiol* 44, no. 6 (June 1978): 918-25. Topics discussed include: heart rate, oxygen consumption, gender effects, sweat glands physiology, acclimation, thermoregulation, physical work, humidity.

2380. ---. "Wind speed limits to work under hot environments for clothed men." *J Appl Physiol* 46, no. 2 (February 1979): 340-5. Topics discussed include: acclimation, body temperature, oxygen consumption, air movements, thermoregulation, physical work, heat, wind.

2381. ---. "Heart rate and rectal temperature relationships during work in hot humid environments." *J Appl Physiol* 31, no. 3 (September 1971): 472-7. Topics discussed include: acclimation, aerobiosis, air, body regulation, clothing, oxygen consumption, partial pressure, spirometry, physical work.

2382. Kazantseva, L. et al. "Physiological and hygienic properties of clothing according to data from a subjective assessment of the human thermal state." *Gig Sanit no.* 10 (October 1978): 98-100. Topics discussed include: climates, USSR, thermoregulation.

2383. Kenney, W. L. et al. "Physiological evaluation of liquid-barrier, vapor-permeable protective clothing ensembles for work in hot environments." *American Industrial Hygiene Association Journal* 54 (July 1993): 397-402. Topics discussed include: permeability, hot weather, heat, physiology.

2384. Kessler, P. "Clothing and nutrition in the tropics." *Z Allgemeinmed* 48, no. 27 (30 September 1972): 1241-3. Topics discussed include: travel, tropical climates.

2385. Korenevskaia, E. et al. "Seasonal changes in the thermal state of school children in Ashkhabad." *Gig Sanit* 35, no. 8 (August 1970): 46-50. Topics discussed include: clothing, desert climates, Turkmenistan, body temperature, health services.

2386. Koshcheev, V. et al. "A method of studying physiologic hygienic problems concerning use of winter clothing under windy conditions." *Gig Sanit no.* 1 (January 1975): 62-4. Topics discussed include: cold, air movements, thermoregulation, wind.

2387. Kurpita, P. "The heat insulation effect of quilted clothing with a wind protection interlayer of paper." *Gig Sanit* 32, no. 3 (March 1967): 26-30. Topics discussed include: body temperature, paper, weather.

2388. Lamb, G. E. R. et al. "Heat loss from a ventilated clothed body." *Textile Research Journal* 60 (July 1990): 378-83. Topics discussed include: polyester fabrics, textiles, heat transmission.

2389. Lee, Y. et al. "Seasonal adaptation of thermal and metabolic responses in men wearing different clothing at 10 degrees C." *Int J Biometeorol* 37, no. 1 (February 1993): 36-41. Topics discussed include: skin temperature, physiology, acclimation, thermoregulation.

2390. Li, Y. et al. "Physiological responses and psychological sensations in wearer trials with knitted sportswear. Special Issue: Sport, leisure and ergonomics." *Ergonomics* 31, no. 11 (November 1988): 1709-1721. Topics discussed include: hot climates, cold environment, exercise, psychology, clothing fibers, physiological correlates, physical comfort.

2391. Lind, A. "The lack of heatstroke in European miners working in hot climates." *Am Ind Hyg Assoc J* 31, no. 4 (July 1970): 460-5. Topics discussed include: acclimation, clothing, environment, exposure, Europe, work-related diseases epidemiology, South Africa, heat exhaustion, mining.

2392. Lotens, W. "The actual insulation of multilayer clothing." *Scand J Work Environ Health 1989* 1 (1915): 66-75. Topics discussed include: air movements, heat, manikins, perspiration, ventilation, thermoregulation physiology.

2393. ---. "Clothing design for work in the cold." *Arctic Med Res* 46, no. 1 (1987): 3-12. Topics discussed include: cold climate.

2394. Mairiaux, P. et al. "Evaluation of the effects of cooling clothes on the adaptation to prolonged exertion in high temperatures by miners." *Rev Inst Hyg Mines Hasselt* 32, no. 3 (1977): 99-122. Topics discussed include: body temperature, body weight, environment, fatigue, heart rate, humidity, coal mining, cold climates, physical work, heat, protective clothing.

2395. Makarov, V. et al. "The effect of local heat on the nature of physiological response reactions and the possibility of its utilization in designing heated winter clothing." *Gig Sanit* 34, no. 12 (December 1969): 90-2. Topics discussed include: USSR, thermoregulation, cold climates.

2396. Maloiy, G. et al. "Thermoregulation, water and electrolyte metabolism of man in the desert." *East Afr Med J* 52, no. 2 (February 1975): 97-112. Topics discussed include: adaptation, clothing, drinking behavior, housing, kidneys, perspiration, desert climates.

2397. McLellan, T. et al. "Influence of a new vapor protective clothing layer on physical work tolerance times at 40 degrees C." *Aviat Space Environ Med* 63, no. 2 (February 1992): 107-13. Topics discussed include: body temperature, climatic effects, heart rate, military personnel, motor skills, perspiration physiology, humidity, work capacity evaluation.

2398. Mecheels, J. "Local heat therapy, climate and thermo economy of the human body." *Schriftenr Zentralbl Arbeitsmed Arbeitsschutz Prophyl Ergonomie* 6 (1981): 157-64. Topics discussed include: body temperature, clothing, rheumatic diseases, treatment.

2399. Milroy, W. "Medical operational problems in cold environment, the role of dehydration, caloric intake, and clothing." *Mil Med* 152, no. 8 (August 1987): 397-8. Topics discussed include: acclimation, caloric intake, cold climates, dehydration, military personnel.

2400. Mizunashi, S. et al. "Seasonal changes of clothing weight and skin temperature of infants." *J Hum Ergol Tokyo* 6, no. 2 (December 1977): 203-6. Topics discussed include: Japan, skin.

2401. Muminov, B. "Studying the properties of textiles in relation to hot, dry climatic conditions." *Gig Sanit* 35, no. 2 (February 1970): 113-5. Topics discussed include: Asia, clothing, desert climates, textiles.

2402. Muminov, B. et al. "Scientifically based sizes for clothing suitable for summer use in Central Asian regions." *Gig Sanit* 36, no. 5 (May 1971): 101-2. Topics discussed include: Siberia, desert climate.

2403. Nakahashi, M. "Study on air circulation in paper and cloth diapers: a comparison of temperature and humidity inside the diapers in summer." *Kango* 34, no. 9 (August 1982): 106-16. Topics discussed include: humidity, seasonality, cotton, infant care, paper.

2404. Nakahashi, M. et al. "Researches on the clothing of adult males classified by the season comparison between the student group and the father group." *Ann Physiol Anthropol* 6, no. 1 (January 1987): 11-20. Topics discussed include: students, temperature.

2405. Natsume, K. et al. "Field studies of clothing microclimate temperatures in human subjects during normal daily life." *J Hum Ergol Tokyo* 17, no. 1 (September 1988): 13-9. Topics discussed include: body temperature, seasonality, climatic effects, microclimates, skin temperature.

2406. Nishi, Y. et al. "Clothing insulation as a biometeorological parameter during rest and exercise." *Int J Biometeorol* 22, no. 3 (September 1978): 177-89. Topics discussed include: body temperature, environment, humidity, skin temperature, temperature, physical work, rest.

2407. Nunneley, S. "Heat stress in protective clothing. Interactions among physical and physiological factors." *Scand J Work Environ Health* 15 Supp, no. 1 (1989): 52-7. Topics discussed include: adaptation, computer simulation, movement, perspiration, task performance.

2408. Obendorf, S. K. et al. "Heat transfer characteristics of nonwoven insulating materials." *Textile Research Journal* 56 (November 1986): 691-6. Topics discussed include: nonwoven fabrics, textiles, fabrics, heat transmission, clothing, winter.

2409. Olumide, Y. et al. "Cutaneous implications of excessive heat in the work place." *Contact Dermatitis* 9, no. 5 (September 1983): 360-3. Topics discussed include: clothing, work-related diagnosis, dermatitis, humidity, Nigeria, posture, tropical climates, ventilation.

2410. Raikhman, S. et al. "Effect of the hygienic properties of special clothing on thermoregulations in man under conditions of inhibited heat loss." *Gig Sanit no.* 5 (May 1984): 16-9. Topics discussed include: absorption, microclimates, temperature, textiles, thermal conductivity, thermoregulation.

2411. Reinberg, A. et al. "Interindividual differences in a set of biological rhythms documented during the high arctic summer (79 degrees N) in three healthy subjects." *Chronobiol Int* 1, no. 2 (1984): 127-38. Topics discussed include: cycles, arctic regions, body temperature, clothing, fatigue, France, heart rate, muscle contraction, seasonality.

2412. Rensburg, A. V. et al. "Physiological reactions of men using microclimate cooling in hot humid environments." *Br J Ind Med* 29, no. 4 (October 1972): 387-93. Topics discussed include: acclimation, body temperature, heart rate, ice, mining, work-related medicine, refrigeration, stress, perspiration, heat, humidity, protective clothing.

2413. Rodahl, K. et al. "Protection against cold. A physiological comparison of the protective value of clothing made of nylon pile and wool pile." *Tidsskr Nor Laegeforen* 93, no. 13 (10 May 1973): 907-10. Topics discussed include: body temperature, physical work, nylons, oxygen consumption, skin temperature, wool, cold climates.

2414. Rogers, A. et al. "Non acclimatization of man to cold proceedings." *J Physiol Lond* 277 (April 1978): 50P-51P. Topics discussed include: clothing, cold climate.

2415. ---. "Proceedings: Clo values of Polar clothing and their relation to 'total number of layers' counts." *J Physiol Lond* 240, no. 2 (July 1974): 22P-24P. Topics discussed include: wind, cold climates, protective standards.

2416. Rogers, W. H. et al. "Manual performance in the cold with gloves and bare hands." *Perceptual and Motor Skills* 59, no. 1 (August 1984): 3-13. Topics discussed include: temperature, gloves, tasks, marines, cold effects, hand anatomy, motor performance, marine personnel.

2417. Salvosa, CB. et al. "Environmental conditions and body temperatures of elderly women living alone or in local authority home." *Br Med J* 4, no. 788 (11 December 1971): 656-9. Topics discussed include: clothing, diet, housing, London, nutrition, seasonality, skin temperature, urine, body temperature, heating.

2418. Sato, M. et al. "Effects of moderate changes in air temperature and clothing on pulmonary exchange and energy metabolism." *J Hum Ergol Tokyo* 2, no. 2 (December 1973): 159-67. Topics discussed include: adaptation, exposure, environment, heart rate, seasonality, respiration.

2419. Schrader, G. et al. "Investigations on the clothing of preschool children during winter." *Arztl Jugendkd* 69, no. 2 (April 1978): 82-9. Topics discussed include: environment, Germany, housing, temperature, child care, seasons.

2420. Schulz, G. et al. "Human requirement conditions in work in extremely warm climate and the effect on their protective clothing." *Int Arch Arbeitsmed* 24, no. 1 (1967): 74-115. Topics discussed include: adaptation, body temperature, environment, exposure, heart rate, infrared rays, occupations, perspiration, tropical climates.

2421. Shimizu, H. et al. "Changes in temperature and humidity at different layers inside clothing during rest, exercise, and recovery." *Nippon Eiseigaku Zasshi* 43, no. 6 (February 1989): 1130-9. Topics discussed include: perspiration, physical work, humidity, rest.

2422. Shvartz, E. et al. "Heat acclimatization while wearing vapor barrier clothing." *Aerosp Med* 44, no. 6 (June 1973): 609-12. Topics discussed include: body temperature, heart rate, humidity, oxygen consumption, skin temperature, perspiration, temperature, water.

2423. Sleggs, G. A. "Thermal comfort and physiological responses during exercise in a warm-humid environment among young men who wore selected upper body garments." Ph.D. diss., Pennsylvania State University, 1986. Topics discussed include: hot climates, weather, clothing, physiology, heat, humidity, thermoregulation.

2424. Smith, J. et al. "Development of weather protective clothing for building site workers." *Proc R Soc Med* 63, no. 10 (October 1970): 1008-9. Topics discussed include: bronchitis, work-related diseases, rheumatic diseases, textiles, architecture, work-related medicine.

2425. Smolander, J. et al. "Cardiorespiratory and thermal effects of wearing gas protective clothing." *Int Arch Occup Environ Health* 54, no. 3 (1984): 261-70. Topics discussed include: exercise, physical work, oxygen consumption, body temperature, heart rate.

2426. ---. "Physiological strain in work with gas protective clothing at low ambient temperature." *Am Ind Hyg Assoc J* 46, no. 12 (December 1985): 720-3. Topics discussed include: aging, chemical industry, heart rate, oxygen consumption, cold climates, gases, work-related diseases, physiopathology, stress.

2427. Steegmann, A. J. "Finger temperatures during work in natural cold: the Northern Ojibwa." *Hum Biol* 49, no. 3 (September 1977): 349-62. Topics discussed include: Canada, diet, physical work, Indians, North America, protective clothing, body temperature, cold climate.

2428. Stuart, I. M. et al. "Wind induced transfer of water vapor and heat through clothing." *Textile Research Journal* 53 (November 1983): 655-60. Topics discussed include: textiles, fabrics, permeability, heat transmission, winter.

2429. Sulli, E. "Clothing hygiene with special reference to space flight. II. Clothing in relation to climate." *Riv Med Aeronaut Spaz* 29, no. 4 (October 1966): 681-742. Topics discussed include: arctic regions, tropical climates, seasonality, space flight.

2430. Tanaka, M. "Experimental studies on human reaction to cold. Differences in the vascular hunting reaction to cold according to sex, season, and environmental temperature." *Bull Tokyo Med Dent Univ* 18, no. 4 (December 1971): 269-80. Topics discussed include: acclimation, body regulation, clothing, exposure, fingers, blood supply, occupations, gender effects, swimming, blood vessels.

2431. Tanaka, M. et al. "Effects on the manual function of men wearing cold protective clothing to cold stress." *Sangyo Igaku* 23, no. 1 (January 1981): 72-8. Topics discussed include: skin temperature, stress etiology, cold climates, physiology, physiopathology.

2432. ---. "Thermal reaction and manual performance during cold exposure while wearing cold protective clothing." *Ergonomics* 26, no. 2 (February 1983): 141-149. Topics discussed include: exposure, cold climates, skin temperature, task performance.

2433. Teitlebaum, A. et al. "Increased energy cost with multiple clothing layers." *J Appl Physiol* 32, no. 6 (June 1972): 743-4. Topics discussed include: cold climates, locomotion, basal metabolism, physical work.

2434. Terent'Eva, G. et al. "Role of the design of children's outer winter clothing in the total insulating effect." *Gig Sanit 1987* no. 7 (July 1987): 86-8. Topics discussed include: cold climates, skin temperature, thermoregulation, seasons.

2435. Tochihara, Y. et al. "A survey on physiological strains of asbestos abatement work wearing protective clothing in summer." *Ann Physiol Anthropol* 12, no. 1 (January 1993): 31-8. Topics discussed include: body temperature, heart rate, heat, humidity, seasonality, skin temperature, weight loss, work-related health.

2436. Utidjian, H. "Criteria for a recommended standard occupational exposure to hot environments. I. Recommendations for a standard for work in hot environments." *J Occup Med* 15, no. 6 (June 1973): 519-23. Topics discussed include: acclimation, cold climates, emergency medical services, health education, physiologic standards, protective clothing, shock treatment, timing, work.

2437. Vaughan, E. "Adequately insulated gloves, boots vital for working in extreme cold." *Occup Health Saf* 59, no. 13 (December 1990): 34. Topics discussed include: equipment design, seasonality, shoes, work-related diseases, protective clothing.

2438. Weiner, J. S. "Biological Assessment of Clothing for Tropical Use." *Med Bull* 5 (1947): 1016. Topics discussed include: biology, climatic effects, temperature, environment.

2439. White, M. K. et al. "The effectiveness of ice- and freon-based personal cooling systems during work in fully encapsulating suits in the heat." *American Industrial Hygiene Association Journal* 52 (March 1991): 127-35. Topics discussed include: protective clothing.

2440. Williams, G. "Foul-weather fabrics: clothing for outdoor enthusiasts who haven't the sense to come in from the rain." *Sierra* 73 (September 1988): 110-11. Topics discussed include: textiles, fibers, protection.

2441. Winslow, C. E. A. et al. "Reactions of the Clothed Human Body to Variation in Atmospheric Humidity." *Am J Physiol* 124 (1938): 692-704. Topics discussed include: biology, environment, climatic effects, health, behavior.

2442. Witkowski, S. R. et al. "Climate, clothing, and body-part nomenclature." *Ethnology* 24 (July 1985): 197-214. Topics discussed include: anatomy, terminology, language, culture, anatomy, dress.

2443. Woodbury, R. "Clothing, its evolution and development by the inhabitants of the Artic." *Arch Environ Health* 17, no. 4 (October 1968): 586-91. Topics discussed include: arctic regions, cotton, medicine, wool, history, cold climates, Eskimos, Indians.

2444. Woods, R. "Cooling of motorcyclists in various clothing during winter in Britain." *Ergonomics* 26, no. 9 (September 1983): 847-61. Topics discussed include: body temperature, United Kingdom, seasonality, thermoregulation, transportation.

2445. Wyon, D. P. et al. "The mental performance of subjects clothed for comfort at two different air temperatures." *Ergonomics* 18, no. 4 (July 1975): 359-374. Topics discussed include: air temperature, light clothing, heavy clothing, mental performance, effort, arousal, fatigue, fashion.

2446. Yagloglou, C. B. et al. "Effective Temperature with Clothing." *J Am Soc of Heat Vent Eng* 31 (1925): 69-70. Topics discussed include: behavior.

2447. Yaglou, C. P. "The Summer Comfort Zone, Climate and Clothing." *J Indus Hyg* 10 (1928): 350. Topics discussed include: seasonality, temperature, behavior.

2448. Yokura, H. et al. "Durability of fabric handle and shape retention during wear of men's summer suits." *Textile Research Journal* 60 (April 1990): 194-202. Topics discussed include: textiles, fabrics, clothing, wool.

2449. Yu, Y. "Study of the thermal insulation criteria of winter clothes." *Chung Hua Yu Fang I Hsueh Tsa Chih* 14, no. 1 (February 1980): 4-9. Topics discussed include: seasonality, body temperature, cold weather.

2450. Zerkal', L. et al. "Physiological and hygienic evaluation of special summer clothing for female workers in the mining industry." *Gig Tr Prof Zabol no.* 10 (October 1981): 19-22. Topics discussed include: hygiene, microclimates, Ukraine, mining, seasonality.

Energy

2451. Archibald, R. B. et al. "Seasonal Variation in Residential Electricity Demand: Evidence from Survey Data." *Applied Economics* 14, no. 2 (April 1982): 167-81. Topics discussed include: communication, information, services, energy, consumer behavior, expenditures, cycles, consumption.

2452. Bach, W. *Carbon dioxide : current views and developments in energy, climate research: 2nd course of the International School of Climatology, Ettore Majorana Centre for Scientific Culture, Erice, Italy, July 16-26, 1982.* 525 pages. International School of Climatology. Dordrecht; Boston; Hingham, MA: D. Reidel; Sold and distributed in the U.S.A. and Canada by Kluwer Academic Publishers, 1983. Topics discussed include: atmospheric carbon dioxide.

2453. ---. *Our threatened climate : ways of averting the CO2 problem through rational energy use.* 368 pages. Dordrecht; Boston; Hingham, MA: D. Reidel Pub. Co.; Sold and distributed in the U.S.A. and Canada by Kluwer Academic Publishers, 1984. Topics discussed include: atmospheric effects, carbon dioxide, climatic changes, climatology, sociology, consumption.

2454. Bahel, V. et al. "A computer study for estimation of seasonal energy consumption and selection of air-conditioner in six Saudi Arabian cities." *Energy Conversion and Management* 28, no. 1 (1988): 27-33. Topics discussed include: electricity, air conditioning, buildings, Saudi Arabia, climate.

2455. Bayliss-Smith, T. P. "Energy and Economic Development." In *Subsistence and Survival: Rural Ecology in the Pacific*, editors T. P. Bayliss-Smith, and R. G. A. Feachem. London: Academic Press, 1977. Topics discussed include: environment, economy.

2456. Blocker, T. J. et al. "Household income, electricity use, and rate structure preferences." *Environment and Behavior* 16, no. 5 (September 1984): 551-572. Topics discussed include: households, income, preferences, conservation, ecology, consumer behavior.

2457. Bopp, A. E. et al. "A Comparison of the Effect of Seasonal Adjustment on Parameter Values, Forecasts and Policy Analysis." *Journal of Economic Studies* 5, no. 1 (May 1978): 20-30. Topics discussed include: forecasting, econometrics, methodology, time series analysis, spectral analysis models, energy.

2458. Bower, R. S. et al. "Weather Normalization and Natural Gas Regulation." *Energy Journal* 6, no. 2 (April 1985): 101-15. Topics discussed include: consumer behavior, expenditures, cycles, consumption.

2459. Collins, R. A. et al. "Derivation of a Theoretical Form of a Production Function for Home Temperature - A Comment The Economics of House Heating." *Energy Economics* 5, no. 4 (October 1983): 273-75. Topics discussed include: consumer behavior, expenditures, cycles, consumption.

2460. Cramer, J. C. et al. "Structural-behavioral determinants of residential energy use: summer electricity use in Davis." *Energy (Oxford, England)* 9 (March 1984): 207-16. Topics discussed include: houses, heating, cooling, power consumption, electricity, air conditioning.

2461. De, R. C. et al. "Solar housing potential in the climates of the Andean region in Argentina." *Energy Sources* 13, no. 1 (1991): 19-38. Topics discussed include: solar houses, Argentina, climate, sociology.

2462. Dubin, J. A. et al. "An Engineering, Econometric Analysis of Seasonal Energy Demand and Conservation in the Pacific Northwest." *Journal of Business and Economic Statistics* 6, no. 1 (January 1988): 121-34.

2463. Edelstein, R. "Cost Requirements for Solar Thermal Electric and Industrial Process Heat." *Economic analysis of solar thermal energy systems Solar Heat Technologies: Fundamentals and Applications series, vol 3,* 337-71. Ronald E. Kreith, Frank West. Cambridge, Mass. and London : MIT Press, 1988. Topics discussed include: development.

2464. Engle, R. F. et al. "Merging Short- and Long-run Forecasts: An Application of Seasonal Cointegration to Monthly Electricity Sales Forecasting." *Journal of Econometrics* 40, no. 1 (January 1989): 45-62. Topics discussed include: construction, analysis, models, forecasting.

2465. Engle, R. F. et al. "Semiparametric Estimates of the Relation between Weather and Electricity Sales." *Journal of the American Statistical Association* 81, no. 394 (June 1986): 310-20. Topics discussed include: gas, communication, information, services, energy, consumer behavior, expenditures, cycles, consumption.

2466. "Federal efforts to aid low-income and elderly individuals affected by life-threatening heat conditions : hearing before the Subcommittee on Human Resources of the Committee on Education and Labor, House of Representatives, Ninety-sixth Congress, second session, hearing held in Washington, D.C., on July 30, 1980." United States. Congress. House. Committee on Education and Labor. Subcommittee on Human Resources. U.S. G.P.O., Washington, January 1980. Topics discussed include: heat, physiology, poverty, energy, public assistance.

2467. Ferrier, G. D. et al. "Climate Control Efficiency." *Energy Journal* 13, no. 1 (1992): 37-54. Topics discussed include: resources, conservation, environment, management, water, air, pollution.

2468. Garbacz, C. "A National Micro-Data Based Model of Residential Electricity Demand: New Evidence on Seasonal Variation." *Southern Economic Journal* 51, no. 1 (July 1984): 235-49. Topics discussed include: energy, consumer behavior, expenditures, cycles, consumption.

2469. ---. "Seasonal and Regional Residential Electricity Demand." *Energy Journal* 7, no. 2 (April 1986): 121-34. Topics discussed include: electricity, gas, consumer behavior, expenditures, cycles, consumption.

2470. Gilles, J. L. et al. "Farm Size, Farm Structure, Energy and Climate: An Alternate Ecological Analysis of United States Agriculture." *Rural Sociology* 45, no. 2 (June 1980): 332-339. Topics discussed include: farming, ecology.

2471. Guzzi, R. et al. *Physical Climatology for Solar & Wind Energy*. River Edge, NJ: World Scientific Publishing Company, 1988. Topics discussed include: climate.

2472. Herbert, J. H. "Weather, the Estimation of Residential Energy Demand Relationships and the Reporting of Data." *Energy Economics* 10, no. 4 (October 1988): 324-25. Topics discussed include: consumer behavior, expenditures, cycles, consumption, urban housing, nonurban housing.

2473. Hourcade, J. C. et al. "Energy Politics and Climatic Risks. How to Manage Uncertainty?" *Futuribles* 135 (September 1989): 35-60. Topics discussed include: weather, environment, politics, sociology, ecology, climatic effects.

2474. Hurst, C. "Establishing New Markets for Mature Energy Equipment in Developing Countries - Experience with Windmills, Hydro-Powered Mills and Solar Water Heaters." *World Development* 18, no. 4 (April 1990): 605-15. Topics discussed include: technology change, innovation.

2475. Klein, Y. L. "An Econometric Model of the Joint Production and Consumption of Residential Space Heat." *Southern Economic Journal* 55, no. 2 (October 1988): 351-59. Topics discussed include: energy, consumer behavior, expenditures, cycles, consumption.

2476. Kushman, J. E. et al. "A Model of Individual Household Temperature Demand and Energy-Related Welfare Changes Using Satiety." *Energy Economics* 8, no. 3 (July 1986): 147-54. Topics discussed include: consumer behavior, expenditures, cycles, consumption.

2477. Lawrence, A. et al. "Modelling and Forecasting Time of Day and Seasonal Electricity Demand." *Journal of Econometrics* 9 (1979): 1-237. Topics discussed include: economy, energy.

2478. Lillard, L. A. et al. "Seasonal Electricity Demand and Pricing Analysis with a Variable Response Model." *Bell Journal of Economics* 12, no. 1 (March 1981): 71-92. Topics discussed include: public utilities, consumer behavior, expenditures, cycles, consumption.

2479. ---. "Time-of-Day Electricity Consumption Response to Temperature and the Ownership of Air Conditioning Appliances." *Journal of Business and Economic Statistics* 2, no. 1 (January 1984): 40-53. Topics discussed include: public utilities, electricity, gas, consumer economics, living standards, consumption, savings.

2480. Markham, S. F. *Climate and the energy of nations.* 236 pages. New York: AMS Press, 1982.

2481. McGregor, R. et al. "Radon and radon daughter levels in energy efficient housing." *Sci Total Environ* 45 (October 1985): 271-8. Topics discussed include: environment, exposure, microclimates, Ontario, radiation, seasonality.

2482. Minor, J. E. *Overview of Seasonal Thermal Energy Storage in the United States*, Performer: Battelle Pacific Northwest Labs., Richland, WA. Department of Energy, Washington, DC, September 1981. Topics discussed include: storage, equipment, aquifers, economics.

2483. Molnar, A. "Statistics on the Production and Use of Low-Temperature Thermal Water in Hungary." *Statistical Journal* 2, no. 1 (January 1984): 73-83. Topics discussed include: energy.

2484. Monts, K. et al. "Weather Normalization of Electricity Sales to the School Sector: Potential Model Misspecification." *Energy Economics* 11, no. 2 (April 1989): 127-32.

2485. Reddy, T. A. "Statistical analyses of electricity use during the hottest and coolest days of summer for groups of residences with and without air-conditioning." *Energy (Oxford, England)* 15 (January 1990): 45-61. Topics discussed include: power plants, management, houses, heating, cooling.

2486. Reilly, J. et al. "Changing Climate and Energy Modeling: A Review." *Energy Journal* 6, no. 3 (July 1985): 137-54.

2487. Ritchie, J. B. et al. "Complexities of household energy consumption and conservation." *Journal of Consumer Research* 8, no. 3 (December 1981): 233-242. Topics discussed include: climate, possessions, demography, conservation, attitudes, Canada, consumer behavior, attitudes, temperature.

2488. Rushdi, A. A. "Weather, the Estimation of Residential Energy Demand Relationships and the Reporting of Data: Reply." *Energy Economics* 10, no. 4 (October 1988): 325. Topics discussed include: consumer behavior, expenditures, cycles, consumption, urban housing, nonurban housing.

2489. Sebald, A. V. et al. "On eliminating peak load auxiliary energy consumption in passive solar residences during winter." *Solar Energy* 39, no. 4 (1987): 307-19. Topics discussed include: solar houses, power systems, plants, peak loads.

2490. Seligman, C. et al. "Predicting summer energy consumption from homeowners' attitudes." *Journal of Applied Social Psychology* 9, no. 1 (January 1979): 70-90. Topics discussed include: attitudes, summer consumption, married couples, homeowners, spouses, consumer behavior.

2491. Shore, H. "Corrections Summer Time and Electricity Conservation: The Israeli Case." *Energy Journal* 6, no. 1 (January 1985): 169-70. Topics discussed include: electricity, gas, consumer behavior, expenditures, cycles, consumption.

2492. ---. "Summer Time and Electricity Conservation: The Israeli Case." *Energy Journal* 5, no. 2 (April 1984): 53-70. Topics discussed include: electricity, gas, consumer behavior, expenditures, cycles, consumption.

2493. Spewak, P. C. "Analyzing the Effect of Economic Policy on Solar Markets." *Economic analysis of solar thermal energy systems Solar Heat Technologies: Fundamentals and Applications series*, 167-204. eds Ronald E. West, and Frank Kreith. Cambridge, MA and London: MIT Press, 1988. Topics discussed include: planning, policy.

2494. Terjung, W. H. "Urban energy balance climatology: A prleminary inverstigation of the city man system in downtown Los Angeles." *Geographical R* 60, no. 1 (1970): 31-53.

2495. Tiplitz, C. "Optimal Seasonal Distillate Inventory." *Energy Journal* 7, no. 3 (July 1986): 83-97. Topics discussed include: mining industry, oil, gas, fuels.

2496. "USSR : seasonality in energy production and foreign trade : a research paper." United States. Central Intelligence Agency, and National Foreign Assessment Center (U.S.). Central Intelligence Agency : Document Expeditions (DOCEX) Project, Exchange and Gift Division, Library of Congress [distributor], Washington, D.C., 1981. Topics discussed include: Soviet Union, seasonality, economics, business.

2497. Warren, M. L. "Cost Requirements for Active Solar Heating and Cooling." *Economic analysis of solar thermal energy systems Solar Heat Technologies: Fundamentals and Applications series*, 274-308. eds Ronald E. West, and Frank Kreith. Cambridge, MA and London: MIT Press, 1988.

2498. Watson, D. *Climatic Building Design: Energy Efficient Building Principles & Practice*. New York, NY: McGraw-Hill, 1993. Topics discussed include: building, environment, engineering.

2499. Wenders, J. T. "The Welfare Economics of Optional Seasonal Time-of-Day Electricity Tariffs." *Energy Economics* 3, no. 2 (April 1981): 102-04. Topics discussed include: regulation, public utilities.

2500. Wenders, J. T. et al. "Experiments in Seasonal-Time-of-Day Pricing of Electricity to Residential Users." *Bell Journal of Economics* 7, no. 2 (1976): 531-52. Topics discussed include: regulation, public utilities.

Food

2501. Aall, C. et al. "The Sahelian drought. Proposals for a supporting programme in Niger for food provision, nutrition rehabilitation and malnutrition prevention." *J Trop Pediatr Environ Child Health* 20, no. 6 (December 1974): 303-29. Topics discussed include: costs, health care, services, pregnancy, weather.

2502. Abdullah, M. et al. "Seasonal variations, and the intra household distribution of food in a Bangladeshi village." *Am J Clin Nutr* 41, no. 6 (June 1985): 1305-13. Topics discussed include: body weight, caloric consumption, rural areas, population, gender effects, socioeconomics.

2503. Abdullah, M. "The Effect of Seasonality on Intrahousehold Food Distribution and Nutrition in Bangladesh." *Seasonal variability in third world agriculture: The consequences for food security*, 57-65. ed. David E. Sahn. Baltimore and London: Johns Hopkins University Press for the International Food Policy Research Institute, 1989. Topics discussed include: supply, health, medical programs, poverty, rural areas, economics.

2504. Agarwal, B. "Social Security and the Family: Coping with Seasonality and Calamity in Rural India." *Journal of Peasant Studies* 17, no. 3 (April 1990): 341-412. Topics discussed include: rural areas, communities, family farms, agriculture, production, food, division of labor by gender.

2505. Aleksiejczyk, Z. et al. "Structure and seasonal variations of food consumption by 4 member families of workers in the Olsztyn region 1980-1982." *Rocz Panstw Zakl Hig* 36, no. 3 (1985): 221-7. Topics discussed include: family composition, social status, family health.

2506. Annegers, J. F. "Seasonal Food Shortages in West Africa." *Ecol Food Nutr* 2 (1973): 251-257. Topics discussed include: nutrition, food, agriculture.

2507. Bach, W. *Interactions of Food & Climate.* Norwell, MA: Kluwer Academic Publishers, 1981.

2508. Behrman, J. R. et al. "Agricultural Wages in India: The Role of Health, Nutrition, and Seasonality." *Sahn, David E, ed Seasonal variability in third world agriculture: The consequences for food security,* 107-117. Baltimore, London: John Hopkins University Press for the International Food Policy Research, 1989. Topics discussed include: supply, rural areas, economics, health issues, medical programs, poverty.

2509. ---. "Seasonal Demands for Nutrient Intakes and Health Status in Rural South India." *Sahn, David E, ed Seasonal variability in third world agriculture: The consequences for food security,* 66-78. Baltimore and London: John Hopkins University Press for the International Food Policy Research Institute, 1989. Topics discussed include: supply, health issues, medical programs, poverty, rural areas, economics.

2510. Benhura, M. et al. "Seasonal variation in the food consumption patterns of the people of Mutambara district of Zimbabwe." *Cent Afr J Med* 38, no. 1 (January 1992): 8-13. Topics discussed include: cereals, food preferences, fruit, legumes, meat, rural areas, population, timing, vegetables, diet.

2511. Biswas, M. R. *Food, Climate, & Man.* Ann Arbor, MI: Wiley, 1979. Topics discussed include: science, meteorology, agriculture.

2512. Blake, D. et al. "The Demand for Cider in the United Kingdom." *Oxford Bulletin of Economics and Statistics* 54, no. 1 (February 1992): 73-86. Topics discussed include: consumer expenditures, seasonal patterns, consumption, living standards, savings.

2513. Blaxter, K. *Food, Nutrition & Climate.* New York: Elsevier Science Publishing Company, 1982. Topics discussed include: food, climate.

2514. Boiko, E. "The vitamin C supply of personnel of seagoing ships." *Vopr Pitan* 27, no. 3 (May 1968): 46-52. Topics discussed include: climates, physical labor, food, seasonality, ascorbic acid, naval medicine, nutrition.

2515. Bouis, H. E. "Seasonal Rice Price Variation in the Philippines: Measuring the Effects of Government Intervention." *Food Research Institute Studies* 19, no. 1 (1983): 81-92. Topics discussed include: policy, agriculture, markets, marketing, cooperatives.

2516. Bryson, R. A. et al. "Climatic Variation and Implications for World Food Production." *World Development* 5, no. 5-7 (May 1977): 507-18. Topics discussed include: agriculture, natural resources.

2517. Cabezas, G. M. et al. "The nutritional status of the Kung population (the Kalahari Desert)." *Rev Enferm* 16, no. 175 (March 1993): 81-4. Topics discussed include: Angola, Botswana, diet, food, supply, namibia, water supply, desert climates, ethnic groups, nutrition.

2518. Callaway, J. M. et al. *Analysis of Methods and Models for Assessing the Direct and Indirect Economic Impacts of CO sub 2 Induced Environmental Changes in the Agricultural Sector of the US Economy,* Performer: Battelle Pacific Northwest Labs., Richland, WA. Department of Energy, Washington, DC, August 1982. Topics discussed include: greenhouse effect, climatic effects, productivity, economics, crops, behavior, sociology, food, business, marketing.

2519. CFNPP Staff. *Short Term Consumption Behavior, Seasonality, & Labor Market Uncertainty in Rural India,* R. Sudarshan Canagarajah, and S. E. Pudney. Ithaca, NY: Cornell Food & Nutrition Policy Programs, 1993. Topics discussed include: consumption, consumer behavior, labor, agriculture.

2520. Chowdhury, N. "Seasonality of Foodgrain Price and Procurement Programme in Bangladesh since Liberation - A Rejoinder." *Bangladesh Development Studies* 20, no. 1 (March 1992): 121-27. Topics discussed include: agriculture, natural resources, food, policy, supply, economics, development.

2521. Cooper, P. J. M. et al. "Climatic Variability and Food Production in West Asia and North Africa." *Climate and food security: Papers presented at the International Symposium on Climate Variability and Food Security in Developing Countries, 5 9 February 1987 New Delhi, India*, 69-83. P. A. Oram. et al Manila: International Rice Research Institute in collaboration with the American Association for the Advancement of Science, 1989. Topics discussed include: agriculture, supply, demand.

2522. Dadhwal, V. K. "Effect of Temperature on Wheat in India." *Climate and food security: Papers presented at the International Symposium on Climate Variability and Food Security in Developing Countries, 5 9 February 1987, New Delhi, India*, 137-44. P. A. Oram et al.. Manila: International Rice Research Institute in collaboration with the American Association for the Advancement of Science, 1989. Topics discussed include: agriculture, supply, demand, economics.

2523. *Seasonal variability in third world agriculture : The consequences for food security*, ed David E. Sahn. Baltimore: Johns Hopkins University Press, 1989.

2524. Desai, P. "Weather and grain yields in the Soviet Union." *Research Report series*, 94 Washington, DC: International Food Policy Research Institute, 1986. Topics discussed include: agriculture, supply, demand, economics.

2525. Edwards, J. et al. "The influence of a calorie supplement on the consumption of the meal, ready to eat in a cold environment." *Mil Med* 156, no. 9 (September 1991): 466-71. Topics discussed include: Alaska, cold climates, diet, urine chemistry, caloric consumption, eating physiology, food, military personnel.

2526. Emerson, R. D. "Seasonal agricultural labor markets in the United States: final report." University of Florida. Agricultural Experiment Station. Food and Resource Economics Department, Florida Agricultural Experiment Stations, Institute of Food and Agricultural Sciences, University of Florida, Gainesville, 1981. Topics discussed include: seasonality, labor, migrants.

2527. Ferguson, A. "Some aspects of birth seasonality in Kenya." *Soc Sci Med* 25, no. 7 (1987): 793-801. Topics discussed include: climates, food, supply, health, policy, holidays, Kenya.

2528. Fox, R. H. "Heat Stress and Nutrition." *Proc Nutrition Soc* 17 (1958): 173-179. Topics discussed include: temperature, food, environment.

2529. Gao, L. et al. "Climatic Variation and Food Production in Jiangsu, China." *Climate and food security: Papers presented at the International Symposium on Climate Variability and Food Security in Developing Countries, 5 9 February 1987, New Delhi, India*, 557-62. P. A. Oram et al.. Manila: International Rice Research Institute in collaboration with the American Association for the Advancement of Science, 1989. Topics discussed include: agriculture, supply, demand, economics.

2530. Gassaway, A. "Natural and economic events influencing arctic food consumption data: per capita results in northernmost Norway." *Arch Environ Health* 26, no. 1 (January 1973): 41-7. Topics discussed include: agriculture, arctic regions, climatic effects, commerce, economics, environment, fisheries, geography, margarine, milk, seasonality, migrants, diet, food, supply.

2531. Geuns, M. et al. "Child nutrition in the pre-harvest season in Kenya." *East Afr Med J* 68, no. 2 (February 1991): 93-105. Topics discussed include: body height, body weight, breast feeding, diet, eating, Kenya, agriculture.

2532. Gleick, P. H. "Global Climatic Change, Water Resources, and Food Security." *Climate and food security: Papers presented at the International Symposium on Climate Variability and Food Security in Developing Countries, 5 9 February 1987, New Delhi, India*, 415-27. P. A. Oram et al.. Manila: International Rice Research Institute in collaboration with the American Association for the Advancement of Science, 1989. Topics discussed include: agriculture, supply, demand, economics, conservation, pollution, natural resources.

2533. Goldman, R. H. "Seasonal Rice Prices in Indonesia, 1953-69: An Anticipatory Price Analysis." *Food Research Institute Studies* 13, no. 2 (1974): 99-143. Topics discussed include: agriculture, supply, demand, economics.

2534. Guyer, J. "From Seasonal Income to Daily Diet in a Partially Commercialized Rural Economy (Southern Cameroon)." *Seasonal variability in third world agriculture: The consequences for food security*, 137-50. ed David E. Sahn. Baltimore and London: Johns Hopkins University Press for the International Food Policy Research Institute, 1989. Topics discussed include: supply, demand, rural areas, economics, health issues, medical programs, poverty.

2535. Herdt, R. W. "The Impact of Technology and Policy on Seasonal Household Food Security in Asia." *Seasonal variability in third world agriculture: The consequences for food security*, 235-45. ed David E. Sahn. Baltimore and London: Johns Hopkins University Press for the International Food Policy Research Institute, 1989. Topics discussed include: rural areas, economics, policy, technology, change, innovation.

2536. Hill, K. et al. "Seasonal variance in the diet of Ache hunter-gatherers in Eastern Paraguay." *Human Ecology* 12 (June 1984): 101-35. Topics discussed include: hunters, gatherers, Paraguay, food, caloric content, Indians.

2537. Hussain, M. "Seasonal variation and nutrition in developing countries." *Food Nutr Roma* 11, no. 2 (1985): 23-7. Topics discussed include: Africa, Bangladesh, development, food, supply.

2538. International Food Policy Research Institute. "Seasonal variability in Third World agriculture." *The Journal of Development Studies* 28 (January 1992): 353-5.

2539. Janssen, M. et al. "Season and Regional Differences in Agricultural Supply Response in Benin." *European Review of Agricultural Economics* 17, no. 4 (1990): 407-420. Topics discussed include: production, Germany, family farms, food, crops, supply, prices.

2540. Johnson, D. G. *World Grain Economy and Climate Change to the Year 2000: Implications for Policy Final rept*, Performer: Chicago Univ., IL., 1983. Topics discussed include: agriculture, food, farms, crops, economics, production, international trade, atmospheric effects, meteorology.

2541. Kauffman, H. E. et al. "Climatic Vulnerability of Major Food Crops: Soybeans." *Climate and food security: Papers presented at the International Symposium on Climate Variability and Food Security in Developing Countries, 5 9 February 1987, New Delhi, India*, 155-64. P. A. Oram et al. Manila: International Rice Research Institute in collaboration with the American Association for the Advancement of Science, 1989. Topics discussed include: agriculture, supply, demand, economics.

2542. Khanna Chopra, R. et al. "Impact of Climate Variation on Production of Pulses." *Climate and food security: Papers presented at the International Symposium on Climate Variability and Food Security in Developing Countries, 5 9 February 1987, New Delhi, India*, 219-36. P. A. Oram et al.. Manila: International Rice Research Institute in collaboration with the American Association for the Advancement of Science, 1989. Topics discussed include: agriculture, supply, demand, economics.

2543. King, K. et al. "Food patterns from dietary surveys in rural Haiti." *J Am Diet Assoc* 53, no. 2 (August 1968): 114-8. Topics discussed include: diet, food, supply, Haiti, seasonality, socioeconomics, nutrition, rural areas, health.

2544. Kulesza, W. et al. "Dietary survey of children and adolescents in Warsaw schools." *Rocz Panstw Zakl Hig* 24, no. 3 (1973): 353-64. Topics discussed include: food, consumption behavior, nutrition, Poland, health services, seasonality.

2545. Kumar, S. K. "Effect of seasonal food shortage on agricultural production in Zambia." *World Development* 16 (September 1988): 1051-63. Topics discussed include: food, supply, Zambia, seasonality, economics.

2546. Lander, F. et al. "Chronic subclinical intake of dietary anticholinesterase agents during the spraying season." *Food Chem Toxicol* 30, no. 1 (January 1992): 37-40. Topics discussed include: seasonality, blood, diet, food, contamination.

2547. Lawrence, M. et al. "Seasonal Pattern of Activity and Its Nutritional Consequence in Gambia." *Seasonal variability in third world agriculture: The consequences for food security*, 47-56. editor David

E. Sahn. Baltimore and London: Johns Hopkins University Press for the International Food Policy Research Institute, 1989. Topics discussed include: supply, demand, rural areas, economics, health issues, medical programs, poverty.

2548. Leonard, W. R. "Household-Level Strategies for Protecting Children from Seasonal Food Scarcity." *Social Science and Medicine* 33, no. 10 (1991): 1127-1133. Topics discussed include: children, welfare services, strategies, Peru, diet, family farms, agriculture, work.

2549. Leonard, W. et al. "Biosocial responses to seasonal food stress in highland Peru." *Hum Biol* 61, no. 1 (February 1989): 65-85. Topics discussed include: caloric consumption, family composition, Peru, social adjustment, food, supply, stress.

2550. Levin, C. E. "The Effect of Seasonal Fluctuations on Household Food Security and Intertemporal Decision-Making - A Study of Rural Households in Indonesia." Ph.D. diss., Cornell University, 1992. Topics discussed include: farm households, input markets, agriculture management; available from UMI.

2551. Li, J. "Climatic Vulnerability of Maize in China." *Climate and food security: Papers presented at the International Symposium on Climate Variability and Food Security in Developing Countries, 5-9 February 1987, New Delhi, India*, 145-54. P. A. Oram et al. Manila: International Rice Research Institute in collaboration with the American Association for the Advancement of Science, 1989. Topics discussed include: agriculture, supply, demand, economics.

2552. "Living with climatic change: proceedings of a Seminar on Food and Climatic Change, Winnipeg, Canada 13 January 1977." 90 pages. Seminar on Food and Climatic Change, Winnipeg, Man., 1977, Science Council of Canada, and Canadian Meteorological Society. Ottawa: Science Council of Canada, 1977. Topics discussed include: crops, climatic effects.

2553. Maunder, W. J. "Climatic Variability and Agricultural Production in Temperate Regions: Six Case Studies." *Climate and food security: Papers presented at the International Symposium on Climate Variability and Food Security in Developing Countries, 5 9 February 1987, New Delhi, India*, 45-55. P. A. Oram et al. Manila: International Rice Research Institute in collaboration with the American Association for the Advancement of Science, 1989. Topics discussed include: supply, demand, capital markets, regulation.

2554. McCabe, J. T. "Success and failure: the breakdown of traditional drought coping institutions among the pastoral Turkana of Kenya." *Journal of Asian and African Studies* 25 (July 1990): 146-60. Topics discussed include: Kenya, Africa, sociology, customs, food, economics, development, pastoralists, herdsmen, culture.

2555. Messer, E. "Seasonality in Food Systems: An Anthropological Perspective on Household Food Security." *Seasonal variability in third world agriculture: The consequences for food security*, 151-75. ed David E. Sahn. Baltimore and London: Johns Hopkins University Press for the International Food Policy Research Institute, 1989. Topics discussed include: supply, demand, rural areas, economics, health issues, medical programs, poverty.

2556. Miller, S. E. "Seasonality and Stock-Adjustment Models of Retail Food Store Inventory Behavior." *American Journal of Agricultural Economics* 72, no. 4 (November 1990): 1018-27. Topics discussed include: distribution, retailing.

2557. Mongkolsmai, D. et al. "The Effect of Irrigation on Seasonal Rice Prices, Farm Income, and Labor Demand in Thailand." *Seasonal variability in third world agriculture: The consequences for food security*, 246-63. ed David E. Sahn. Baltimore and London: Johns Hopkins University Press for the International Food Policy Research Institute, 1989. Topics discussed include: supply, demand, rural areas, economics, development, irrigation, policy.

2558. Mortimore, M. *Adapting to Drought: Farmers, Famine & Desertification in West Africa*. New York, NY: Cambridge University Press, 1989. Topics discussed include: economy, agriculture, climatic effects, Africa, food.

2559. Narojek, L. et al. "Effects of environmental factors on the dietary patterns of children and adolescents." *Rocz Panstw Zakl Hig* 26, no. 5 (1975): 589-98. Topics discussed include: Poland, seasonality, sociology, environment, food, consumption behavior, nutrition.

2560. Nasolodin, V. et al. "Comparative effectiveness of microelements added to the diet of athletes during training in different seasons." *Vopr Pitan no.* 6 (November 1984): 33-8. Topics discussed include: copper, iron, manganese, physical education, training, food, running.

2561. National Academy of Sciences. *Climate and Food Climatic Fluctuation and U S Agricultural Production A Report of the Committee on Climate and Weather Fluctuations and Agricultural Production*, National Academy and Sciences, Washington, DC, 1976. Topics discussed include: nutrition.

2562. N'Zi, G. et al. "Health of school children on the Ivory Coast. Survey of food consumption." *Acta Trop Basel* 37, no. 4 Suppl 11 (December 1980): 34-40. Topics discussed include: caloric consumption, climatic effects, carbohydrates, fats, proteins, environment, feeding behavior, Ivory Coast, food, consumption behavior.

2563. Obasi, G. O. P. "Climate and Food Crop Production." *Climate and food security: Papers presented at the International Symposium on Climate Variability and Food Security in Developing Countries, 5 9 February 1987, New Delhi, India*, 503-14. P. A. Oram et al.. Manila: International Rice Research Institute in collaboration with the American Association for the Advancement of Science, 1989. Topics discussed include: agriculture, supply, demand, economics, conservation, pollution.

2564. Ono, T. et al. "Seasonal changes in food consumption." *Hokkaido Igaku Zasshi* 51, no. 6 (November 1976): 549-50. Topics discussed include: caloric consumption, diet.

2565. Oram, P. A. *Climate and food security: Papers presented at the International Symposium on Climate Variability and Food Security in Developing Countries, 5-9 February 1987, New Delhi, India.* 602 pages. Manila: International Rice Research Institute in collaboration with the American Association for the Advancement of Science, 1989. Topics discussed include: agriculture.

2566. ---. "Sensitivity of Agricultural Production to Climatic Change, an Update." *Climate and food security: Papers presented at the International Symposium on Climate Variability and Food Security in Developing Countries, 5 9 February 1987, New Delhi, India*, 25-44. P. A. Oram et al. Manila: International Rice Research Institute in collaboration with the American Association for the Advancement of Science, 1989. Topics discussed include: supply, demand, economics.

2567. Ostberg, O. "Circadian rhythms of food intake and oral temperature in 'morning' and 'evening' groups of individuals." *Ergonomics* 16, no. 2 (0173): 203-209. Topics discussed include: biology, cycles, food intake.

2568. Ouma, D. E. "Small Farmer Food Production and Security: A Response to Seasonal Food Supply Shortages in Kenya." *Dissertation Abstracts International, A: The Humanities and Social Sciences* 49, no. 5 (November 1988): 1289-A. Topics discussed include: Kenya, farms, agriculture, development, food, self sufficiency.

2569. Parry, M. L. et al. "The Effect of Climatic Variations of Food Production." *Climate and food security: Papers presented at the International Symposium on Climate Variability and Food Security in Developing Countries, 5 9 February 1987, New Delhi, India*, 429-49. P. A. Oram et al. Manila: International Rice Research Institute in collaboration with the American Association for the Advancement of Science, 1989. Topics discussed include: agriculture, supply, demand, economics.

2570. ---. "IIASA, UNEP climate impacts project: the vulnerability of food production in climate-sensitive areas." *Bulletin of the American Meteorological Society* 66 (October 1985): 1310-11. Topics discussed include: atmosphere, temperature, agricultural.

2571. Pasricha, S. et al. "Seasonal changes in the cost of balanced diets." *Indian J Med Res* 77 (March 1983):359-65. Topics discussed include: food, supply, economics, India, diet.

2572. Pasternak, B. "Seasons of Birth and Marriage in Two Chinese Localities." *Human Ecology* 6, no. 3 (September 1978): 299-323. Topics discussed include: seasonality, Taiwan, food, supply, marriage, timing.

2573. Payne, P. "Public Health and Functional Consequences of Seasonal Hunger and Malnutrition." *Seasonal variability in third world agriculture: The consequences for food security*, 19-46. David E. Sahn ed. Baltimore and London: Johns Hopkins University Press for the International Food Policy Research Institute, 1989. Topics discussed include: supply, demand, health issues, medical programs, poverty, rural areas, economics.

2574. Pellerin, P. et al. "1st report on 7 years' research on contamination levels in the environment and food supply from radioactive fallout on French territory." *Bull Fed Soc Gynecol Obstet Lang Fr* 19, no. 3 (June 1967): 357-82. Topics discussed include: climates, environment, exposure, France, milk, food, contamination.

2575. Penning de Vries, F. W. T. et al. "Simulated Yields of Wheat and Rice in Current Weather and When Ambient CO2 Has Doubled." *Climate and food security: Papers presented at the International Symposium on Climate Variability and Food Security in Developing Countries, 5 9 February 1987, New Delhi, India*, 347-57. P. A. Oram et al. Manila: International Rice Research Institute in collaboration with the American Association for the Advancement of Science, 1989. Topics discussed include: agriculture, supply, demand, economics.

2576. Pinstrup Andersen, P. et al. "The Impact of Drought and Technological Change in Rice Production on Intrayear Fluctuations in Food Consumption: The Case of North Arcot, India." *Seasonal variability in third world agriculture: The consequences for food security*, 264-84. ed David E. Sahn. Baltimore and London: Johns Hopkins University Press for the International Food Policy Research Institute, 1989. Topics discussed include: rural areas, economics, supply, demand, technology, change, innovation.

2577. Qudrat i Elahi, K. "Seasonality of Foodgrain Price and Procurement Programme in Bangladesh since Liberation; An Exploratory Study; Comments." *Bangladesh Development Studies* 20, no. 1 (March 1992): 117-19. Topics discussed include: economics, development, agriculture, natural resources, food, policy, supply, demand.

2578. Reardon, T. et al. "Coping with Household-Level Food Insecurity in Drought-Affected Areas of Burkina Faso." *World Development* 16, no. 9 (September 1988): 1065-74. Topics discussed include: rural areas, economics, agriculture farm, management.

2579. ---. "Seasonal Food Insecurity and Vulnerability in Drought-Affected Regions of Burkina Faso." *Seasonal variability in third world agriculture: The consequences for food security*, 118-36. ed. David E. Sahn. Baltimore and London: Johns Hopkins University Press for the International Food Policy Research Institute, 1989. Topics discussed include: supply, demand, rural areas, economics, health issues, medical programs, poverty.

2580. Richardson, G. et al. "Estimating fish consumption rates for Ontario Amerindians." *J Expo Anal Environ Epidemiol* 3, no. 1 (January 1993): 23-38. Topics discussed include: age effects, Ontario, seasonality, gender effects, fish, contamination, food, consumption behavior, Indians, North America.

2581. Sahn, D. E. "A Conceptual Framework for Examining the Seasonal Aspects of Household Food Security." *Seasonal variability in third world agriculture: The consequences for food security*, 3-16. ed David E. Sahn. Baltimore and London: Johns Hopkins University Press for the International Food Policy Research Institute, 1989. Topics discussed include: development, supply, demand, poverty, rural areas, economics.

2582. ---. "Seasonal fluctuations in factor and products markets : Is there a nutritional risk?" 30 pages. Pew, Cornell lecture series on food and nutrition policy, Ithaca, NY: Cornell Food and Nutrition Policy Program, Division of Nutritional Sciences, Cornell University, 1989. Topics discussed include: food, supply, seasonality, malnutrition, development.

2583. ---. *Seasonal variability in third world agriculture : The consequences for food security*. 366 pages. International Food Policy Research Institute. Baltimore: Johns Hopkins University Press, 1989. Topics discussed include: food, supply, development, seasonality, economics.

2584. Sahn, D. E. et al. "The Nature and Implications for Market Interventions of Seasonal Food Price Variability." *Seasonal variability in third world agriculture: The consequences for food security*, 179-95. ed. David E. Sahn. Baltimore and London: Johns Hopkins University Press for the International Food Policy Research Institute, 1989. Topics discussed include: markets, marketing, cooperatives, policy.

2585. ---. "Seasonal variability in Third World agriculture." *American Anthropologist* 93 (March 1991): xiv, 366 214-15.

2586. Sai, F. "Ecology and environment. 2. Nutrition in industrial populations in the tropics." *J Trop Med Hyg* 73, no. 11 (November 1970): 294-8. Topics discussed include: agriculture, proteins, food, consumption behavior, preservation, nutrition, growth, rural areas, unemployment, urbanization, development.

2587. Schneider, S. H. "Climate and Food: Signs of Hope, Despair, and Opportunity." *The Cassandra Conference: Resources and the human predicament*, 17-51. eds Paul R. Ehrlich, and John P. Holdren. College Station: Texas A&M University Press, 1988. Topics discussed include: agriculture, natural resources.

2588. Seeber, E. "Health hazards of food consumption during the warmer season." *Ther Ggw* 110, no. 7 (July 1971): 992 passim. Topics discussed include: heat, salmonella, food, poisoning.

2589. Seshu, D. V. et al. "Effect of Weather and Climate on Production and Vulnerability of Rice." *Climate and food security: Papers presented at the International Symposium on Climate Variability and Food Security in Developing Countries, 5 9 February 1987, New Delhi, India*, 93-113. P. A. Oram et al. Manila: International Rice Research Institute in collaboration with the American Association for the Advancement of Science, 1989. Topics discussed include: agriculture, supply, demand, economics.

2590. Sinha, S. K. et al. "Food Security in the Changing Global Climate." *Climate and food security: Papers presented at the International Symposium on Climate Variability and Food Security in Developing Countries, 5 9 February 1987, New Delhi, India*, 579-97. P. A. Oram et al. Manila: International Rice Research Institute in collaboration with the American Association for the Advancement of Science, 1989. Topics discussed include: agriculture, supply, demand, economics, agriculture.

2591. Sivakumar, M. V. K. et al. "Climate and Production in the Semi-arid and the Humid Tropics." *Global aspects of food production Natural Resources and the Environment Series*, 129-69. eds. M. S. Swaminathan, and S. K. Sinha. London; Riverton, NJ and Dehra Dun, India: Tycooly International,: 1986. Topics discussed include: agriculture, supply, demand, economics.

2592. Slater, L. E. *Food & Climate Review 1980-81*. Lanham, MD: University Press of America, 1981. Topics discussed include: food, nutrition, agriculture, tropics.

2593. Swaminathan, M. S. et al. "Global aspects of food production." *Natural Resources and the Environment Series, vol 20*, 449 pages London; Riverton, NJ and Dehra Dun, India: Tycooly International: 1986. Topics discussed include: agriculture, supply, demand, economics.

2594. Szlyk, P. C. et al. "Patterns of human drinking: Effects of exercise, water temperature, and food consumption." *Aviation, Space, and Environmental Medicine* 61, no. 1 (January 1990): 43-48. Topics discussed include: exercise, food consumption, drinking behavior, water consumption.

2595. Takahashi, K. *Climatic Change & Food Production*. New York: Columbia University Press, 1978. Topics discussed include: climates, food, agriculture.

2596. Tanner, M. et al. "Food consumption patterns in a rural Tanzanian community (Kikwawila village, Kilombero District, Morogoro Region) during lean and post harvest season." *Acta Trop Basel* 44, no. 2

(June 1987): 229-44. Topics discussed include: age effects, diet surveys, rural areas, population, gender effects, Tanzania, eating, food, supply.

2597. Teklu, T. et al. "Drought and famine relationships in Sudan: Policy implications." *IFPRI Research Report,* 135 Washington, DC: International Food Policy Research Institute, 1991. Topics discussed include: agriculture.

2598. Thomas, R. et al. "Household coping strategies and contradictions in response to seasonal food shortage." *Eur J Clin Nutr* 44 Supp, no. 1 (1990): 103-11. Topics discussed include: adaptation, gender effects, social status, socioeconomics, food, supply.

2599. Ulbricht, G. et al. "The informativeness of 2 month surveys of food consumption in private households." *Nahrung* 28, no. 2 (1984): 185-90. Topics discussed include: carbohydrates, fruit, seasonality, vegetables, diet, nutrition.

2600. Van, S. W. et al. "Seasonal variation in food intake, pattern of physical activity and change in body weight in a group of young adult Dutch women consuming self selected diets." *Int J Obes* 10, no. 2 (1986): 133-45. Topics discussed include: caloric consumption, fats, fiber, Netherlands, body weight, diet, eating, physical labor.

2601. Varela, G. et al. "Changes in the quantitative and qualitative composition of fat from fish, due to seasonality and industrial and culinary processing." *Bibl Nutr Dieta* no. 46 (1990): 104-9. Topics discussed include: food, preservation, fatty acids, fish, food, handling.

2602. Wandel, M. et al. "Food availability and nutrition in a seasonal perspective: a study from the Rukwa region in Tanzania." *Human Ecology* 20 (March 1992): 89-107. Topics discussed include: food, supply, economics, children, corn.

2603. Wittwer, S. *Climate and Food.* Chelsea, MI: Lewis Publishers, 1994. Topics discussed include: food.

2604. Young, A. "Climatic-change and food production - international symposium on recent climatic-change and food production, held October 4-8, 1976, Tsukuba and Tokyo - Takahashi, K., Yoshino, M. M." *Geographical Journal* 145 (July 1979): 332.

Health Care

2605. Aizenshtat, B. et al. "Thermal condition of mechanics in the cotton growing industry in the hot season of the year." *Gig Tr Prof Zabol* 17, no. 10 (October 1973): 8-11. Topics discussed include: environment, health issues, work-related medicine.

2606. Avery, K. "The dental health of children of migrant and seasonal agricultural workers." *J Sch Health* 45, no. 1 (January 1975): 24-6. Topics discussed include: Florida, gingivitis, epidemiology, oral health, transients.

2607. Bruce, N. et al. "The contribution of environmental temperature and humidity to geographic variations in blood pressure." *J Hypertens* 9, no. 9 (September 1991): 851-8. Topics discussed include: body physiology, United Kingdom, humidity, blood pressure, climate.

2608. Buettner, K. J. "Our Present Knowledge on Correlations between Weather Changes, Atmospheric and Air Electric Space Changes, and Human Health and Behavior." *Federation Proc* 16 (1957): 631-637. Topics discussed include: climatic effects, environment, health issues, behavior, biology.

2609. Chambers, R. et al. "Seasonal dimensions to rural porverty: analysis and practical implications." *J Trop Med Hyg* 82, no. 8 (August 1979): 156-72. Topics discussed include: agriculture, Bangladesh, Gambia, India, Kenya, Mali, Nigeria, Tanzania, tropical climates, poverty, health issues, rural areas.

2610. Chambers, R. "Health, Agriculture, and Rural Poverty: Why Seasons Matter." *Journal of Development Studies* 18, no. 2 (January 1982): 217-238. Topics discussed include: tropics, environment, health issues, wet season.

2611. Christensen, R. "Anchorage, Alaska and Honolulu, Hawaii: seasonal differences in demand for counseling services." *Alaska Med* 25, no. 3 (July 1983): 79-80. Topics discussed include: Alaska, Hawaii, climatic effects, mental health, services, demand.

2612. ---. "Seasonal variations in demand for mental health services." *Ariz Med* 40, no. 4 (April 1983): 252. Topics discussed include: Arizona, mental health, services.

2613. Crow, R. "Health care in the Turkana desert." *Nurs Mirror Midwives J* 144, no. 13 (31 March 1977): 62-5. Topics discussed include: Kenya, rural areas, health issues, desert climates, services, supply, distribution.

2614. Decker, S. et al. "Functional health pattern assessment: a seasonal migrant farmworker community." *J Community Health Nurs* 7, no. 3 (1990): 141-51. Topics discussed include: mass screening, Oregon, health care, agriculture, nursing, health services, demand, transients.

2615. Diehl, A. et al. "Use of calendar and weather data to predict walk in attendance." *South Med J* 74, no. 6 (June 1981): 709-12. Topics discussed include: forecasting, timing, health centers, demand, emergency service, hospital demand, seasonality.

2616. Donle, W. "Dependence of Outbreaks of Influenza on Season and Weather." *Infection* 3 (1975): 23-27. Topics discussed include: health issues, illness.

2617. Evarett, J. T. "Studies in Relation to the Production of Pain by Weather." *Chicago Med, J and Exam* 38 (1879): 253-260. Topics discussed include: health issues, illness, physiology.

2618. Fouts, J. R. et al. "Human health effects and global climate change." *Marine Technology Society Journal* 25 (December 1991): 38-44. Topics discussed include: air pollution, physiology, diseases, ozone layer depletion.

2619. Grima, F. "Industrial city in the desert." *J Soc Occup Med* 36, no. 2 (June 1986): 45-50. Topics discussed include: health services, Saudi Arabia, desert climates, industry, work-related diseases.

2620. Guntheroth, W. et al. "A seasonal association between SIDS deaths and kindergarten absences." *Public Health Rep* 107, no. 3 (May 1992): 319-23. Topics discussed include: Oregon, epidemiology, absenteeism, schools, nursery.

2621. Huntington, E. "Influenza and the Weather in the United States in 1919." *Sci Monthly* 17 (1923): 462-472. Topics discussed include: illness, health.

2622. Hygge, S. "Heat and performance." *Handbook of human performance, Vol 1: The physical environment; Vol 2: Health and performance; Vol 3: State and trait*, pp. 79-104. Academic Press, Inc, London, England. eds Dylan M. Jones, Andrew P. Smith. 1992. Topics discussed include: performance.

2623. Ikeda, M. et al. "A field survey on the health status of workers in dye producing factories." *Int Arch Occup Environ Health* 39, no. 4 (15 September 1977): 219-35. Topics discussed include: Japan, work-related diseases, seasonality, chemical industry, dyes, health.

2624. Jonkers, G. H. "Seasonal Variations in the Occurence of Retinal Detachments (From the Nederlandsch Gasthuis voor OOglijders, Utrecht)." *Ophthalmologica, Z Augenheilk* 115, no. 5 (1948): 308-315. Topics discussed include: illness, health.

2625. Kao, L. t. c. *Health through the four seasons.* 2 v. Ya-chou min su, she hui sheng huo chuan k'an; Asian folklore & social life monographs; v. 153, Chung-hua min kuo T'ai-pei shih shih-lin ch'u: Chung-kuo min su hsueh hui : Tung fang wen hua shu chu, 1985. Topics discussed include: diseases, seasonality, medicine, China.

2626. Lathrop, B. M. "Seasonal Variations in Incidence of Dental Caries: A Preliminary Report." *Am J Mental Deficiency* 46 (1941): 33-36. Topics discussed include: illness, health issues, dental care.

2627. Lutalo, S. et al. "Complications of seasonal adult malaria at a central hospital." *Cent Afr J Med* 36, no. 11 (November 1990): 268-73. Topics discussed include: hospitals, quality, health care, Zimbabwe.

2628. Martin, H. W. et al. "Texas snowbirds: Seasonal migrants to the Rio Grande Valley." *Research on Aging* 9, no. 1 (March 1987): 134-147. Topics discussed include: sociodemographics, lifestyle, health issues.

2629. Miller, M. "Industrialization, ecology and health in the tropics." *J Occup Med* 15, no. 4 (April 1973): 365-8. Topics discussed include: Africa, agriculture, Asia, culture, pollution, malaria, work-related medicine, South America, urbanization, water supply, developing countries.

2630. Mills, C. A. et al. "Climatic Basis for Susceptibility to Heat Stroke or Exhaustion." *American Journal of Hygiene* 17 (1933): 686-696. Topics discussed include: behavior, biology, weather, health issues, temperature.

2631. Nikolaeva, E. "Seasonal changes in the bactericidal function of the skin in persons working under various conditions of industrial microclimate." *Gig Sanit* 36, no. 11 (November 1971): 25-8. Topics discussed include: environment, exposure, temperature, USSR, bacterial infections, health issues, work-related medicine.

2632. Nimmo, A. et al. "A survey of in patients from the oil related industries in north east Scotland." *Health Bull Edinb* 37, no. 1 (January 1979): 20-3. Topics discussed include: patient admissions, patient discharges, Scotland, seasonality, hospitals, work-related diseases, petroleum.

2633. Palerm, J. "A season in the life of a migrant farm worker in California." *West J Med* 157, no. 3 (September 1992): 362-6. Topics discussed include: attitudes, health, California, emigration, immigration, health services, supply, agriculture, transients, migrants.

2634. Petersen, W. F. *The Patient and the Weather.* Ann Arbor, MI: Edwards Bros, 1936. Topics discussed include: illness, health.

2635. Pocock, S. J. "Harmonic Analysis Applied to Seasonal Variations in Sickness Absence." *Appl Stat* 23 (1974): 103-120. Topics discussed include: health issues, illness.

2636. Radsma, W. "The Influence of a Temporary Sojourn in a Cool Environment on Various Vegetative Functions in Inhabitants of the Tropics." *Acta Physiol et Pharmacol Neerl* I (1950): 112-169. Topics discussed include: environment, temperature, physiology, climatic effects, biology.

2637. Ricketts, T. et al. "An evaluation of subsidized rural primary care programs: II. The environmental contexts." *Am J Public Health* 73, no. 4 (April 1983): 406-13. Topics discussed include: health, services, economics, ethnic groups, labor, supply, distribution, rural areas.

2638. Russell, C. S. *Drought & Water Supply: Implications of the Massachusetts Experience for Municipal Planning.* Washington, D.C.: Resources for the Future, 1970. Topics discussed include: policy, public health.

2639. Sala, C. "Solvents and climate in the viscous textile industry." *Med Lav* 66, no. 4 (July 1975): 348-56. Topics discussed include: air pollution, environment, health issues, Italy, textile industry.

2640. Schenker, M. et al. "Occupational health among migrant and seasonal farmworkers: the specific case of dermatitis." *Am J Ind Med* 18, no. 3 (1990): 345-51. Topics discussed include: migrants, agricultural labor, diseases, dermatitis, epidemiology.

2641. Shulman, J. et al. "The prevalence of recurrent herpes labialis during an army hot weather exercise." *J Public Health Dent* 52, no. 4 (June 1992): 198-203. Topics discussed include: California, diseases, pathology, timing, desert climates, military personnel.

2642. Smiley, D. F. "Seasonal Factors in the Incidence of the Acute Respiratory Infections." *Amp J Hyg* 6 (1926): 621. Topics discussed include: illness, health.

2643. Smith, E. F. "Climate and Health in the Royal Navy." *Brit J Ind Med* 15 (1958): 197-203. Topics discussed include: health issues, illness.

2644. Turkat, D. et al. "Seasonal demands in community mental health." *Psychological Reports* 43, no. 2 (October 1978): 486. Topics discussed include: seasonality, service consumption, mental health.

2645. Walter, S. D. "A Study of Seasonal Incidence of Congenital Malformation in the United States." *Br J Prev Soc Med* 24 (1970): 24-32. Topics discussed include: demography, health issues, illness.

2646. Weihe, W. H. "Climate Change and Human Health." *Climate and development: Climatic change and variability and the resulting social, economic and technological implications*, 409-32. eds H. J. Karpe, D. Otten, and S. C. Trinidade. New York; Berlin; London and Tokyo: Springer in cooperation with the United Nations, 1990. Topics discussed include: resources, health issues, medical policy.

2647. Weiner, J. S. et al. "A Classification of Heat Illness. A Memorandum Prepared for the Climatic Physiology Committee of the Medical Research Council." *Brit Med J* I (1958): 1533-1535. Topics discussed include: temperature, environment, health issues, illness, climate.

2648. Wernsdorfer, W. "Quinine in health care in the tropics." *Acta Leiden* 55 (1987): 197-208. Topics discussed include: history, medicine, falciparum, tropical climates, malaria.

2649. Wheatley, B. et al. "Methylmercury in the Canadian Arctic environment past and present natural or industrial?" *Arctic Med Res* 47 Supp, no. 1 (1988): 163-7. Topics discussed include: arctic regions, environment, Ontario, cold climates, health.

Household Demand, General

2650. Beck, R. J. "Economic growth, low prices to hike U.S. winter fuels demand." *Oil & Gas Journal* 86 (7 November 1988): 14-16. Topics discussed include: fuel oil, supply, demand, gas.

2651. Beck, R. J. et al. "Cold weather, economic growth to hike U.S. winter fuel demand." *Oil & Gas Journal* 81 (5 December 1983): 57-60. Topics discussed include: petroleum refineries, fuel oil, supply, demand.

2652. ---. "U.S. winter fuel demand to top volume of last two seasons." *Oil & Gas Journal* 85 (2 November 1987): 14-16. Topics discussed include: fuel oil, supply, demand.

2653. Birchenhall, C. R. et al. "Modelling Seasonality in Economic Variables, with Special Reference to Consumption." paper presented at the Conference on Dynamic Modelling and Seasonality, Manchester, 13 July 1987. 1987. Topics discussed include: economy, consumption.

2654. Brown, K. et al. "Milk consumption and hydration status of exclusively breast fed infants in a warm climate." *J Pediatr* 108, no. 5 Pt 1 (May 1986): 677-80. Topics discussed include: Peru, urine analysis, breast feeding, tropical water.

2655. Dahl, C. et al. "Analysing Gasoline Demand Elasticities: A Survey." *Energy Economics* 13, no. 3 (July 1991): 203-10. Topics discussed include: demand, supply, mining, refining, consumer behavior, extractive industries, oil, gas, consumer expenditures, cycles, consumption.

2656. Engle, R. F. et al. "Seasonal cointegration: the Japanese consumption function." *J Econometrics* 55, no. 1-2 (1993): 275-303. Topics discussed include: numerical statistics.

2657. Felton, J. R. "Regulatory Rigidity and Seasonal Variations in Demand." *Regulation and deregulation of the motor carrier industry*, 82-98. eds John Richard Felton, and Dale G. Anderson. Ames: Iowa State University Press, 1989. Topics discussed include: transportation economics, regulatory policy.

2658. Ferson, W. E. et al. "Seasonality and Consumption-Based Asset Pricing." *Journal of Finance* 47, no. 2 (June 1992): 511-52. Topics discussed include: capital and asset pricing, financial markets, regulatory policies, portfolio composition.

2659. Heien, D. "Seasonality in U.S. Consumer Demand." *Journal of Business and Economic Statistics* 1, no. 4 (October 1983): 280-84. Topics discussed include: consumer expenditure cycles, consumption.

2660. Herrmann, M. et al. "Applying Almon-Type Polynomials in Modelling Seasonality of the Japanese Demand for Salmon." *Marine Resource Economics* 7, no. 1 (March 1992): 3-13. Topics discussed include: natural resources, conservation, environmental management, fishery industry, consumption.

2661. Ingemann, H. T. et al. "Seasonal variation of maximal oxygen consumption rate in humans." *Eur J Appl Physiol* 49, no. 2 (1982): 151-7. Topics discussed include: adaptation, oxygen consumption.

2662. Jain, R. K. "The Seasonal Adjustment Procedures for the Consumer Price Indexes: Some Empirical Results." *Journal of Business and Economic Statistics* 7, no. 4 (October 1989): 461-69.

2663. Katsaitis, O. "Seasonality and the Estimation of Dynamic Demand Systems: Some Exploratory Results." *Economic Letters* 26, no. 2 (1988): 129-32. Topics discussed include: consumer expenditure cycles, consumption.

2664. Kelly, J. P. et al. "Increasing the Efficiency of Forecasting Seasonal Demand for Individual Products." *Marketing channels: Relationships and performance Advances in Retailing Series*, 183-95. Luca Pellegrini and Srinivas K.Reddy eds., Lexington, Mass., and Toronto,: Heath, Lexington Books, 1986. Topics discussed include: forecasting, markets, industrial strategy, marketing, advertising, distribution, retailing.

2665. Kemmerer, E. W. et al. *Seasonal Variations in the Relative Demand for Money & Capital in the United States*. Salem, NH: Ayer Company Publishers, 1981. Topics discussed include: supply, demand, interest rates, money.

2666. Lawrence, A. G. et al. *Electric Appliance Energy Consumption Survey: Analysis and Revision of the MRI Data Final Report*, Performer: Cambridge Systematics, Inc., Berkeley, CA., September 1982. Topics discussed include: consumption, electric appliances, residential sector, demand, electric power, refrigerators, television, water heaters, microwave ovens, seasonality, food.

2667. Ogg, C. W. et al. "Western Irrigation Response to Pumping Costs: A Water Demand Analysis Using Climatic Regions." *Water Resources Research* 25, no. 5 (May 1989): 767-73. Topics discussed include: natural resources.

2668. Osborn, D. R. et al. "Seasonality and the Order of Integration for Consumption." *Oxford Bulletin of Economics and Statistics* 50, no. 4 (November 1988): 361-77. Topics discussed include: econometrics, time series analysis, spectral analysis models.

2669. Petty, M. T. "Weather and Consumer Sales." *Bulletin American Meteorological Society* 44 (1963): 68-71. Topics discussed include: climate, economy, society, behaviour.

2670. Planting, R. "Prices and Seasonality-The Case of Motor Gasoline." *Business Economics* 17, no. 1 (January 1982): 56-58. Topics discussed include: prices, consumer expenditure cycles, consumption.

2671. Russell, J. D. "Seasonal and time-of-day pricing." *American Water Works Association Journal* 76 (September 1984): 63-5. Topics discussed include: water prices, water consumption.

2672. Taylor, L. D. "Time-of-Day and Seasonal Demand for Electric Power." *Growth and Change* 10, no. 1 (January 1979): 105-10. Topics discussed include: forecasting, regulation, public utilities, consumer expenditure cycles, consumption.

2673. Waldrop, J. "The seasons of business." *American Demographics* 14 (May 1992): 40-4. Topics discussed include: consumption, economics, seasonality, cycles.

2674. Waldrop, J. et al. *The seasons of business : the marketer's guide to consumer behavior*. 269 pages. Ithaca, N.Y.: American Demographics, 1992. Topics discussed include: consumer behavior, seasonal cycles, economics.

Manufacturing, General

2675. Barnett, G. A. "Seasonality in Television Viewing: A Mathematical Model of Cultural Processes." *Communication Research* 18, no. 6 (December 1991): 755-772. Topics discussed include: programming, production, sociocultural factors, television ratings.

2676. Barsky, R. B. "The Seasonal Cycle and the Business Cycle." *Journal of Political Economy* 97, no. 3 (June 1989): 503-34. Topics discussed include: economic cycles.

2677. Beaulieu, J. J. "The seasonal cycle in U.S. manufacturing." 35 pages. National Bureau of Economic Research. NBER working paper series; working paper no. 3450 4003562772, Cambridge, MA: National Bureau of Economic Research, 1990. Topics discussed include: business cycles, manufacturing.

2678. ---. "Why do countries and industries with large seasonal cycles also have large business cycles?" 44 pages. National Bureau of Economic Research. NBER working papers series; working paper no. 3635, Cambridge, MA: National Bureau of Economic Research, 1991. Topics discussed include: business cycles, economics.

2679. Blanco, M. "Desestacionalizacion del producto bruto manufacturero." 41 pages. Banco Central de la Republica Argentina. Gerencia de Investigaciones y Estadisticas Economicas. Serie de trabajos metodologicos y sectoriales, Buenos Aires: Banco Central de la Republica Argentina, 1987. Topics discussed include: gross national product, Argentina, econometrics, seasonality, economics, time series analysis.

2680. Britton, A. "Seasonal Patterns in the British Economy." *National Institute Economic Review* 0, no. 117 (August 1986): 33-42. Topics discussed include: Europe, income.

2681. Coelho, C. H. M. "Trading Days, Seasonal Unit Root, and Variance Change." *International Journal of Forecasting* 8, no. 1 (June 1992): 61-67. Topics discussed include: econometrics, methodology, manufacturing.

2682. Dufour, J. M. "The Importance of Seasonality in Inventory Models: Discussion." *Journal of Econometrics* 55, no. 1-2 (January 1993): 129-33. Topics discussed include: intertemporal choice by firms, growth, investment, finance, microeconomics, production theory.

2683. Froggatt, P. "Short term absence from industry. 3. The interfence of 'proneness' and a search for causes." *Br J Ind Med* 27, no. 4 (October 1970): 297-312. Topics discussed include: age effects, probability, seasonality, absenteeism, psychology, industry.

2684. Ghali, M. A. "Seasonality, aggregation and the testing of the production smoothing hypothesis." *The American Economic Review* 77 (June 1987): 464-9. Topics discussed include: seasonality, economics, production, inventories, cement industry, production smoothing.

2685. Gunn, P. A. "Productive Cycles and the Season of Marriage: A Critical Test." *Journal of Interdisciplinary History* 21, no. 2 (1990): 217-243. Topics discussed include: marriage timing, Australia, United Kingdom, nineteenth century, production cycles.

2686. Haas, J. "Industrial medicine and industrial physiology in the tropics." *Med Klin* 63, no. 25 (21 June 1968): 1001-4. Topics discussed include: adaptation, environment, exposure, humidity, perspiration, temperature, Venezuela, deafness, epidemiology, work-related diseases.

2687. Halkier, S. L. "The relationship between skin surface temperature, transepidermal water loss and electrical capacitance among workers in the fish processing industry: comparison with other occupations. A field study." *Contact Dermatitis* 24, no. 5 (May 1991): 345-55. Topics discussed include: physiology, work-related exposure, seasonality, fish products, occupations, skin physiology.

2688. Hall, A. "Induced Seasonality and Production-Smoothing Models of Inventory Behavior: Discussion." *Journal of Econometrics* 55, no. 1-2 (January 1993): 169-72. Topics discussed include: intertemporal choice of firms, growth, investment, finance, microeconomics, production theory.

2689. Hayes, J. "Emphysema in a non industrialized tropical island." *Thorax* 24, no. 5 (September 1969): 623-5. Topics discussed include: age effects, air pollution, ethnic groups, Jamaica, lung pathology, pulmonary emphysema, pathology, epidemiology, gender effects.

2690. Hickish, D. "Thermal Sensations of Workers in Light Industry in Southern England." *Journal pf Hygiene* 53 (1955): 112-123. Topics discussed include: temperature, behavior, biology.

2691. Hindy, K. T. "Monthly and seasonal trends of total suspended particulate matter and smoke concentration in industrial and residential areas in Cairo." *Atmospheric Environment Part B, Urban Atmosphere* 24B, no. 2 (1990): 343-53. Topics discussed include: air pollution, Cairo, Egypt, atmosphere, smoke.

2692. Hutchings, R. *Seasonal influences in Soviet industry.* 321 pages. Royal Institute of International Affairs. London, New York: Oxford University Press for the Royal Institute of International Affairs, 1971. Topics discussed include: seasonality, economics, Soviet Union, industry.

2693. Ion, N. N. "Efficacy of influenza vaccination in an industrial collective in the municipality of Bucharest during the 1985-1986 influenza season." *Rev Ig Bacteriol* 32, no. 1 (January 1987): 47-52. Topics discussed include: costs, benefits, risks, Romania, textile industry, vaccination economics, influenza prevention.

2694. Krane, S. D. "Induced Seasonality and Production-Smoothing Models of Inventory Behavior." *Journal of Econometrics* 55, no. 1-2 (January 1993): 135-68. Topics discussed include: intertemporal choice of firms, growth, investment, finance, microeconomics, production theory.

2695. Kuznets, S. S. *Seasonal variations in industry and trade.* 455 pages. National Bureau of Economic Research publications in reprint, New York: Arno Press, 1975. Topics discussed include: economics, industry.

2696. Legostaeva, E. "The hygienic characteristics of the manufacturing environment in modern lead production and of the concomitant production processes." *Gig Tr Prof Zabol no.* 12 (1989): 4-7. Topics discussed include: air pollutants, dust, Russia, climate, lead.

2697. McHugh, J. "Offshore seasonal gas supply." *Pipeline & Gas Journal* 217 (October 1990): 20-2+. Topics discussed include: offshore gas production, North Sea, natural gas.

2698. Mikhailuts, A. "Hygienic evaluation of the combined effects of ammonia and low temperatures on chemical industry workers." *Gig Tr Prof Zabol no.* 4 (1989): 31-4. Topics discussed include: Russia, seasonality, air pollutants, cold climates, work-related diseases.

2699. Miron, J. A. "Seasonality, Cost Shocks, and the Production Smoothing Models of Inventories." *Econometrica* 56, no. 4 (July 1988): 877-908. Topics discussed include: business investment, economic cycles, manufacturing.

2700. Monteith, J. L. "Solar Radiation and Productivity in Tropical Ecosystems." *Journal of Applied Ecology* 9 (1972): 747-766. Topics discussed include: environment, climate.

2701. Murota, T. "Heat Economy of the Water Planet Earth: Part II, Revision and Some New Results." *Hitotsubashi Journal of Economics* 26, no. 2 (December 1985): 181-85. Topics discussed include: economic history, Europe, product prices, markets, Asia, natural resources.

2702. Nerlove, M. "The Importance of Seasonality in Inventory Models: Evidence from Business Survey Data." *Journal of Econometrics* 55, no. 1-2 (January 1993): 105-28. Topics discussed include: intertemporal choice of firms, growth, finance, investment, microeconomics, production.

2703. Russo, J. A. Jr. "The economic Impact of Weather on the Constuction Industry of the United States." *Bulletin of the American Meteorological Society* 47 (1966): 967-972. Topics discussed include: economy, climate.

2704. *Seasonal fluctuations in Maryland business; war and postwar adjustments, industry and area differences.* 10 pages. University of Maryland. Bureau of Business and Economic Research Studies, v. 3. no. 1, College Park: 1949. Topics discussed include: seasonal variations, industry.

2705. Wilson, A. "The Impact of Climate on Industrial Growth: A Case Study." *Human Dimensions of Weather Modification*, editor W. R. D. Sewell. Research Paper, 105. Chicago: University of Chicago, Departement of Geography, 1966. Topics discussed include: economy, society.

Mining

2706. Amundsen, E. S. "Seasonal Fluctuations of Demand and Optimal Inventories of a Non-renewable Resource Such as Natural Gas." *Resources and Energy* 13, no. 3 (September 1991): 285-306. Topics discussed include: demand, supply, mining, extraction, refining, hydrocarbon fuels, conservation, pollution, natural resources.

2707. Andrlikova, J. et al. "The levels of immunoglobulins (IgG, IgA, IgM) in the four seasons in uranium miners." *Czech Med* 8, no. 2 (1985): 112-8.

2708. Barkhudarian, M. et al. "Physiological and hygienic substantiation of the resting time for transportation workers in the mining industry in relation to the workload and meteorological conditions." *Gig Tr Prof Zabol no.* 6 (1992): 7-10. Topics discussed include: microclimates, seasonality, work capacity, thermoregulation, physiology, rest, transportation.

2709. Berezovskii, V. et al. "Effect of artificial mountain climate on respiration and circulation in coal miners with chronic dust bronchitis." *Fiziol Zh* 31, no. 5 (September 1985): 619-23. Topics discussed include: altitude, bronchitis therapy, mining, dust, work-related diseases, treatment.

2710. Conwell, C. N. "Mercury mining revival in Alaska despite poor access and climate." *World Min (US Ed)* 7, no. 8 (1971): 38-40. Topics discussed include: economics, geology, mercury, history, mineral exploration.

2711. Cosio, G. "Mining work in high altitude." *Arch Environ Health* 19, no. 4 (October 1969): 540-7. Topics discussed include: adaptation, anoxia complications, cardiac volume, electrocardiography, pulmonary radiography, work-related diseases, etiology, Peru.

2712. Cross, R. et al. "A simple field test for assessing salt balance in heat stressed miners." *J Occup Med* 31, no. 8 (August 1989): 668-73. Topics discussed include: work-related diseases, stress, diagnosis, mining, work-related diseases, metabolism.

2713. Dedenko, I. "Study of human heat exchange in the Far North under different industrial microclimate conditions." *Gig Sanit no.* 3 (March 1979): 75-8. Topics discussed include: arctic regions, drinking, mining, temperature, USSR, thermoregulation, climatic effects, microclimates, work-related medicine.

2714. Dedenko, I. et al. "Heat exchange characteristics of miners during open pit mining in the Far North." *Gig Tr Prof Zabol no.* 7 (July 1980): 18-21. Topics discussed include: air movements, energy metabolism, temperature, USSR, thermoregulation, cold climates, mining.

2715. Deinega, V. et al. "Effect of a mountain climate on respiratory hemodynamic function in miners with the initial stage of pneumoconiosis." *Fiziol Zh* 28, no. 5 (September 1982): 542-7. Topics discussed include: adaptation, anthracosilicosis therapy, hemodynamics, pulmonary gas exchange, timing, Ukraine, altitude, respiration.

2716. ---. "Assessment of peripheral thermoregulation in vibration disease patients by dynamic heat calorimetry." *Gig Tr Prof Zabol no.* 3 (March 1985): 29-31. Topics discussed include: coal mining, blood supply, skin temperature, Ukraine, thermoregulation, work-related diagnosis, vibration.

2717. Ellis, F. "Heat illness. I. Epidemiology." *Trans R Soc Trop Med Hyg* 70, no. 5-6 (1977): 402-11. Topics discussed include: acclimation, climatic effects, United Kingdom, military medicine, mining, work-related diseases, ships, South Africa, exhaustion.

2718. Fabry, H. "Mine climate and work disability caused by dermatoses in the Ruhr mining district." *Berufsdermatosen* 15, no. 5 (October 1967): 270-9. Topics discussed include: dermatitis, work-related, etiology, epidemiology, disability evaluation, environment, exposure, Germany, socioeconomics, coal mining.

2719. Gaebelein, H. et al. "Climatic stress in selected daily activities in potash mining in East Germany." *Z Gesamte Hyg* 29, no. 3 (March 1983): 156-8. Topics discussed include: Germany, heart rate, thermoregulation, heat, hydroxides, mining, potassium.

2720. Gardner, J. S. "Snow as a resource and hazard in early-twentieth-century mining, Selkirk Mountains, British Columbia." *The Canadian Geographer* 30 (September 1986): 217-28. Topics discussed include: hazards, geography, environment, mineral resources, Canada.

2721. Gedymin, M. et al. "Cold vasodilation in persons exposed to the chronic action of low positive temperatures." *Fiziol Cheloveka* 10, no. 1 (January 1984): 52-8. Topics discussed include: thermoregulation, climatic effects, environment, exposure, mining, skin temperature, timing, USSR.

2722. German, G. "Influence of industrial occupational, climatic and geographic factors on the health of miners in the Northeastern part of the country." *Vestn Akad Med Nauk USSR* no. 9 (1989): 48-51. Topics discussed include: acclimation, geography, diseases, etiology, occupations, Siberia, mining.

2723. Goedecke, M. et al. "Unfavorable mine climates and their effects on miners." *Z Gesamte Hyg* 25, no. 8 (August 1979): 575-81. Topics discussed include: environment, exposure, humidity, temperature, microclimates, mining, work-related medicine.

2724. Gottlieb, J. "Episodes of illness and medical service in a geographically isolated mine village in Greenland." *Arctic Med Res* 49, no. 3 (July 1990): 128-31. Topics discussed include: occupations, cold climates, mining, morbidity, work, health services, consumption.

2725. Haas, J. "Studies on human physical capability and gross energy transfer during industrial and traditional work in tropical climate." *Z Tropenmed Parasitol* 21, no. 3 (September 1970): 267-306. Topics discussed include: agriculture, psychology, Europe, physical labor, heat, intestinal diseases, parasitic problems, Liberia, mining, Negroid race.

2726. Hallmeyer, A. et al. "Conclusions from field studies of stress on miners caused by dry hot climate conditions." *Z Gesamte Hyg* 36, no. 6 (June 1990): 302-4. Topics discussed include: Germany, risk, heat exhaustion, mining, work-related diseases, work capacity.

2727. Harada, N. "Esthesiometry, nail compression and other function tests used in Japan for evaluating the hand arm vibration syndrome." *Scand J Work Environ Health* 13, no. 4 (August 1987): 330-3. Topics discussed include: aging, arm physiopathology, cold climates, mining, work-related diseases, diagnosis, seasonality, skin temperature.

2728. Harshbarger, J. W. "Source, movement and development of water in arid regions, in mining and ecology in the arid environment - Conf., Tucson, Ariz., 1970, proc. (W. C. Peters, editor)." *Tucson, arizona, university of arizona, college of mines, p 122 134, illus (19701* (1970): 122-134. Topics discussed include: arid lands, ground water, hydrogeology, resource development.

2729. Ivanov, A. et al. "Sanatorium climate therapy for miners with pneumoconiosis as a method of preventing progression of the disease." *Vopr Kurortol Fizioter Lech Fiz Kult* 37, no. 2 (March 1972): 129-30. Topics discussed include: mining, anthracosilicosis treatment.

2730. Ivanov, I. et al. "Physiologo hygienic aspects of artificial heat adaptation (review of the literature)." *Voen Med Zh no.* 4 (April 1983): 46-8. Topics discussed include: acclimation, diet, drinking, physical labor, military medicine, mining, potassium, tropical climates, adaptation.

2731. Joutz, F. et al. "Using Stochastic Simulation to Test the Effect of Seasonal Adjustment on Forecast Standard Errors of Motor Gasoline Demand." *International Journal of Forecasting* 8, no. 2 (October 1992): 219-31. Topics discussed include: resources, conservation, demand, supply, mining, extraction, refining, econometrics, coal.

2732. Kharaberiush, V. et al. "Incidence of proctologic diseases with temporary disability." *Sov Zdravookhr* no. 4 (1986): 39-41. Topics discussed include: seasonality, Ukraine, coal mining, disabilities, epidemiology.

2733. Knapik, Z. et al. "Impact of high temperature and relative humidity on thermoregulation in miners subjected to controlled physical loads." *Gig Tr Prof Zabol no.* 12 (December 1987): 27-30. Topics discussed include: body regulation, physical labor, heat, humidity, mining.

2734. Krashchenko, V. "The seasonality of pyoderma in workers of the Don Basin coal mines." *Vestn Dermatol Venerol no.* 10 (1989): 55-8. Topics discussed include: absenteeism, meteorology, Ukraine, mining, dermatitis, work-related epidemiology.

2735. Lembke, D. et al. "The stress process of hot dry climatic conditions in a simulation chamber." *Z Gesamte Hyg* 36, no. 6 (June 1990): 294-302. Topics discussed include: thermoregulation, Germany, heart rate, heat exhaustion, mining, work-related diseases, potassium, work.

2736. Macfarlane, W. et al. "Content and turnover of water in Bantu miners acclimatizing to humid heat." *J Appl Physiol* 21, no. 3 (May 1966): 978-84. Topics discussed include: Africa, Angola, body weight, humidity, urine, acclimation, mining, tropical climates, water.

2737. Mambetaliev, B. "Use of some indices of external respiration in the evaluation of work adaptation of miners under conditions of the foothill climate of Kirghizia." *Zdravookhr Kirg* no. 5 (September 1974): 38-41. Topics discussed include: Kyrgyzstan, disabilities, mining, respiratory functions, work capacity.

2738. ---. "Age characteristics of the reactions of miners to production work at various altitudes in the Tien Shan." *Zdravookhr Kirg* no. 2 (March 1979): 18-21. Topics discussed include: age effects, climatic effects, energy metabolism, Kyrgyzstan, oxygen consumption, physical labor, mining.

2739. Mambetaliev, B. et al. "Utilization of the gas metabolism indicators of the efficiency of blood circulation in autonomic shifts in the body for the evaluation of adaptation to work in miners in different climatic zones." *Sov Zdravookhr Kirg* 1, no. 0 (January 1974): 26-30. Topics discussed include: exertion, Kazakhstan, oximetry, rest, Siberia, spirometry, adaptation, blood circulation, mining, oxygen consumption.

2740. Miura, T. "Human performance under hot thermal condition of work environment." *J Hum Ergol Tokyo* 7, no. 2 (December 1978): 157-68. Topics discussed include: acclimation, coal mining, humidity, Japan, temperature, physiology, heat, work.

2741. Odinaev, F. "The characteristics of the development and course of pneumoconiosis under the conditions of a mountain climate." *Gig Tr Prof Zabol no.* 7 (1992): 13-4. Topics discussed include: mining, epidemiology, silicosis, ethnology, etiology, Tajikistan epidemiology, altitude, pneumoconiosis.

2742. Peters, W. C. "Mining and ecology in the Arid Environment - conference: tucson, arizona, 1970, Proceedings." *College of Mining, Univeristy of Arizona* (1970): 229 pages. Topics discussed include: ecology, mining.

2743. Schutte, P. et al. "Heat acclimatization by a method utilizing microclimate cooling." *Aviat Space Environ Med* 49, no. 5 (May 1978): 710-4. Topics discussed include: body temperature, heart rate, South Africa, perspiration, climatic effects, microclimates, mining, work-related medicine.

2744. Senay, L. J. "Early response of plasma contents on exposure of working men to heat." *J Appl Physiol* 44, no. 2 (February 1978): 166-70. Topics discussed include: physiology, environment, physical labor, hematocrit, mining, timing, acclimation, blood, protein.

2745. Shearer, S. "Intestinal parasites in novice black miners. Prevalence and possible effect on heat acclimatization." *S Afr Med J* 73, no. 4 (20 February 1988): 214-7. Topics discussed include: gold, intestinal diseases, parasites, physiopathology, mining, physical fitness, South Africa, intestinal diseases, epidemiology.

2746. Strydom, N. "The prevention of heat stroke." *Proc Mine Med Off Assoc* 46, no. 396 (November 1966): 63-6. Topics discussed include: acclimation, exhaustion, mining, work-related diseases.

2747. Strydom, N. et al. "Effect of ascorbic acid on rate of heat acclimatization." *J Appl Physiol* 41, no. 2 (August 1976): 202-5. Topics discussed include: environment, heart rate, mining, perspiration, timing, drug effects, thermoregulation.

2748. Wenzel, H. "Occupational medicine significance of climatic chamber experiments for the assessment and improvement of unfavorable climatic and occupational conditions." *Z Gesamte Hyg* 34, no. 2 (February 1988): 97-104. Topics discussed include: risks, thermoregulation, mining, diseases.

2749. Williams, C. et al. "Rate of loss of acclimatization in summer and winter." *J Appl Physiol* 22, no. 1 (January 1967): 21-6. Topics discussed include: body temperature, physical labor, heart rate physiology, mining, perspiration, cold, heat, seasons.

2750. Wyndham, C. "Effect of acclimatization on the sweat rate rectal temperature relationship." *J Appl Physiol* 22, no. 1 (January 1967): 27-30. Topics discussed include: climates, hypothalamus, mining, South Africa, physiology, body regulation, physical labor, heat, rectum physiology.

2751. ---. "The effects of heat stress upon human productivity. pp. A491 7." *In: Prevision quantitative des effets physiologiques et psychologiques de l'environnement thermique chez l'homme Paris, Centre National de la Recherche Scientifique* (1975): A491 7. Topics discussed include: acclimation, air movements, body temperature, gold, heart rate, humidity, oxygen consumption, stress, mining.

2752. Wyndham, C. et al. "Seasonal variation in acute respiratory diseases and meningitis in Black miners living in hostels." *S Afr Med J* 54, no. 9 (26 August 1978): 353-8. Topics discussed include: bronchitis, crowding, mining, pneumonia epidemiology, respiratory tract transmission, South Africa, temperature, timing, meningitis.

2753. ---. "Limiting rates of work for acclimatization at high wet bulb temperatures." *J Appl Physiol* 35, no. 4 (October 1973): 454-8. Topics discussed include: body temperature, humidity, timing, physical labor, heat, mining, oxygen consumption.

2754. ---. "Relation between VO2 max and body temperature in hot humid air conditions." *J Appl Physiol* 29, no. 1 (July 1970): 45-50. Topics discussed include: body weight, environment, exposure, physical labor, metabolism, mining, acclimation, body regulation, heat, humidity, oxygen, respiration, work.

2755. Young, G. J. "Physiological Effects of High Temperature and Humidity." *Eng and Mining J* 88 (1909): 1155-11156. Topics discussed include: environment, biology.

2756. Yu, Y. "Investigation on the hygienic standard of environmental meteorology for underground work in coal mines." *Chung Hua Yu Fang I Hsueh Tsa Chih* 18, no. 1 (January 1984): 1-4. Topics discussed include: heart rate, diseases, etiology, perspiration, temperature, mining, work-related medicine.

Savings, Debt, Income & Finance

2757. Ball, R. "Daily Seasonals in Equity and Fixed-Interest Returns: Australian Evidence and Tests of Plausible Hypotheses." *Dimson, Elroy, ed Stock market anomalies*, 74-90. Cambridge, New York,

Melbourne: Cambridge University Press, 1988. Topics discussed include: capital and financial markets, regulatory policy.

2758. Birchenhall, C. R. et al. "A Seasonal Model of Consumption." *Economic Journal* 99, no. 397 (September 1989): 837-43. Topics discussed include: consumer behavior, economics, living standards, composition of expenditures, consumption, savings.

2759. Bo, T. "Seasonality in Risk Premia and Tests of APT--Estimation Using Kalman Filtering and Evaluation Using Multi-valued Logic." Ph.D. diss., SUNY, Binghamton, 1991. Topics discussed include: capital and financial markets.

2760. Butter, F. A. G. "Seasonal Adjustment of the Causes of Changes in the Domestic Money Supply." *De Nederlandsche Bank NV, Quarterly Statistics* no. 3 (1980): 98-121. Topics discussed include: economy, finance.

2761. Calvet, A. L. "Risk and Return on Canadian Capital Markets: Seasonality and Size Effect." *Finance* 10, no. 1 (June 1989): 21-39. Topics discussed include: capital and financial markets, regulatory policy.

2762. Chan, K. C. "Can Tax-Loss Selling Explain the January Seasonal in Stock Returns?" *Journal of Finance* 41, no. 5 (December 1986): 1115-28. Topics discussed include: capital and financial markets, regulatory policy.

2763. Chang, E. C. "A Fundamental Study of the Seasonal Risk-Return Relationship: A Note." *Journal of Finance* 43, no. 4 (September 1988): 1035-39. Topics discussed include: capital and financial markets, regulatory policy.

2764. Cho, D. C. "The Seasonal Stability of the Factor Structure of Stock Returns." *Journal of Finance* 42, no. 5 (December 1987): 1195-1211. Topics discussed include: capital and financial markets, regulatory policy.

2765. Clark, R. A. "Seasonalities in NYSE Bid-Ask Spreads and Stock Returns in January." *Journal of Finance* 47, no. 5 (December 1992): 1999-2014. Topics discussed include: asset prices, capital and financial markets, regulatory policy, portfolio composition.

2766. Corhay, A. "Seasonality in the Risk-Return Relationship: Some International Evidence." *Journal of Finance* 42, no. 1 (March 1987): 49-68. Topics discussed include: capital and financial markets, regulatory policy.

2767. Cornioley, C. "CAPM, Risk Premium Seasonality and Size Anomaly: The Swiss Case." *Finance* 12, no. 1 (June 1991): 23-44. Topics discussed include: asset prices, capital and financial markets, regulatory policy, portfolio composition.

2768. De, L. F. "The measurement and significance of the cyclically adjusted federal budget and debt." *Journal of Money, Credit and Banking* 17 (May 1985): 232-42. Topics discussed include: budget, debts, seasonality, economics, gross national product.

2769. De Bondt, W. F. M. "Further Evidence on Investor Overreaction and Stock Market Seasonality." *Journal of Finance* 42, no. 3 (July 1987): 557-81. Topics discussed include: capital and financial markets, regulatory policy.

2770. Dickey, D. A. "Seasonal Unit Roots in Aggregate U.S. Data: Discussion." *Journal of Econometrics* 55, no. 1-2 (January 1993): 329-31. Topics discussed include: wealth, income.

2771. Dickinson, A. "Seasonality in the Option Market." *Financial Review* 24, no. 4 (November 1989): 529-40. Topics discussed include: capital and financial markets, regulatory policy.

2772. Flannery, M. J. "From T-Bills to Common Stocks: Investigating the Generality of Intra-Week Return Seasonality." *Journal of Finance* 43, no. 2 (June 1988): 431-50. Topics discussed include: capital and financial markets, regulatory policy.

2773. Ghosh, R. N. "The quantification of seasonality of money supply in Ghana." Institute of Statistical, Social and Economic Research, University of Ghana, Legon, 1972. Topics discussed include: Ghana, money supply, seasonality.

2774. Givoly, D. "Year-End Tax-Induced Sales and Stock Market Seasonality." *Journal of Finance* 38, no. 1 (March 1983): 171-85. Topics discussed include: capital and financial markets, regulatory policy, taxation, income, subsidies.

2775. Hamon, J. "Monthly Return Seasonality on the Paris Stock Exchange." *Finance* 7, no. 1 (June 1986): 57-74. Topics discussed include: capital and financial markets, regulatory policy.

2776. Jegadeesh, N. "Seasonality in Stock Price Mean Reversion: Evidence from the U.S. and the U.K." *Journal of Finance* 46, no. 4 (September 1991): 1427-44. Topics discussed include: asset prices, capital and financial markets, regulatory policy, portfolio composition, theory.

2777. Kozak, J. "An Attempt of Mathematical Formulation of Seasonal Variations of Inputs, Outputs and Prices in the Definitional Equation of Income." *Ekonomicko Matematicky Obzor* 5, no. 2 (1969): 209-23. Topics discussed include: communist systems, finance, capital.

2778. Larsen, G. A., Jr. "Seasonality in Firm-Size Portfolio Returns: A Nonparametric Analysis." *Journal of Economics and Finance* 16, no. 3 (September 1992): 121-30. Topics discussed include: regulatory policy, capital and financial markets, portfolio composition.

2779. Lee, H. S. "The influence of seasonal adjustment on the Canadian consumption function, 1947-1991." *Canadian Journal of Economics* 26 (August 1993): 575-89. Topics discussed include: Canada, seasonality, income.

2780. Lucken, J. A. "Seasonal Income and Consumption in Japan and the United States, 1961, 4 to 1969, 1." *Hitotsubashi Journal of Economics* 15, no. 1 (June 1974): 19-25. Topics discussed include: macroeconomic demand, consumption.

2781. Meric, I. "Potential Gains from International Portfolio Diversification and Inter-temporal Stability and Seasonality in International Stock Market Relationships." *Journal of Banking and Finance* 13, no. 4-5 (September 1989): 627-40. Topics discussed include: capital and financial markets, regulatory policy.

2782. Miron, J. A. *Seasonal fluctuations and the life cycle-permanent income model of consumption*. 19 pages. NBER working paper series; working paper no. 1845, Cambridge, MA (1050 Massachusetts Ave., Cambridge 02138): National Bureau of Economic Research, 1986. Topics discussed include: consumption, economics, models, seasonality.

2783. Mishra, P. "Household Consumer Expenditure Inequalities in India: A Decomposition Analysis." *Review of Income and Wealth* 38, no. 2 (June 1992): 225-36. Topics discussed include: living standards, composition of expenditures, consumption, savings.

2784. Neusser, K. "Intertemporal Nonseparability, Liquidity Constraints, and Seasonality of Aggregate Consumer Expenditures: An Empirical Investigation." *Empirical Economics* 17, no. 3 (1992): 363-82. Topics discussed include: consumption, savings, living standards, composition of expenditures.

2785. Osborn, D. R. "Seasonality and Habit Persistence in a Life Cycle Model of Consumption." *Journal of Applied Econometrics* 3, no. 4 (October 1988): 255-66. Topics discussed include: consumer economics, living standards, consumption, savings.

2786. ---. "The Performance of Periodic Autoregressive Models in Forecasting Seasonal U. K. Consumption." *Journal of Business and Economic Statistics* 7, no. 1 (January 1989): 117-27. Topics discussed include: forecasting, time series analysis, spectral analysis models, consumer behavior, economics, living standards, savings.

2787. Pesando, J. E. "Seasonal Variability in Distributed Lag Models." *Journal of the American Statistical Association* 67, no. 338 (June 1972): 311-12. Topics discussed include: inferential problems, single equation systems, insurance.

2788. Theobald, M. "Seasonality Estimation in Thin Markets." *Journal of Finance* 39, no. 2 (June 1984): 377-92. Topics discussed include: capital and financial markets, regulatory policy.

2789. ul Haq, M. "Redefining wealth and progress: New ways to measure economic, social and environmental change. The Caracas report on alternative development indicators." *Second edition, Foreword by Carlos Andres Perez, A TOES (The Other Economic Summit) Book,* 99 pages. New York, Indianapolis: International Technology Development Group of North America, Bootstrap Press; Knowledge Systems, 1990. Topics discussed include: growth, productivity, development, theories, developing countries, development models.

2790. Williamson, P. "Degrees of national wealth." *Nature* 362 (29 April 1993): 782. Topics discussed include: climate, temperature, economic conditions.

2791. Wolpin, K. I. "A New Test of the Permanent Income Hypothesis: The Impact of Weather on the Income and Consumption of Farm Households in India." *International Economic Review* 23, no. 3 (October 1982): 583-94. Topics discussed include: rural areas, economics, consumer economics, living standards, composition of expenditures, consumption, savings.

2792. Yadav, P. K. "Intraweek and Intraday Seasonalities in Stock Market Risk Premia: Cash and Futures." *Journal of Banking and Finance* 16, no. 1 (February 1992): 233-70. Topics discussed include: asset prices, regulatory policy, capital and financial markets, portfolio composition.

Tourism, Leisure & Transportation

2793. Dannenberg, A. et al. "Leisure time physical activity in the Framingham Offspring Study. Description, seasonal variation, and risk factor correlates." *Am J Epidemiol* 129, no. 1 (January 1989): 76-88. Topics discussed include: coronary disease, physiopathology, Massachusetts, risks, leisure activities.

2794. Fritz, A. "Health hazards of modern tropic tourism." *Ther Ggw* 108, no. 9 (September 1969): 1286-8 passim. Topics discussed include: Africa, travel, tropical climates, tropical medicine.

2795. Greene, D. L. *Transportation & Global Climate Change*. Berkeley, CA: American Council for an Energy Efficient Economy, 1993. Topics discussed include: meteorology, greenhouse effect.

2796. Greene, D. L. et al. *Transportation and global climate change*. 357 pages. Washington, DC: American Council for an Energy-Efficient Economy, 1993. Topics discussed include: greenhouse gases, energy conservation, energy consumption, synthetic fuels.

2797. Gu, Z. et al. "Terrorism, Seasonality, and International Air Tourist Arrivals in Central Florida: An Empirical Analysis." *Journal of Travel and Tourism Marketing* 1, no. 1 (1992): 3-17. Topics discussed include: tourism, air transportation, Florida.

2798. Hartmann, R. "Tourism, Seasonality and Social Change." *Leisure Studies* 5, no. 1 (January 1986): 25-33. Topics discussed include: travel, environmental factors, sociology, recreation, seasonality.

2799. Kerin, P. D. "Some Implications of Sunk, Congestion and Seasonal Opening Costs within a Regional Grain Handling and Transport System." *Journal of Transport Economics and Policy* 22, no. 2 (May 1988): 175-96. Topics discussed include: transportation, agriculture, markets, marketing, cooperatives.

2800. Magnus, K. et al. "Walking, cycling, or gardening, with or without seasonal interruption, in relation to acute coronary events." *Am J Epidemiol* 110, no. 6 (December 1979): 724-33. Topics discussed include: Netherlands, timing, coronary diseases, exertion, leisure.

2801. Mechkuev, R. et al. "Hygienic evaluation of the effectiveness of short term suburban recreation from the viewpoint of natural climatic and demographic factors." *Probl Khig* 6 (1981): 135-42. Topics discussed include: Bulgaria, leisure, timing, health, recreation, population.

2802. Mieczkowski, Z. "The tourism climatic index: a method of evaluating world climates for tourism." *The Canadian Geographer* 29 (September 1985): 220-33. Topics discussed include: climatology, tourism.

2803. Mourdoukoutas, P. "Seasonal Employment, Seasonal Unemployment and Unemployment Compensation: The Case of the Tourist Industry of the Greek Islands." *American Journal of Economics and Sociology* 47, no. 3 (July 1988): 315-29. Topics discussed include: employment, unemployment level, services, entertainment, recreation, tourism.

2804. Nappi, G. et al. "Hot springs, tourism and climatology." *Ann Ig* 3, no. 3-4 (May 1991): 175-9. Topics discussed include: Italy, health resorts, travel.

2805. Orlov, A. S. "Seasonal Variations in the Quantity and Structure of a Recreational Zone Population." *Sotsiologicheskie Issledovaniya* 13, no. 3 (July 1986): 173-177. Topics discussed include: USSR, distribution, tourism, social planning, seasonality, tourism planning.

2806. Perry, A. H. "Weather, Climate, and Tourism." *Weather* 27 (1972): 199-203. Topics discussed include: economy, society, culture, behavior.

2807. Pertsovskii, M. et al. "Value of local tourism in complex sanatorium climatic treatment of coronary and cerebral arteriosclerosis." *Vopr Kurortol Fizioter Lech Fiz Kult no.* 2 (March 1977): 81-3. Topics discussed include: health resorts, Ukraine, coronary disease therapy, exercise.

2808. Rose, L. S. et al. "Seasonal Migration of Retired Persons: Estimating Its Extent and Its Implications for the State of Florida." *Journal of Economic and Social Measurement* 15, no. 1 (1989): 91-104. Topics discussed include: Florida, tourism, elderly, retirement.

2809. Seldon, Z. A. "Seasonality in Movement Patterns on the Upper Mississippi." *Logistics and Transportation Review* 25, no. 1 (March 1989): 23-38. Topics discussed include: economics, transportation.

2810. Sharma, S. C. et al. "Duration and frequency of seasonal traffic counts." *Journal of Transportation Engineering* 119 (May 1993): 344-59. Topics discussed include: highway engineering.

2811. Sutcliffe, C. M. S. et al. "The Measurement of Seasonality within the Tourist Industry: An Application to Tourist Arrivals in Spain." *Applied Economics* 12, no. 4 (December 1980): 429-41. Topics discussed include: services, entertainment, recreation, tourism.

2812. Tepper, P. A. "Rheumatism and Its Relation to Cold, Especially as it Affects Workers in Transportation Industry." *Klin Med* 14 (1936): 1607-1614. Topics discussed include: weather, climate, temperature, disease, industry.

2813. Tuppen, J. N. "Tourism in France: recent trends in winter sports." *Geography* 73 (October 1988): 359-63. Topics discussed include: tourist trade, France.

2814. Uitenbroek, D. "Seasonal variation in leisure time physical activity." *Med Sci Sports Exerc* 25, no. 6 (June 1993): 755-60. Topics discussed include: age effects, golf, jogging, gender effects, exercise, leisure activities, sports.

2815. Van, D. L. F. et al. "Risk groups in travel to tropical countries within the frame of mass tourism." *Soz Praventivmed* 22, no. 4 (July 1977): 159-60. Topics discussed include: Africa, Asia, diarrhea, South America, tropical medicine, travel.

2816. Vargha, G. "Medical geography of health tourism in Hungary considering the seasonal combinations." *Geogr Med* 11 (1981): 141-50. Topics discussed include: Hungary, therapy, travel.

2817. Woo, M. et al. "Impacts of climate warming on transportation and the regional economy of Canada's North." *The Canadian Geographer* 36 (March 1992): 73-5. Topics discussed include: transportation, global temperature changes, climate.

2818. Zimmermann, F. "Austria: Contrasting Tourist Seasons and Contrasting Regions." *Tourism and economic development: Western European experiences*, 145-61. ed Allan M. Williams, and Gareth Shaw. London and New York: Pinter, Belhaven Press; distributed by Columbia University Press, 1988. Topics discussed include: services, entertainment, recreation, tourism.

2819. Zuidema, P. "Tourism and imported diseases from the tropics." *Tijdschr Ziekenverpl* 30, no. 5 (1 March 1977): 203-9. Topics discussed include: Africa, Asia, Mediterranean islands, vaccinations, travel, tropical climate.

MACROECONOMIC POLICY & METHODOLOGIES

Climatic Change Policy

2820. Bach, W. "Modelling the Climatic Effects of Trace Gases: Reduction Strategy and Options for a Low Risk Policy." *Climate and development: Climatic change and variability and the resulting social, economic and technological implications*, 193-218. eds H. J. Karpe, D. Otten, and S. C. Trinidade. New York, Berlin, London, Tokyo: Springer in cooperation with the United Nations, 1990. Topics discussed include: conservation, pollution.

2821. Baehr, L. "Climate Change--Its Political Challenge to the International Community." *Climate and development: Climatic change and variability and the resulting social, economic and technological implications*, 60-63. eds H. J. Karpe, D. Otten, and S. C. Trinidade. New York, Berlin, London, Tokyo: Springer in cooperation with the United Nations, 1990. Topics discussed include: conservation, pollution.

2822. Baker, D. N. et al. *Effects of Increased CO sub 2 on Photosynthesis and Agricultural Productivity Volume II, Part 6 Environmental and Societal Consequences of a Possible CO sub 2 Induced Climate Change*, Performer: Mississippi State Univ., Mississippi State. University of Florida, Gainesville. Clemson Univ., Clemson, SC. Department of Energy, Washington, DC, October 1982. Topics discussed include: carbon dioxide, greenhouse effects, atmosphere, meteorology.

2823. Barrett, S. "Convention on climate change : economic aspects of negotiations." Organisation for Economic Co-operation and Development. Environment Committee. Organisation for Economic Co-operation and Development ; OECD Publications and Information Centre [distributor], Paris, Washington, DC, 1992. Topics discussed include: negotiation, global warming, legal aspects.

2824. Batie, S. S. et al. "The Biological Consequences of Climate Changes: An Ecological and Economic Assessment." *Greenhouse warming: Abatement and adaptation Proceedings of a workshop held in Washington, D.C., June 14-15, 1988*, 121-131. eds Norman J. Rosenberg et al., Washington, DC: Resources for the Future, 1989. Topics discussed include: conservation, pollution.

2825. Beltzner, K. P. *Living with climatic change : proceedings, Toronto conference workshop, November 17-22, 1975*. 105 pages. American Meteorological Society. Ottawa: Science Council of Canada, 1976.

2826. Bentley, C. R. "Environmental and societal consequences of a possible CO2-induced climate change." U.S. Dept. of Energy. Office of Health and Environmental Research, and U.S. Dept. of Energy. Office of Basic Energy Sciences. Carbon Dioxide Research Division. U.S. Dept. of Energy, Office of Energy Research, Office of Health and Environmental Research; Available from National Technical Information Service, Washington, DC, Springfield, VA, 1982. Topics discussed include: atmosphere, carbon dioxide.

2827. Bohme, W. "Programmes and Co-operation in Climate Research--A Means to the Understanding of the Climate System as a Requirement to Cope with the Climate Issue." *Karpe, H J; Otten, D; Trinidade, S C, eds Climate and development: Climatic change and variability and the resulting social, economic and technological implications*, 123-140. New York, Berlin, London, Tokyo: Springer in cooperation with the United Nations, 1990. Topics discussed include: conservation, pollution.

2828. Brightwell, H. "Parliamentary Forum on Global Climate Change: proceedings." Parliamentary Forum on Global Climate Change (1990: Ottawa, Ont.). Queen's Printer for Canada, Ottawa, 1990. Topics discussed include: global warming, environment, policy, greenhouse effect.

2829. Brunner, R. D. "Global Climate Change: Defining the Policy Problem." *Policy Sciences* 24, no. 3 (August 1991): 291-311. Topics discussed include: conservation, environment, management, policy, pollution, resources.

2830. Bundesverband der Deutschen Industrie. "Climate and Development: Climatic Change and Variability and the Resulting Social, Economic and Technological Implications: Industry's Attitude." *Karpe, H J; Otten, D; Trinidade, S. C., eds Climate and development: Climatic change and variability and the resulting social, economic and technological implications*, 295-99. New York; Berlin; London and Tokyo: Springer in cooperation with the United Nations, 1990. Topics discussed include: resources.

2831. Burton, I. "To Limit and to Adjust: Some Human Dimensions of Global Climate Change." *Climate and development: Climatic change and variability and the resulting social, economic and technological implications*, 141-51. eds H. J. Karpe, D. Otten, and S. C. Trinidade. New York; Berlin; London; Tokyo: Springer in cooperation with the United Nations, 1990. Topics discussed include: conservation, pollution.

2832. "Carbon dioxide effects research and assessment program : Workshop on Environmental and Societal Consequences of a Possible CO2-Induced Climate Change, Annapolis, Maryland, April 2-6, 1979." Workshop on Environmental and Societal Consequences of a Possible CO2-Induced Climate Change (1979: Annapolis, MD), and American Association for the Advancement of Science. U.S. Dept. of Energy, Assistant Secretary for Environment, Office of Health and Environmental Research; Available from National Technical Information Service, U.S. Dept. of Commerce, Washington, DC, Springfield, VA, 1980. Topics discussed include: sociology, atmosphere, carbon dioxide.

2833. Changnon, S. A. et al. "Shifts in perceptions of climate change: a Delphi experiment revisited." *Bulletin of the American Meteorological Society* 73 (October 1992): 1623-7. Topics discussed include: sociology.

2834. Chen, R. S. et al. *Social science research and climate change : an interdisciplinary appraisal.* 255 pages. Dordrecht, Holland; Boston, Hingham, MA: D. Reidel Pub. Co.; Sold and distributed in the U.S.A. and Canada by Kluwer Boston, 1983. Topics discussed include: sociology.

2835. Churchill, R. R. et al. *International law and global climate change.* 447 pages. International environmental law and policy series, London; Boston: Graham & Trotman, 1991. Topics discussed include: law, global warming.

2836. "Climate change and transnational corporations: analysis and trends." Centre on Transnational Corporations (United Nations). United Nations, New York, 1992. Topics discussed include: international aspects, business, environment, policy.

2837. "Climate change: designing a tradeable permit system." Organisation for Economic Co-operation and Development. Organisation for Economic Co-operation and Development, Paris, 1992. Topics discussed include: environment, policy, greenhouse gases.

2838. "Climate change: evaluating the socio-economic impacts." Organisation for Economic Co-operation and Development ; OECD Publications and Information Centre, distributor, Paris, Washington, DC, 1991. Topics discussed include: sociology evaluation, issues evaluation.

2839. "Climate change, policy initiatives." International Energy Agency. International Energy Agency, Paris, 1992. Topics discussed include: policy, Europe.

2840. Climate Impact Assessment Program. *Economic and Social Measures of Biological and Climatic Change*. Washington, DC: Monograph 6, U. S. Department of Transportation, 1975. Topics discussed include: economy, sociology, biology.

2841. "Climatic change: impacts on New Zealand: implications for the environment, economy, and society." New Zealand Climate Change Programme, and New Zealand. Ministry for the Environment.

Ministry for the Environment, Wellington, N.Z., 1990. Topics discussed include: New Zealand, economics, sociology.

2842. Clinton, B. et al. "The Climate change action plan." United States. Executive Office of the President. Executive Office of the President, Washington, DC, 1993. Topics discussed include: global warming, policy, pollution.

2843. Cogan D. *The greenhouse gambit : business and investment responses to climate change,* Washington, DC: Investor Responsibility Research Center, 1992. Topics discussed include: economics, greenhouse effects, atmosphere.

2844. "A Compendium of options for government policy to encourage private sector responses to potential climate changes : report to the Congress of the United States." U.S. Dept. of Energy. Assistant Secretary for Environment, Safety, and Health. U.S. Dept. of Energy, Office of Environmental Analysis, Assistant Secretary for Environment, Safety and Health; Available from the National Technical Information Service, U.S. Dept. of Commerce, Washington, DC, Springfield, VA, 1989. Topics discussed include: greenhouse effects, atmospheric effects, policy, industries.

2845. "Confronting climate change : strategies for energy research and development." National Research Council (U.S.). Committee on Alternative Energy Research and Development Strategies. National Academy Press, Washington, DC, 1990. Topics discussed include: policy, greenhouse effect.

2846. Craig, P. P. et al. "Energy policies to address global climate change: research needs and recommendations: a University of California workshop executive summary." University of California Global Climate Change Workshop (2nd: 1989: Davis, California), and University of California, Davis. Public Service Research and Dissemination Program. University of California, Public Service Research and Dissemination Program, Davis, 1989. Topics discussed include: policy, California, global warming, greenhouse effect.

2847. Crosson, P. R. "Climate Change: Problems of Limits and Policy Responses." *Greenhouse warming: Abatement and adaptation Proceedings of a workshop held in Washington, D.C., June 14-15, 1988*, 69-82. eds Norman J. Rosenberg et al., Washington, DC: Resources for the Future, 1989. Topics discussed include: conservation, pollution.

2848. D'Arge, R. C. "Economic and social measures of biologic and climatic change: final report." Panel on Economic and Social Measures of Biologic and Climatic Change, U.S. Dept. of Transportation. Climatic Impact Assessment Program Office, and Institute for Defense Analyses. Science and Technology Division. Dept. of Transportation, Climatic Impact Assessment Program, Office of the Secretary of Transportation, Washington, 1975. Topics discussed include: stratosphere, air pollution, bioclimatology.

2849. Diaz, H. F. "Global climate change." *Bulletin of the American Meteorological Society* 71 (July 1990): 424 1034-6.

2850. Department of Energy (US). *Energy & Climate Changes.* Chelsea, MI: Lewis Publishers, 1990. Topics discussed include: climate.

2851. "DOE's national energy plan and global warming : hearing before the Committee on Energy and Natural Resources, United States Senate, One Hundred First Congress, first session, on the formulation of a national energy plan and related policies which affect global climate change, July 26, 1989." U.S. Senate, Committee on Energy and Natural Resources. U.S. G.P.O.: For sale by the Supt. of Docs., Congressional Sales Office, U.S. G.P.O., Washington, 1989. Topics discussed include: dept, policy, global temperature changes.

2852. Donahue, M. J. "The Great Lakes Basin climate change & policy issues." *Geotimes* 37 (December 1992): 19-20. Topics discussed include: water level, great lakes region, environment, management.

2853. Dooge, J. C. I. "Sensitivity of runoff to climate change: a Hortonian approach." *Bulletin of the American Meteorological Society* 73 (December 1992): 2013-24. Topics discussed include: runoff, hydrology.

2854. Drennen, T. E. et al. "Negotiating a Response to Climate Change: Role of Biological Emissions." *Contemporary Policy Issues* 10, no. 3 (July 1992): 49-58. Topics discussed include: conservation, environment, management, water, air, policy, pollution, resources.

2855. Easterling, W. E. "Social science research and climate change." *Bulletin of the American Meteorological Society* 66 (June 1985): 699-71.

2856. "The Economics of long-term global climate change: a preliminary assessment: report of an interagency task force." U.S. Dept. of Energy, Office of Policy, Planning, and Analysis; Available from the National Technical Information Service, U.S. Dept. of Commerce, Washington, DC, Springfield, VA, 1990. Topics discussed include: economics, global warming, greenhouse gases.

2857. "Energy policies to address global climate change: a University of California Workshop executive summary: research needs and recommendations." University of California Global Climate Change Workshop (2nd: 1989: Davis, CA), and University of California, Davis. Public Service Research and Dissemination Program. University of California, Davis, 1989. Topics discussed include: policy, California, global warming, greenhouse effects.

2858. Ewert, A. W. "Outdoor Recreation and Global Climate Change: Resource Management Implications for Behaviors, Planning, and Management." *Society and Natural Resources* 4, no. 4 (October 1991): 365-377. Topics discussed include: environment, resources, management, weather, recreation, behavior.

2859. Fairbridge, R. W. *Solar variations, climatic change, and related geophysical problems: papers.* 740 pages. New York Academy of Sciences, and American Meteorological Society. Annals of the New York Academy of Sciences; v. 95, New York: New York Academy of Sciences, 1961. Topics discussed include: solar effects.

2860. Feldman, D. L. *Democratic theory, science, and decision making in global climate change : culpability, uncertainty, and fairness.* 13 pages. American Political Science Association Annual Meeting, August 30-September 2, 1990. Topics discussed include: environment, policy.

2861. Feldman, D. *Global Climate Change & Public Policy.* Urbana, IL: Policy Studies Organization, 1991. Topics discussed include: history, environment, policy.

2862. Feldman, D. L. et al. "Policy analysis and the management of climate change: institutional adaptability in the face of scientific uncertainty; introducing the symposium." *Policy Studies Journal* 19, no. 2 (1991): 43-9. Topics discussed include: greenhouse effects, environment, policy.

2863. Firth, P. et al. *Global climate change and freshwater ecosystems.* 321 pages. United States. National Aeronautics and Space Administration, United States. Environmental Protection Agency, and North American Benthological Society. New York: Springer-Verlag, 1992. Topics discussed include: freshwater, ecology, bioclimatology.

2864. Gates, D. M. *Climate Change & Its Biological Consequences.* Sunderland, MA: Sinauer Associates, 1993. Topics discussed include: biology, meteorology.

2865. Glantz, M. H. "Climate Variability, Climate Change and the Development Process in Sub-Saharan Africa." *Climate and development: Climatic change and variability and the resulting social, economic and technological implications*, 173-92. eds H. J. Karpe, D. Otten, and S. C. Trinidade. New York; Berlin; London and Tokyo: Springer in cooperation with the United Nations, 1990. Topics discussed include: conservation, pollution.

2866. ---. *Societal responses to regional climatic change : forecasting by analogy.* 428 pages. Boulder, CO: Westview Press, 1988. Topics discussed include: sociology, North America, environment.

2867. Gleick, P. H. "The Colorado River Basin and the greenhouse effect : water resources and water management : proceedings of a Workshop on the Implications of Climatic Change for the Colorado River Basin, May 17-18, 1990, Denver, Colorado." 76 pages. Pacific Institute for Studies in Development, Environment, and Security, and Workshop on the Implications of Climatic Change for the Colorado

River Basin (1990: Denver, CO). Denver, CO: Pacific Institute for Studies in Development, Environment, and Security, 1990. Topics discussed include: Colorado river, Mexico, sociology, environment, policy, greenhouse effects, water supply, climate.

2868. "Global climate change: hearing before a subcommittee of the Committee on Appropriations, United States Senate, One Hundred First Congress, second session: special hearing." U.S. Senate. Committee on Appropriations. Subcommittee on VA-HUD-Independent Agencies. For sale by the Supt. of Docs., Congressional Sales Office, U.S. G.P.O., Washington, 1990. Topics discussed include: policy, global warming.

2869. "Global climate change: potential impacts & policy recommendation." California Energy Commission. Technology Evaluation Office, Sacramento, CA, 1991. Topics discussed include: greenhouse effects, atmosphere, global warming, policy, California.

2870. "Global climate change and the Pacific islands: hearing before the Committee on Energy and Natural Resources, United States Senate, One Hundred Second Congress, second session, on the scientific evidence and policy implications of sea level rise ... Honolulu, HI, May 26, 1992." U.S. Senate Committee on Energy and Natural Resources. For sale by the U.S. G.P.O., Supt. of Docs., Congressional Sales Office, Washington, 1992. Topics discussed include: global warming, Pacific islands.

2871. "Global environmental change research: hearing before the Subcommittee on Science, Technology, and Space and the National Ocean Policy Study of the Committee on Commerce, Science, and Transportation, United States Senate, One Hundredth Congress, first session, on global climate change due to manmade changes in the earth's atmosphere, July 16, 1987." U.S. Senate Committee on Commerce, Science, and Transportation. Subcommittee on Science, Technology, and Space, and U.S. Senate National Ocean Policy Study. For sale by the Supt. of Docs., Congressional Sales Office, U.S. G.P.O., Washington, 1987. Topics discussed include: temperature changes, greenhouse effects, atmospheric ozone.

2872. Gray, J. E. et al. *Global climate change: U.S.-Japan cooperative leadership for environmental protection.* 96 pages. Resources for the Future, Atlantic Council of the United States, Nihon Enerugi Keizei Kenkyujo, Institute of Energy Economics (Japan), Enerugi Sogo Suishin Iinkai (Japan), Committee for Energy Policy Promotion (Japan), Chikyu Sangyo Bunka Kenkyujo, and Global Industrial and Social Progress Research Institute. Energy & Environment Series, The Atlantic Council of the United States, Washington, DC (1616 H St., NW, Washington 20006): Atlantic Council of the United States, 1991.

2873. Grobecker, A. J. "Economic and social measures of biologic and climatic change." Panel on Economic and Social Measures of Biologic and Climatic Change. Dept. of Transportation, Washington, 1975. Topics discussed include: climatic effects, bioclimatology, air pollution.

2874. *Climate and development : climate change and variability and the resulting social, economic and technological implicationse*, eds H. J. Karpe, D. Otten, S.C. Trindade. Berlin; New York: Springer-Verlag, 1990. Topics discussed include: sociology, environment, policy.

2875. Haas, J. E. "Social Impact of Induced Climatic Change." In *Climate Impact Assessment Program (CIAP), Monograph 6: Economic and Social Measures of Biologic and Climatic Change*, editors R. d'Arge et al. Washington, DC: U.S. Department of Transportation, 1975. Topics discussed include: society, economy, biology.

2876. Hayes, P. et al. *The Global greenhouse regime: who pays?: science, economics and North-South politics in the Climate Change Convention.* 382 pages. London, Tokyo, New York: Earthscan Publications; United Nations University Press, 1993. Topics discussed include: greenhouse effects, global warming.

2877. Haynes, D. L. *Effects of Climate Change on Agricultural Plant Pests Volume II, Part 10 of Environmental and Societal Consequences of a Possible CO sub 2 Induced Climate Change*, Performer: Michigan State Univ., East Lansing. Dept. of Entomology. Department of Energy, Washington, DC, October 1982. Topics discussed include: crops, greenhouse effects, weeds, pests, productivity, medicine, biology.

2878. Hoeller, P. et al. "Energy taxation and price distortions in fossil fuel markets: some implications for climate change policy." Organisation for Economic Co-operation and Development, Paris, 1992. Topics discussed include: environment, policy, econometrics, energy, resources, taxation.

2879. Hourcade, J. C. et al. "International economic instruments and climate change." Organisation for Economic Co-operation and Development. Environment Committee. Organisation for Economic Co-operation and Development; OECD Publications and Information Centre, distributor, Paris; Washington, DC, 1993. Topics discussed include: greenhouse gases, policy, environment.

2880. *Climate change and energy policy: proceedings*, International conference on global climate change: its mitigation through improved production and use of energy (1991: Los Alamos National Laboratory). New York, NY: American Institute of Physics, 1992.

2881. "International Conference on Global Climate Change: Its Mitigation through Improved Production and Use of Energy (1991 : Los Alamos National Laboratory)." *Climate change and energy policy : proceedings of the International Conference on Global Climate Change, Its Mitigation Through Improved Production and Use of Energy, Los Alamos National Laboratory, October 21-24, 1991, Los Alamos, New Mexico, USA*, eds Louis Rosen, Robert Glasser. New York: American Institute of Physics, 1992. Topics discussed include: resources, policy.

2882. Jager, J. "Developing policies for responding to climatic change: a summary of the discussions and recommendations of the workshops held in Villach (28 September-2 October 1987) and Bellagio (9-13 November 1987), under the auspices of the Beijer Institute." World Meteorological Organization, World Climate Programme, United Nations Environment Programme, and Beijer Institute. WCIP; 1. WMO-TD; no. 225, Geneva, Switzerland: World Meteorological Organization Secretariat, 1988. Topics discussed include: climatology, greenhouse effects, atmosphere.

2883. ---. *Responding to climate change: tools for policy development: (summary report)*. 26 pages. Stockholm Environment Institute. Report. Stockholm: Stockholm Environment Institute, 1990. Topics discussed include: economics, global warming.

2884. Jager, J. et al. *Climate change: science, impacts and policy: proceedings of the Second World Climate Conference.* 578 pages. World Climate Conference (2nd : 1990 : Geneva, Switzerland). Cambridge; New York: Cambridge University Press, 1991. Topics discussed include: climatic effects.

2885. *Global climate change: linking energy, environment, economy, and equity*, editors James C. White, William R. Wagner, and Carole N. Beal. Environmental science research; v. 45, New York: Plenum Press, 1992. Topics discussed include: development, air quality, management, policy.

2886. Jeftic, L. et al. *Climatic change and the Mediterranean: environmental and societal impacts of climate change and sea-level rise in the Mediterranean region*. 673 pages. United Nations Environment Programme. London; New York: E. Arnold, 1992. Topics discussed include: Mediterranean region, sociology, sea level.

2887. Johnson, S. et al. "An Approach to Estimating the Economic Impact of Climatic Change on a Regional Economy." *Environment and Planning A* 23, no. 8 (August 1991): 1197-1208. Topics discussed include: spatial economic models, growth, development, economics, urban areas, economics, policy.

2888. Johnson, W. C. et al. *Ecological Consequences of a CO sub 2 Induced Climatic Change on Forest Ecosystems Volume II, Part 13 of Environmental and Societal Consequences of a Possible CO sub 2 Induced Climate Change*, Performer: Virginia Polytechnic Inst. and State Univ., Blacksburg. Dept. of Biology. Southern Illinois Univ., Carbondale. Dept. of Geography. Department of Energy, Washington, DC, October 1982. Topics discussed include: greenhouse effects, carbon dioxide, forests, ecosystems, geography, medicine, biology.

2889. Jones, B. P. "UN Convention on Climate Change: Effects on Australia's Energy Sector." *Agriculture and Resources Quarterly* 4, no. 2 (June 1992): 186-95. Topics discussed include: resources, conservation, environment, management, water, air, demand, supply, policy, pollution.

2890. Jones, B. S. "State responses to global climate change." *Policy Studies Journal* 19, no. 2 (1991): 73-82. Topics discussed include: greenhouse effects, environment, policy.

2891. Justus, J. R. et al. "Global climate change." Congressional Research Service, Library of Congress, Washington, DC, 1993. Topics discussed include: policy.

2892. Kellogg, W. W. et al. *Climate change and society : consequences of increasing atmospheric carbon dioxide.* 178 pages. Boulder, Colo.: Westview Press, 1981. Topics discussed include: atmosphere, carbon dioxide, sociology.

2893. ---. "Society, Science and Climate Change." *Foreign Affairs* 60, no. 5 (June 1982): 1076-1109. Topics discussed include: agriculture, supply, demand, resources, energy.

2894. Kohlmaier, G. H. "Contributions to the Atmospheric CO2 Increase by Changes in the Land Biosphere - Analysis of the Past and Present, Including Possible Future Developments." *Climate and development: Climatic change and variability and the resulting social, economic and technological implications*, 219-54. editors H. J. Karpe, D. Otten, and S. C. Trinidade. New York; Berlin; London and Tokyo: Springer in cooperation with the United Nations, 1990. Topics discussed include: conservation, pollution.

2895. Kokoski, M. F. et al. "A General Equilibrium Analysis of Partial-Equilibrium Welfare Measures: The Case of Climate Change." *American Economic Review* 77, no. 3 (June 1987): 331-41. Topics discussed include: welfare, economics, theory, resources.

2896. Koo, E. H. "Climate Change--Secular Trends in Urban Temperature." *Climate and development: Climatic change and variability and the resulting social, economic and technological implications*, 285-92. eds. H. J. Karpe, D. Otten, and S. C. Trinidade. New York; Berlin; London and Tokyo: Springer in cooperation with the United Nations, 1990. Topics discussed include: resources.

2897. Kuik, O. et al. "Joint implementation to curb climate change: legal and economic aspects, edited by Onno Kuik, Paul Peters and Nico Schrijver." Environment & Policy v. 2, Netherlands: Dordrecht, Boston: Kluwer Academic Publishers, 1994. Topics discussed include: international issues, cooperation, law.

2898. Kullenberg, G. "Climate Change: Roles of UNESCO and IOC with Particular Reference to the Ocean." *Climate and development: Climatic change and variability and the resulting social, economic and technological implications*, 433-47. eds H. J. Karpe, D. Otten, and S. C. Trinidade. New York; Berlin; London and Tokyo: Springer in cooperation with the United Nations, 1990. Topics discussed include: conservation, pollution, resources.

2899. Lambright, W. H. et al. "Governing global climate change: can we learn from the past in designing the future?" *Policy Studies Journal* 19, no. 2 (1991): 50-60. Topics discussed include: greenhouse effects, environment, policy, environment, law.

2900. Le, H. H. N. "Climatic change and desertization." *Impact of Science on Society* no. 166 (1992): 183-201. Topics discussed include: desertification.

2901. Levine, M. D. et al. "Strategies for addressing climate change: policy perspectives from around the world." *Energy (Oxford, England)* 17 (December 1992): 1121-36. Topics discussed include: sociology.

2902. Liebetrau, A. M. et al. "Strategies for Modeling the Uncertain Impacts of Climate Change." *Journal of Policy Modeling* 13, no. 2 (June 1991): 185-204. Topics discussed include: risk.

2903. MacDonald, G. J. "Climate change: a challenge to the means of technology transfer." University of California Institute on Global Conflict and Cooperation. University of California Institute on Global Conflict and Cooperation, La Jolla, Calif., 1992. Topics discussed include: technology, innovations, transfer, economics, greenhouse gases.

2904. MacKenzie, J. J. *Breathing easier : taking action on climate change, air pollution, and energy insecurity.* 23 pages. Washington, DC: World Resources Institute, 1988. Topics discussed include: policy, greenhouse effects, atmospheric effects.

2905. Magalhaes, A. R. "Social Impacts of Climate Variations, a Developing Region's Viewpoint." *Climate and development: Climatic change and variability and the resulting social, economic and technological implications*, 369-78. eds H. J. Karpe, D. Otten, and S. C. Trinidade. New York; Berlin; London and Tokyo: Springer in cooperation with the United Nations, 1990. Topics discussed include: conservation, pollution.

2906. Markham, A. "Tropical forest conservation and climate change." 8 pages. World Wide Fund for Nature. Gland, Switzerland: Conservation Policy Division, WWF International, 1990. Topics discussed include: forest conservation, tropics, deforestation, rain forests.

2907. Maul, G. A. *Climatic change in the Intra-Americas Sea : implications of future climate on the ecosystems and socio-economic structure in the marine and coastal regions of the Caribbean Sea, Gulf of Mexico, Bahamas, and the northeast coast of South America.* 389 pages. United Nations Environment Programme, and Intergovernmental Oceanographic Commission. London; New York: E. Arnold ; Distributed in the USA by Routledge, Chapman, and Hall, 1993. Topics discussed include: North Atlantic Ocean, Caribbean, global warming.

2908. Meyer Abich, K. M. "Political Perception of Climate Change: Still Chalk on the White Wall?" *Climate and development: Climatic change and variability and the resulting social, economic and technological implications*, 54-59. eds H. J. Karpe, D. Otten, and S. C. Trinidade. New York; Berlin; London and Tokyo: Springer in cooperation with the United Nations, 1990. Topics discussed include: conservation, pollution.

2909. *Societal responses to regional climatic change: forecasting by analogy*, ed Michael H. Glantz. Boulder, CO: Westview Press, 1988. Topics discussed include: sociology, North America, environment.

2910. Michaels, P. J. "Workshop, On Greenhouse-Gas-Induced Climatic Change: A Critical Appraisal Of Simulations And Observations." *Bulletin of the American Meteorological Society* 74 (May 1993): 615-882. Topics discussed include: greenhouse effects, climatic effects.

2911. Morgenstern, R. D. "Towards a Comprehensive Approach to Global Climate Change Mitigation." *American Economic Review* 81, no. 2 (May 1991): 140-45. Topics discussed include: conservation, environment, management, water, air, policy, pollution, resources.

2912. Murdock, S. H. et al. "The Socioeconomic Dimensions of Global Climate Change: Searching for a Theoretical and Empirical Base." *Society and Natural Resources* 4, no. 4 (October 1991): 397-403. Topics discussed include: environment, weather, socioeconomics.

2913. "National action plan for global climate change." U.S. Dept. of State. Office of Global Change. [Dept. of State], Bureau of Oceans and International Environmental and Scientific Affairs, Office of Global Change : For sale by the U.S. G.P.O., Supt. of Docs., Washington, DC, 1992. Topics discussed include: environment, policy, greenhouse gases.

2914. "New Jersey global climate change initiative." New Jersey. Office of the Governor, and New Jersey. Board of Public Utilities. Office of the Governor, Trenton, NJ, 1989. Topics discussed include: policy, New Jersey, global temperature changes.

2915. Nordhaus, W. D. "The Cost of Slowing Climate Change: A Survey." *Energy Journal* 12, no. 1 (1991): 37-65. Topics discussed include: resouces, water, air, conservation, environment, management, forestry, policy, conservation, pollution.

2916. ---. *Managing the global commons: the economics of climate change.* Cambridge, MA: MIT Press, 1994. Topics discussed include: economics, models.

2917. "Ocean and global climate change : bibliography for the second Scientific Meeting, the Oceanography Society." Oceanography Society (Washington, D.C.). Scientific Meeting (2nd: 1991). U.S.

Dept. of Commerce, National Oceanic and Atmospheric Administration, National Environmental Satellite, Data, and Information Service, National Oceanographic Data Center, Library and Information Services Division, Washington, DC, 1991. Topics discussed include: oceanography, atmosphere, interactions.

2918. OECD Staff. *Cars & Climate Change*. Washington, DC: Organization for Economic Cooperation & Development, 1993. Topics discussed include: economy, industry.

2919. ---. *Convention on Climate Change: Economic Aspects of Negotiations*. Washington, DC: Organization for Economic Cooperation & Development, 1992. Topics discussed include: economics, environment, policy.

2920. Ominde, S. H. et al. *A Change in the weather: African perspectives on climatic change*. 210 pages. African Centre for Technology Studies, and International Conference on Global Warming and Climatic Change: African Perspectives (1990: United Nations Environment Programme Headquarters). African Centre for Technology Studies (ACTS) environmental policy series; no. 1, Nairobi, Kenya: ACTS Press, African Centre for Technology Studies, 1991. Topics discussed include: Africa, global warming, economics, policy.

2921. Oppenheimer, M. "Responding to Climate Change: The Crucial Role of the NGO's." *Climate and development: Climatic change and variability and the resulting social, economic and technological implications*, 339-46. eds H. J. Karpe, D. Otten, and S. C. Trinidade. New York; Berlin; London and Tokyo: Springer in cooperation with the United Nations, 1990. Topics discussed include: conservation, pollution, public firms.

2922. "Pacific Rim research development strategies related to climate change: October 22-27, 1989 at University of California, Davis." University of California, Davis. Public Service Research and Dissemination Program, 1989. Topics discussed include: California, Pacific Ocean.

2923. Parry, M. L. et al. "Assessing the impact of climatic change in cold regions : report of a workshop held at Villach, Austria, 19-23 September 1983 as part of the International Study Conference on the Sensitivity of Ecosystems and Society to Climatic Change." 42 pages. International Institute for Applied Systems Analysis, and World Meteorological Organization. Summary report; SR-84-1, Laxenburg, Austria: International Institute for Applied Systems Analysis, 1984. Topics discussed include: models, greenhouse effects, atmospheric effects.

2924. Patterson, F. *Energy policies to address climate change: proceedings of a workshop, 1989 September 6-8, University of California, Davis.* 485 pages. U.S. Dept. of Energy. Office of Environmental Analysis, University of California (System). Universitywide Energy Research Group, and Conference on Energy Policies to Address Global Climate Change (1989: University of California, Davis). UER; 257, Berkeley, CA: Universitywide Energy Research Group, 1991. Topics discussed include: policy.

2925. Pittock, A. B. *Climatic change and variability: a Southern perspective: based on a conference at Monash University, Melbourne, Australia, 7-12 December, 1975*, co-sponsored by the Australian Academy of Science. 455 pages. Royal Meteorological Society (Great Britain). Australian Branch, and Australasian Conference on Climate and Climatic Change, Monash University, 1975. Cambridge; New York: Cambridge University Press, 1978. Topics discussed include: southern hemisphere.

2926. ---. "Societal responses to regional climatic change." *Bulletin of the American Meteorological Society* 70 (September 1989): 428 1150-2.

2927. Potter, T. D. et al. "Climate and Development: Climatic Change and Variability and the Resulting Social, Economic and Technological Implications: An Overview of International Developments, Trends and Actions." *Climate and development: Climatic change and variability and the resulting social, economic and technological implications,* 448-70. eds H. J. Karpe, D. Otten, and S. C. Trinidade. New York; Berlin; London and Tokyo: Springer in cooperation with the United Nations, 1990. Topics discussed include: conservation, pollution.

2928. Rayner, S. et al. "Managing the global commons decision making and conflict resolution in response to climate change : August 1-4, 1989: Knoxville Hilton, Knoxville, Tennessee." Oak Ridge National Laboratory; Available from the National Technical Information Service, Oak Ridge, TN; Springfield, VA, 1990. Topics discussed include: decision making, environment, policy.

2929. "Report of the first U.S.-Canada Symposium on Impacts of Climate Change on the Great Lakes Basin." U.S. & Canada Symposium on Impacts of Climate Change on the Great Lakes Basin (1st : 1988 : Oak Brook, IL), U.S. National Climate Program Office, and Canadian Climate Centre. National Climate Program Office, Rockville, MD, 1989. Topics discussed include: the Great Lakes, watersheds, sociology, policy, environment, Canada, climatic effects.

2930. "Responding to climate change: selected economic issues." Organisation for Economic Co-operation and Development. Organisation for Economic Co-operation and Development; OECD Publications and Information Centre [distributor], Paris; Washington, DC, 1991. Topics discussed include: Europe.

2931. Revelle, R. et al. "Energy Options and Climatic Effects." *Climate and development: Climatic change and variability and the resulting social, economic and technological implications*, 261-69. eds H. J. Karpe, D. Otten, and S. C. Trinidade. New York; Berlin; London and Tokyo: Springer in cooperation with the United Nations, 1990. Topics discussed include: conservation, pollution.

2932. Riesenhuber, H. "Climate and Development: Climatic Change and Variability and the Resulting Social, Economic and Technological Implications: Opening Speech." *Climate and development: Climatic change and variability and the resulting social, economic and technological implications*, 30-39. eds. H. J. Karpe, D. Otten, and S. C. Trinidade. New York; Berlin; London and Tokyo: Springer in cooperation with the United Nations, 1990. Topics discussed include: conservation, pollution, energy.

2933. Rosen, L. G. *Climate Change & Energy Policy*. Woodbury, NY: American Institute of Physics, 1992. Topics discussed include: meteorology.

2934. Rosen, L. et al. *Climate change and energy policy : proceedings of the International Conference on Global Climate Change, Its Mitigation Through Improved Production and Use of Energy, Los Alamos National Laboratory, October 21-24, 1991, Los Alamos, New Mexico, USA*. 554 pages. International Conference on Global Climate Change: Its Mitigation through Improved Production and Use of Energy (1991: Los Alamos National Laboratory). New York: American Institute of Physics, 1992. Topics discussed include: policy.

2935. Rosenberg, N. J. *Towards an integrated impact assessment of climate change : the MINK study*. 173 pages. Dordrecht; Boston: Kluwer Academic, 1993. Topics discussed include: economics.

2936. Schelling, T. C. "Anticipating climate change: implications for welfare and policy." *Environment* 26 (October 1984): 6-9. Topics discussed include: air, carbon dioxide.

2937. Schneider, S. H. "Policymakers' summary of the potential impacts of climate change." *Environment* 33 (February 1991): 25-30.

2938. "Science concerning global climate change : hearing before the Committee on Energy and Natural Resources, United States Senate, One Hundred Third Congress, first session ... March 30, 1993." U.S. Senate Committee on Energy and Natural Resources. For sale by the U.S. G.P.O., Supt. of Docs., Congressional Sales Office, Washington, 1993. Topics discussed include: policy, global warming, greenhouse effects, gases.

2939. Singh, B. "The implications of climate change for natural resources in Quebec." Canada. Atmospheric Environment Service, and Canadian Climate Centre. The Service, Ottawa, 1988. Topics discussed include: resources, Quebec, climatic effects, economics.

2940. Sluyter, D. M. *Living with climatic change Phase II : proceedings of the symposium on ..., Reston, Va., November 9-11, 1976*. 183 pages. American Meteorological Society, Aspen Institute for Humanistic Studies, and Mitre Corporation. Metrek Division. McLean, VA: The Corporation, 1977.

2941. ---. *Proceedings of the Symposium on Living with Climatic Change, Phase II, Reston, Virginia, November 9, 10, 11, 1976.* 183 pages. Symposium on Living with Climatic Change, Phase II, Reston, Va., 1976, Mitre Corporation. Metrek Division, Aspen Institute for Humanistic Studies, and American Meteorological Society. McLean, VA: The Division, 1977.

2942. Stewart, T. R. et al. *Scientists' agreement and disagreement about global climate change : evidence from surveys.* State University of New York at Albany. Graduate School of Public Affairs. Center for Policy Research. Research report. Albany, NY: Nelson A. Rockefeller College of Public Affairs and Policy, State University of New York at Albany, 1992. Topics discussed include: global temperatures, changes, attitudes.

2943. Sullivan, M. et al. "Environmental planning, climate change and potential sea level rise : report on a mission to Kiribati." 55 pages. South Pacific Regional Environment Programme, and Association of South Pacific Environmental Institutions. SPREP reports and studies; no. 50, Noumea, New Caledonia: South Pacific Regional Environment Programme, 1991. Topics discussed include: policy, Kiribati, sea level.

2944. Tegart, W. J. M. et al. "Climate change : the IPCC impacts assessments." Intergovernmental Panel on Climate Change. Working Group II, World Meteorological Organization, United Nations Environment Programme, and Intergovernmental Panel on Climate Change. Australian Govt. Pub. Service, Canberra, 1990. Topics discussed include: policy.

2945. Tickell, C. "Climatic Change and World Affairs." 75 pages. *Harvard Studies in International Affairs*, No.37. Oxford: Pergamon, 1978. Topics discussed include: climate, economy.

2946. ---. "Human effects of climate change: excerpts from a lecture given to the Society on 26 March 1990." *The Geographical Journal* 156 (November 1990): 325-9.

2947. Timmerman, P. et al. "Economic perspectives on the impact of climate variability and change : a summary report." Canada. Atmospheric Environment Service. The Service, Ottawa, 1988. Topics discussed include: Canada.

2948. Tolba, M. "Strategies to Respond to Climate Changes--The Role of International Organizations." *Climate and development: Climatic change and variability and the resulting social, economic and technological implications*, 391-96. eds H. J. Karpe, D. Otten, and S. C. Trinidade. New York; Berlin; London and Tokyo: Springer in cooperation with the United Nations, 1990. Topics discussed include: conservation, pollution.

2949. Vellinga, P. et al. *Climate change policy in the European Community: report of a workshop held at the Royal Institute of International Affairs, October 1992.* 67 pages. Energy and Environmental Programme (Royal Institute of International Affairs). London; Washington, DC: Energy and Environmental Programme, The Royal Institute of International Affairs; distributed by The Brookings Institution, 1993. Topics discussed include: policy, Europe.

2950. Weber, B. "Possible Policies and Options in the European Community to Respond to Climatic Changes." *Climate and development: Climatic change and variability and the resulting social, economic and technological implications*, 50-53. eds H. J. Karpe, D. Otten, and S. C. Trinidade. New York; Berlin; London and Tokyo: Springer in cooperation with the United Nations, 1990. Topics discussed include: conservation, pollution.

2951. White, J. C. *Global climate change : linking energy, environment, economy, and equity.* 242 pages. Air Resources Information Clearinghouse (U.S.). Environmental science research; v. 45, New York: Plenum Press, 1992. Topics discussed include: development, air quality, management, policy.

2952. White, J. C. et al. *Global climate change : the economic costs of mitigation and adaptation : proceedings of a conference.* 438 pages. Center for Environmental Information (U.S.). New York: Elsevier, 1991. Topics discussed include: development, policy, environment, greenhouse effects, atmospheric effects.

2953. ---. *Global Climate Change: Linking Energy, Environment, Economy, & Equity.* New York, NY: Plenum Publishing Corporation, 1993. Topics discussed include: economy, policy, air quality.

2954. "Workshop on Environmental and Societal Consequences of a Possible CO2-Induced Climate Change, Annapolis, Maryland, April 2-6, 1979." Workshop on Environmental and Societal Consequences of a Possible CO2-Induced Climate Change (1979: Annapolis, MD), American Association for the Advancement of Science, and U.S. Dept. of Energy. Office of the Assistant Secretary for Environment. U.S. Dept. of Energy, Assistant Secretary for Environment, Office of Health and Environmental Research ; Available from National Technical Information Service, Washington, DC; Springfield, VA, 1980. Topics discussed include: sociology, atmospheric effects, carbon dioxide.

2955. *Climate change : science, impacts and policy : proceedings of the Second World Climate Conference,* edited by: J. Jager and H.L. Ferguson, World Climate Conference (2nd: 1990: Geneva). Cambridge: Cambridge University Press, 1991. Topics discussed include: climatology.

2956. Wright, H. E. "Societal responses to regional climatic change." *Environment* 33 (April 1991): 428 25-6.

2957. Wyman, R. L. *Global climate change and life on earth.* 282 pages. New York: Routledge, Chapman and Hall, 1991. Topics discussed include: global temperatures, changes, biology, diversity, conservation, energy, policy.

2958. Yohe, G. W. "Uncertainty, Climate Change and the Economic Value of Information: An Economic Methodology for Evaluating the Timing and Relative Efficacy of Alternative Response to Climate Change with Application to Protecting Developed Property from Greenhouse Induced Sea Level Rise." *Policy Sciences* 24, no. 3 (August 1991): 245-69. Topics discussed include: environment, management, conservation, pollution, resources.

2959. Zimmermeyer, G. "Climatic Risks and the Role of Energy Use." *Climate and development: Climatic change and variability and the resulting social, economic and technological implications*, 300-331. eds H. J. Karpe, D. Otten, and S. C. Trinidade. New York; Berlin; London and Tokyo: Springer in cooperation with the United Nations, 1990. Topics discussed include: conservation, pollution.

Drought & Flood Policy

2960. Arthur, L. M. et al. "The Effects of Technological Change on the Economic Impact of Agricultural Drought in Manitoba." *Canadian Journal of Agricultural Economics* 36, no. 2 (July 1988): 221-37. Topics discussed include: supply, development, irrigation, policy, technology, innovation.

2961. Bardsley, P. "A Note on the Viability of Rainfall Insurance--Reply The Economics of Insuring Crops against Drought." *Australian Journal of Agricultural Economics* 30, no. 1 (April 1986): 70-75. Topics discussed include: insurance, finance.

2962. Bardsley, P. et al. "The Economics of Insuring Crops against Drought." *Australian Journal of Agricultural Economics* 28, no. 1 (April 1984): 1-14. Topics discussed include: insurance, finance.

2963. Barker, F. S. "The Macro-economic Consequences of the Drought." *Journal for Studies in Economics and Econometrics* no. 20 (July 1984): 17-27. Topics discussed include: development, economics, Africa, agriculture, supply, demand.

2964. Barnard, J. R. "Externalities from Urban Growth: The Case of Increased Storm Runoff and Flooding." *Land Economics* 54, no. 3 (August 1978): 298-315. Topics discussed include: conservation, pollution, policy.

2965. Baumann, D. D. et al. "Flood Insurance: Some Determinants of Adoption." *Economic Geography* 54, no. 3 (July 1978): 189-96. Topics discussed include: insurance.

2966. Bruvold, W. H. "Residential Response to Urban Drought in Central California." *Water Resources Research* 15, no. 6 (December 1979): 1297-1304. Topics discussed include: conservation, pollution, economics.

2967. Cavalcanti, C. "Drought in the Brazilian Northeast: Some Features of Its Socio-economic Impact." *Climate and development: Climatic change and variability and the resulting social, economic and technological implications*, 379-84. eds H. J. Otten, D. Trinidade, S. C. Karpe. New York; Berlin; London and Tokyo: Springer in cooperation with the United Nations, 1990. Topics discussed include: conservation, pollution.

2968. Changnon, S. "Drought and climate change." *Bulletin of the American Meteorological Society* 72 (November 1991): 1790.

2969. Chen M.A. *Coping with seasonality and drought,* New Delhi ; Newbury Park: Sage Publications, 1991. Topics discussed include: economics, India, development.

2970. Clay, E. "Floods in Bangladesh, 1974 and 1984: From Famine to Flood-Crisis Management." *Preventing famine: Policies and prospects for Africa*, 131-37. Donald Curtis, Michael Hubbard, and Andrew With contributions from Edward Clay et al., Shepherd. New York and London : Routledge, 1988. Topics discussed include: agriculture, policy, poverty.

2971. Day, J. C. "A Linear Programming Approach to Floodplain Land Use Planning in Urban Areas." *American Journal of Agricultural Economics* 55, no. 2 (May 1973): 165-74. Topics discussed include: policy, development, irrigation.

2972. ---. "A Recursive Programing Model for Nonstructural Flood Damage Control." *Water Resources Research* 6, no. 5 (October 1970): 1262-71. Topics discussed include: urban areas, economics, policy, resources.

2973. Elfandy M.G. et al. "Time-series models adoptable for forecasting nile floods and Ethiopian rainfalls." *Bulletin of the American Meteorological Society* 75, no. 1 (January 1994): 83-94. Topics discussed include: atmosphere.

2974. Faulkingham, R. H. et al. "Population Dynamics and Drought: A Village in Niger." *Population Studies* 29, no. 3 (November 1975): 463-77. Topics discussed include: demography, economics.

2975. Forsythe, G. A. "An Assessment of the Role of Insurance and Structural Measures in Flood Mitigation Planning." *Review of Marketing and Agricultural Economics* 43, no. 2 (June 1975): 65-87. Topics discussed include: conservation, pollution, insurance.

2976. Fratkin, E. et al. "Drought and economic differentiation among Ariaal pastoralists of Kenya." *Human Ecology* 18 (December 1990): 385-402. Topics discussed include: pastoralists, herdsmen, Africa, livestock, households.

2977. Freebairn, J. W. "Drought Assistance Policy." *Australian Journal of Agricultural Economics* 27, no. 3 (December 1983): 185-99. Topics discussed include: policy, finance.

2978. Frey, R. S. "The structural context of city adoption of war on poverty and national flood insurance programs." *The Sociological Quarterly* 24 (December 1983): 59-74. Topics discussed include: decision making, policy, economics, assistance, insurance.

2979. Glantz, M. H. "Consequences and Responsibilities in Drought Forecasting: The Case of Yakima, 1977." *Water Resources Research* 18, no. 1 (February 1982): 3-13. Topics discussed include: econometrics, development, irrigation, policy.

2980. ---. "Floods, fires, and famine: is El Nino to blame?" *Oceanus* 27 (June 1984): 14-19. Topics discussed include: El Nino, ocean currents, weather, economics.

2981. Glau, T. E. "The Commonwealth Drought Bond." *Australian Journal of Agricultural Economics* 14, no. 2 (December 1970): 121-30. Topics discussed include: policy, taxation, revenues, subsidies.

2982. Gray, L. et al. "For Whom Is the Rural Economy Resilient? Initial Effects of Drought in Western Sudan." *Development and Change* 24, no. 1 (January 1993): 159-76. Topics discussed include: economics, development, rural areas, agriculture, resources, primary goods, urban areas, policy.

2983. Holm, J. D. et al. "Coping with drought in Botswana: an African success." *The Journal of Modern African Studies* 23 (September 1985): 463-82. Topics discussed include: Botswana, economics.

2984. Jeremy, D. J. "Damming the Flood: British Government Efforts to Check the Outflow of Technicians and Machinery, 1780-1843." *Business History Review* 51, no. 1 (March 1977): 1-34. Topics discussed include: economics, history, Europe, policy.

2985. Karrenbrock, J. D. "The 1988 Drought: Its Impact on District Agriculture." *Federal Reserve Bank of St Louis Review* 71, no. 3 (May 1989): 3-13. Topics discussed include: regional variations, economics.

2986. Kebbede, G. et al. "Drought, famine and the political economy of environmental degradation in Ethiopia." *Geography* 73 (January 1988): 65-70. Topics discussed include: agriculture, famines, Ethiopia, policy.

2987. Krenz, M. E. "Drought and hunger in Africa: denying famine a future." *Africa Today* 32, no. o4 (1985): 55-61. Topics discussed include: Africa, economics, development, famine.

2988. Kunreuther, H. et al. "An Economically Meaningful and Workable System for Calculating Flood Insurance Rates." *Water Resources Research* 6, no. 2 (April 1970): 659-67. Topics discussed include: resources, insurance.

2989. Lacewell, R. D. et al. "A General Model for Evaluating Agricultural Flood Plains." *American Journal of Agricultural Economics* 54, no. 1 (February 1972): 92-101. Topics discussed include: development, irrigation, policy.

2990. Lang, I. "Long-Range Strategic Concepts to Reduce Drought Damages to Agriculture." *Climate and development: Climatic change and variability and the resulting social, economic and technological implications*, 255-60. editors H. J. Karpe, D. Otten, and S. C. Trinidade. New York; Berlin; London and Tokyo: Springer in cooperation with the United Nations, 1990. Topics discussed include: policy, conservation, pollution.

2991. Lant, C. L. et al. "The economic value of riparian corridors in cornbelt floodplains: a research framework." *The Professional Geographer* 41 (August 1989): 337-49. Topics discussed include: spatial analysis, agriculture, geography, ecology, wetlands, environment, quality, public goods.

2992. Le Nay, J. et al. "The Impact of Drought on the National Accounts for Livestock in Sahelian Countries." *Review of Income and Wealth* 35, no. 2 (June 1989): 209-24. Topics discussed include: income.

2993. Lewandowski, A. et al. "Application of DIDASS Methodology to a Flood Control Problem--Numerical Experiments." *Plural Rationality and Interactive Decision Processes: Proceedings of an IIASA Systems Analysis) Summer Study on Plural Rationality and Interactive Decision Processes Held at Sopron, Hungary, August 16 26, 1984 Lecture Notes in Economics and Mathematical Systems series*, ed M. Grauer, M. Thompson, and A. P. Wierzbicki. New York; Berlin and Tokyo,: Springer, 1985. Topics discussed include: conservation, pollution.

2994. Maunder, W. J. "The Economic Consequences of Drought: with Particular Reference to the 1969-70 Drought in New Zealand." *New Zealand Meteorological Service Technical Note* (1972): 208. Topics discussed include: economy, weather, agriculture.

2995. McBride, B. "'If people be killing killing . . . ' Plagued by poachers, drought, and economic instability, Tanzania remains committed to the conservation ethic." *Sierra* 70 (March 1985): 64-8. Topics discussed include: wildlife, conservation, national parks, reserves, Tanzania.

2996. McDonald, G. T. "Agricultural Flood Damage Assessment: A Review and Investigation of a Simulation Method." *Review of Marketing and Agricultural Economics* 38, no. 3 (September 1970): 105-20. Topics discussed include: resources.

2997. Meyer, D. "Does Drought Really Depress Beef Prices in South Africa?" *Journal for Studies in Economics and Econometrics* 12, no. 2 (July 1988): 7-14. Topics discussed include: agriculture, supply, demand, agriculture, markets, marketing, cooperatives.

2998. Mitra, S. et al. "The Sahel: Drought, Desertification and Man-Made Problems." *Africa: Dimensions of the economic crisis An analysis of the problems and constraints of development*, 128-45. eds Shanti Sadiq, Gupta, Anirudha Ali. New Delhi: Sterling; disbributed by Apt Books, New York, 1987. Topics discussed include: development, economics, Africa.

2999. Moncur, J. E. T. "Urban Water Pricing and Drought Management." *Water Resources Research* 23, no. 3 (March 1987): 393-98. Topics discussed include: resources, consumer behavior, expenditures, cycles, consumption.

3000. Morgan, R. "Drought Relief Programmes in Botswana." *Preventing famine: Policies and prospects for Africa With contributions from Edward Clay et al.*, 112-20. Donald Curtis, Michael Hubbard, and Andrew Shepherd. New York and London: Routledge, 1988. Topics discussed include: agriculture, policy, poverty.

3001. Mortimore, M. J. "Adapting to drought." *Economic Geography* 66 (January 1989): 97-100.

3002. Nadkarni, M. V. et al. "Socio-economic conditions in drought-prone areas." *Economic Development and Cultural Change* 36 (April 1988): 609-13.

3003. Pearce; F. et al. "West sinks Bangladesh flood plan." *New Scientist* 139 (21 August 1993): 4. Topics discussed include: Bangladesh, economics, assistance.

3004. Pritchett, S. T. et al. "A Case Study of Flood Losses: Implications for Flood Insurance Product Development." *Journal of Risk and Insurance* 42, no. 1 (March 1975): 105-15. Topics discussed include: taxation, revenues, subsidies, insurance, urban areas, economics, policy.

3005. Quiggin, J. "A Note on the Viability of Rainfall Insurance The Economics of Insuring Crops against Drought." *Australian Journal of Agricultural Economics* 30, no. 1 (April 1986): 63-69. Topics discussed include: insurance, finance.

3006. Quinn, V. et al. "Crisis-Proofing the Economy: The Response of Botswana to Economic Recession and Drought." *Adjustment with a human face Volume 2 Country case studies*, 3-27. eds Giovanni Andrea Cornia, Richard Jolly and Frances Stewart. Oxford; New York; Toronto and Melbourne: Oxford University Press, Clarendon Press, 1988. Topics discussed include: welfare, policy, agriculture, economic stabilization, theory.

3007. Rao, C. H. H. et al. *Unstable agriculture and droughts: Implications for policy*. 192 pages. Institute of Economic Growth Studies in Economic Development and Planning series ed. New Delhi: Vikas, 1988. Topics discussed include: supply, demand, policy.

3008. Rettger, M. J. et al. "Flood Insurance or Disaster Loans: An Economic Evaluation." *American Journal of Agricultural Economics* 61, no. 3 (August 1979): 496-505. Topics discussed include: insurance, welfare, health issues, education.

3009. Robie, R. B. "Pressures Created by a Severe Drought on Water Institutions." *American Journal of Agricultural Economics* 59, no. 5 (December 1977): 938-42. Topics discussed include: resources.

3010. Rosine, J. et al. "Drought, Agriculture, and the Economy." *Federal Reserve Bulletin* 75, no. 1 (January 1989): 1-12.

3011. Ross, M. H. "Kalamazoo River Flood Control Proposal: A Case Study." *Land Economics* 48, no. 1 (February 1972): 57-61. Topics discussed include: conservation, pollution.

3012. Sims, J. H. et al. "The adoption of residential flood mitigation measures: what price success?" *Economic Geography* 63 (July 1987): 259-72. Topics discussed include: hazards, geography, environment, disasters, Illinois.

3013. Smith, H. H. "Drought 1988: Farmers and the Macroeconomy." *Federal Reserve Bank of Dallas Economic Review* (September 1988): 15-22. Topics discussed include: agriculture, supply, demand.

3014. Subrahmanyam, V. P. et al. "A Climatic Study of Droughts in Peninsular India." *Geogr Journ* 129 (1963): 122-125. Topics discussed include: climate, agriculture.

3015. Taylor, J. G. et al. "Perceptions of Drought in the Ogallala Aquifer Region." *Environment and Behavior* 20, no. 2 (March 1988): 150-175. Topics discussed include: Western States, farmers, weather, water supply, perceptions, environment, adaptation.

3016. Thatcher, L. P. et al. "Least-Cost Feed Reserve Using Drought Probabilities Derived from a Grazing Model." *Review of Marketing and Agricultural Economics* 43, no. 3 (September 1975): 113-25. Topics discussed include: agribusiness management.

3017. Thomas, J. Murray et al. *Climates of hunger: mankind and the world's changing weather.* Madison, WI: University of Wisconsin Press, 1977. Topics discussed include: paleoclimatology, droughts.

3018. Toulmin, C. "Livestock Policy in the Sahel: Why It Must Become More Drought Oriented." *Preventing famine: Policies and prospects for Africa With contributions from Edward Clay et al.*, 171-77. Donald Curtis, Michael Hubbard, and Andrew Shepherd. New York and London: Routledge, 1988. Topics discussed include: agriculture, policy, poverty, agriculture, supply, demand analysis.

3019. Uddin, F. "Impact of Drought on Agriculture: Some Observations." *Economic Affairs* 29, no. 2 (April 1984): 121-26.

3020. "U.S. economic and social impacts of the record 1976-77 winter freeze and drought." Center for Environmental Assessment Services (U.S.). U.S. Dept. of Commerce, National Oceanic and Atmospheric Administration, Environmental Data and Information Service, Center for Environmental Assessment Services, Washington, DC, 1982. Topics discussed include: frost, sociology.

3021. Wheaton, E. E. "Frequency and Severity of Drought and Dust Storms." *Canadian Journal of Agricultural Economics* 38, no. 4 (December 1990): 695-700. Topics discussed include: conservation, pollution.

3022. Wonder, B. et al. "The Influence of Season and Property Characteristics upon the Viability of Cattle Producers in Drought Prone Areas: A Case Study of the Alice Springs District." *Quarterly Review of Agricultural Economics* 30, no. 3 (July 1977): 235-53. Topics discussed include: industry, agribusiness management.

3023. Wurbs, R. A. "Economic feasibility of flood control improvements." *Journal of Water Resources Planning and Management* 110 (January 1984): 126-7.

Labor & Unemployment

3024. Alderman, H. et al. "Understanding the Seasonality of Employment, Wages, and Income." *Seasonal variability in third world agriculture: The consequences for food security*, 81-106. ed David E. Sahn. Baltimore and London: Johns Hopkins University Press for the International Food Policy Research Institute, 1989. Topics discussed include: supply, rural areas, economics, policy.

3025. Aringazina, A. et al. "The dependence of physical work capacity on genotype in unpleasant weather." *Fiziol Cheloveka* 15, no. 4 (July 1989): 163-5. Topics discussed include: exercise, phenotype, seasonality, genetics, medicine, work capacity.

3026. Asshauer, E. et al. "On the evaluation of employees returning from the tropics." *Munch Med Wochenschr* 109, no. 22 (2 June 1967): 1209-16. Topics discussed include: dysentery, amebic, bacillary, filariasis, helminthiasis, hepatitis, hookworm infections, gastrointestinal diseases, work-related diseases.

3027. Barro, R. J. "An Efficiency-Wage Theory of the Weather." *Journal of Political Economy* 97, no. 4 (August 1989): 999-1001. Topics discussed include: labor economics.

3028. Becker, F. "Summary of the work of physicians and meterologists in bioclimatologic expert opinions, consultation and information." *Arch Phys Ther Leipz* 19, no. 5 (September 1967): 498-500. Topics discussed include: climates, health resorts.

3029. Bedford, T. "Some Effects of Atmospheric Conditions on the Industrial Worker." *J Ind Hyg* 10 (1928): 364. Topics discussed include: weather, environment, biology, behavior, economy, labor.

3030. Berdyshev, V. "Effect of increased doses of vitamin C on the adaptation and working ability of sailors in the tropical climate." *Voen Med Zh no.* 6 (June 1986): 47-8. Topics discussed include: work capacity, adaptation, physiological drug effects, ascorbic acid, naval medicine.

3031. Beveridge, W. H. "Prices and Wages in England, London, 1939.'Bibliography on Climatic Changes'." in *Meteorological Abstracts and Bibliography*, July 1950. Topics discussed include: climates, economy.

3032. Bloch, M. N. "The effect of seasonal maternal employment on young Senegalese children's behavior." *Journal of Comparative Family Studies* 19, no. 3 (September 1988): 397-417. Topics discussed include: behavior, learning activities, Senegal, social adjustment, learning, work, women, preschool age children, mothers, childhood.

3033. Bonin, J. M. et al. "Estimated Seasonal Components in Recorded Unemployment in 1970." *Industrial and Labor Relations Review* 25, no. 3 (April 1972): 383-98. Topics discussed include: employment, unemployment levels, vacancy levels, retirement patterns.

3034. Bonnet, M. H. "Dealing with shift work: Physical fitness, temperature, and napping. Special Issue: Factors for promoting adjustment to night and shift work." *Work and Stress* 4, no. 3 (July 1990): 261-274. Topics discussed include: physical condition, naps, adjustment to night, workday shifts, fitness, sleep.

3035. Brittain, J. A. "Testing for and Removing Bias in the Seasonally Adjusted Unemployment Series." *Brookings Papers on Economic Activity* no. 1 (1976): 211-24. Topics discussed include: employment, unemployment levels, vacancy levels, retirement patterns.

3036. Brose, H. G. et al. "Temporary Work: Outlines of a New Employment Pattern." *Soziale Welt* 38, no. 3 (1987): 282-308. Topics discussed include: seasonality, labor market segments, Germany, France, temporary work.

3037. Bushov, I. et al. "Effect of moderate high altitude hypoxia on human function and work capacity in relation to environmental temperature." *Fiziol Cheloveka* 13, no. 2 (March 1987): 284-9. Topics discussed include: atmosphere, electroencephalography, fever physiopathology, hemodynamics, timing, anoxia physiopathology, work capacity.

3038. Card D. et al. "Unemployment insurance taxes and the cyclical and seasonal properties of unemployment", *NBER working papers series*; working paper no. 4030, Cambridge, Mass.: National Bureau of Economic Research, 1992. Topics discussed include: insurance, unemployment taxation, econometrics.

3039. Card D. et al. "Unemployment insrurance taxes and the cyclical and seasonal properties of unemployment." *Journal of Public Economics* 53, no. 1 (January 1994): 1-29. Topics discussed include: layoffs.

3040. Carton de Grammont, H. "Some Reflections on the Labor Market in the Latin American Countryside; Algunas reflexiones en torno al mercado de trabajo en el campo latinoamericano." *Revista*

Mexicana de Sociologia 54, no. 1 (January 1992): 49-58. Topics discussed include: Latin America, rural areas, labor market trends.

3041. Chiswick, B. R. "The Effect of Unemployment Compensation on a Seasonal Industry: Agriculture." *Journal of Political Economy* 84, no. 3 (June 1976): 591-602. Topics discussed include: labor markets, unemployment insurance.

3042. Choi, K. E. "Seasonal variations in Louisiana employment 1947-1964." 43 pages. Louisiana. State University in New Orleans. Division of Business and Economic Research Research study no. 5. Employment study no. 2, New Orleans: Division of Business and Economic Research, Louisiana State University in New Orleans, 1967. Topics discussed include: seasonality, labor, Louisiana.

3043. Crouter, A. C. et al. "Temporal rhythms in family life: Seasonal variation in the relation between parental work and family processes." *Developmental Psychology* 29, no. 2 (March 1993): 198-205. Topics discussed include: seasonality, parent involvement, parental monitoring, activities, childrearing practices.

3044. Dales, R. et al. "Does seasonal employment in grain elevators increase nonspecific airways responsiveness?" *J Occup Med* 30, no. 8 (August 1988): 613-6. Topics discussed include: Quebec, smoking, spirometry, dust, lung physiology, work-related diseases.

3045. Datta, S. K. et al. "Seasonality, Differential Access Interlinking of Labour and Credit." *Journal of Development Studies* 24, no. 3 (April 1988): 379-93. Topics discussed include: rural areas, labor, economics, finance, agriculture.

3046. Daum, M. et al. "Seasonal vulnerability of the old and cold: The role of the senior citizen center." *Journal of Gerontological Social Work* 5, no. 4 (June 1983): 81-106. Topics discussed include: cold weather problems, transportation, community services.

3047. Davis, R. E. "Summer weather and employee absenteeism in six United States cities." 55 pages. *Publications in climatology*; 41, no. 2 (1988); Elmer, NJ: C.W. Thornthwaite Associates, Laboratory of Climatology. Topics discussed include: absenteeism, labor, psychology.

3048. de Klerk, M. "Seasons That Will Never Return: The Impact of Farm Mechanization on Employment, Incomes and Population Distribution in the Western Transvaal." *Journal of Southern African Studies* 11, no. 1 (October 1984): 84-105. Topics discussed include: employment, farming, labor markets, South Africa.

3049. Dernedde, E. "A correlation of the wet-bulb globe temperature and Botsball heat stress indexes for industry." *American Industrial Hygiene Association Journal* 53 (March 1992): 169-74. Topics discussed include: heat, physiology, foundries, labor, metal workers, diseases, hygiene.

3050. Diesfeld, H. "Hepatitis infections due to stay in tropical and subtropical developing countries, and their evaluation by an accident insurance carrier." *Med Klin* 69, no. 40 (4 October 1974): 1634-8. Topics discussed include: Germany, health issues, manpower, microbiology, work-related diseases.

3051. Dmytrow, E. "The Impact of the Winter of 1977 on Payroll Employment." *Monthly Labor Review* 100, no. 8 (August 1977): 43-45. Topics discussed include: employment, unemployment levels, vacancy levels, retirement patterns.

3052. Dragel', V. S. "Seasons and Collective Farmers' Labor Rhythms." *Sotsiologicheskie Issledovaniya* 10, no. 3 (July 1983): 126-128. Topics discussed include: USSR, farming efficiency, seasonality, policy.

3053. Dunn, R. "Analysing spatial time series of local unemployment: a graphical approach using principal components analysis and seasonal adjustment procedures." *Environment and Planning A* 19 (February 1987): 225-46. Topics discussed include: time series models, United Kingdom, seasonality, economics.

3054. Edebalk, P. G. et al. "Unemployment Insurance and Seasonal Unemployment." *Economy and History* 21, no. 1 (1978): 3-12. Topics discussed include: employment, vacancy levels, retirement patterns.

3055. Elterich, G. J. "Seasonality of Farm Labor Use Patterns in the United States; Discussion." *Seasonal Agricultural Labor Markets in the United States*, 96-103. ed Robert D. Emerson. Ames: Iowa State University Press, 1984. Topics discussed include: labor markets, unemployment levels, vacancy levels, retirement patterns, agribusiness management.

3056. Emerson, R. D. "Seasonal Agricultural Labor Markets in the United States: Summary." *Seasonal Agricultural Labor Markets in the United States*, 482-523. ed Robert D. Emerson. Ames: Iowa State University Press, 1984. Topics discussed include: labor markets, public policy, government.

3057. "Employment and unemployment." Time Series Research and Analysis. Statistics Canada, Ottawa, 1982. Topics discussed include: labor supply, Canada, unemployment levels, seasonality, economics.

3058. Enander, A. E. "Effects of thermal stress on human performance." *Scandinavian Journal of Work, Environment and Health* 15, no. Suppl 1 (1989): 27-33. Topics discussed include: thermal stress, performance, a literature review, temperature effects.

3059. Engerman, S. L. et al. "Seasonality in nineteenth century labor markets." 28 pages. National Bureau of Economic Research. NBER working papers series on historical factors in long run growth; working paper no. 20, Cambridge, MA: National Bureau of Economic Research, 1991. Topics discussed include: seasonality, labor, history.

3060. Exley, H. J. et al. "Employment and its seasonality in Tasmania." Govt. Printer, Hobart, 1938. Topics discussed include: seasonal industries, unemployment levels, Tasmania.

3061. Fabella, R. V. "Rural Non-Farm Employment in the Philippines: Composition, Growth and Seasonality." *Off farm employment in the development of rural Asia Volume 2 Papers presented at a conference held in Chiang Mai, Thailand, 23 to 26 August 1983*, 69-97. ed R. T. Shand. Canberra: Australian National University, National Centre for Development Studies, 1986. Topics discussed include: labor markets, rural areas, labor, economics, unemployment levels, vacancy levels, retirements.

3062. Fearn, R. M. "Cyclical, Seasonal, and Structural Factors in Area Unemployment Rates." *Industrial and Labor Relations Review* 28, no. 3 (April 1975): 424-31. Topics discussed include: employment, unemployment levels, vacancy levels, retirement patterns, regional economics, public policy, government.

3063. Fehrs, L. J. et al. "Prevalence of hepatitis B markers among pregnant Hispanic women in migrant, seasonal work in Oregon." *American Journal of Public Health* 78 (August 1988): 971-2. Topics discussed include: women, agriculture, labor, diseases.

3064. Ferris, J. S. et al. "Labor Mobility, Seasonal Unemployment Insurance, and the Newfoundland Inshore Fishery." *Canadian Journal of Economics* 15, no. 3 (August 1982): 426-41. Topics discussed include: resources, unemployment insurance.

3065. Fiske, C. N. "The Effects of Exposure to Heat on the Working Organism." Trans 15th Internat cong Hyg Demog, 655-672. 1913. Topics discussed include: temperature, biology, labor, environment.

3066. Florian, H. "Occupational medical experiences with personnel employed overseas with special regard to stress of heat and high altitude." *Munch Med Wochenschr* 109, no. 10 (10 March 1967): 515-22. Topics discussed include: anoxia, environment, exposure, diseases, stress, adaptation, physiology.

3067. Frazen, E. "Medical aspects of work in tropical countries." *Z Arztl Fortbild Jena* 72, no. 19 (1 October 1978): 950-5. Topics discussed include: aerospace medicine, development, Europe, ethnology, malaria, tropical climates, work-related medicine.

3068. Friedland, W. H. "Seasonal Farm Labor and Worker Consciousness." *Research in the Sociology of Work* no. 1 (1981): 351-380.

3069. Fritsch, C. F. "Seasonality of Farm Labor Use Patterns in the United States." *Seasonal Agricultural Labor Markets in the United States*, 64-95. ed Robert D. Emerson. Ames: Iowa State University Press, 1984. Topics discussed include: labor markets, unemployment levels, vacancy levels, retirement patterns, agribusiness management.

3070. Gardner, B. "Seasonal Farm Labor and U.S. Farm Policy." *Emerson, Robert D, ed Seasonal Agricultural Labor Markets in the United States*, 450-76. Ames: Iowa State University Press, 1984. Topics discussed include: policy, labor markets, wages, employment.

3071. Gerster, H. J. "A Statistical Evaluation of the Labor Market-Interpretations Based on Seasonally Adjusted Data." *Jahrbucher fur Nationalokonomie und Statistik* 201, no. 2 (March 1986): 152-74. Topics discussed include: forecasting, unemployment levels, vacancy levels, retirement patterns.

3072. Gertner, A. et al. "Effects of work and motivation on the heart rates of chronic heat exposed workers during their regular work shifts." *Ergonomics* 27, no. 2 (February 1984): 135-146. Topics discussed include: type of work, motivation, heart rate, exposed metal workshop workers, Israel, motivation, heart rate, skilled industrial workers.

3073. Gifford, A. "Growth and the trend in seasonality in employment in California, 1950-1965." 28 pages. Institute of Labor Economics, San Diego State College. Bulletin no. 6, San Diego: Institute of Labor Economics, San Diego State College, 1969. Topics discussed include: seasonality, labor, California, seasonal industries.

3074. Gliner, J. et al. "Man's Physiological Response to Long Term Work During Thermal and Pollutant Stress." *Journal of Applied Physiology* 39 (1975): 628-632. Topics discussed include: biology, behavior, temperature, environment.

3075. Gutmann, S. *Now Hiring! Jobs In Asia: The Insider's Guide to Gaining Seasonal & Year round Employment Throughout Asia.* Seattle, WA: Progressive Media, 1993. Topics discussed include: employment, East Asia.

3076. Hald, H. F. "Seasonal adjustment factors; wholesale price index: selected series, 1948-61." U.S. Dept. of Labor, Bureau of Labor Statistics; for sale by the Superintendent of Documents, U.S. Govt. Print. Off., Washington, 1963. Topics discussed include: wholesale price index, seasonality, economics.

3077. Halperin, R. H. "Ecology and Mode of Production: Seasonal Variation and the Division of Labor by Sex among Hunter-Gatherers." *Journal of Anthropological Research* 36, no. 3 (September 1980): 379-399. Topics discussed include: hunting, variations, ecology.

3078. Halpin, T. C. "The Effect of Unemployment Insurance on Seasonal Fluctuations in Employment." *Industrial and Labor Relations Review* 32, no. 3 (April 1979): 353-362. Topics discussed include: unemployed, highways, construction, outerwear, wood products.

3079. Heatley, R. B. "The impact of winter on the Canadian worker; an examination of seasonal variations in employment and unemployment in Canada." Canada. Dept. of Labour. Economics and Research Branch. Queen's Printer, Ottawa, 1966. Topics discussed include: unemployment, seasonal, Canada.

3080. Hellstrom, R. et al. "Physical working capacity, training and climate." *Acta Med Scand Suppl* 472 (1967): 207-14. Topics discussed include: blood pressure, body weight, hemoglobinometry, physical labor, physiology, humidity, physical education, work.

3081. Hioki, A. et al. "Effect of seasonal and intermittent work at high altitude on health status." *Sangyo Igaku* 34, no. 3 (May 1992): 272-8. Topics discussed include: work-related health, blood pressure physiology, health status, hemoglobins analysis.

3082. Holt, J. S. "Introduction to the Seasonal Farm Labor Problem." *Seasonal Agricultural Labor Markets in the United States*, 3-32. ed Robert D. Emerson. Ames: Iowa State University Press, 1984. Topics discussed include: labor markets, unemployment levels, vacancy levels, retirement patterns, agribusiness management.

3083. Horvath, S. M. *Physiological Adjustments to Hemorrhage, Altitude, and Work Final scientific rept 1 Dec 77 31 Aug 83*, Performer: University of California, Santa Barbara. Inst. of Environmental Stress. Air Force Office of Scientific Research, Bolling AFB, Washington, DC, October 1983. Topics discussed include: adaptation, stress, physiology, psychology, work, exercise, behavior.

3084. Huber, F. "Seasonal workers and unemployment insurance in Great Britain, Germany, and Austria. A survey of legal provisions and administrative practice through 1938." U. S. Govt. Print. Off., Washington, 1940. Topics discussed include: unemployment, United Kingdom, insurance, Germany, Austria, seasonality, labor, Europe.

3085. Inoue, K. "Seasonal change in farmers' working and sleeping hours." *J Hum Ergol Tokyo* 1, no. 1 (September 1973): 19-28. Topics discussed include: environment, Japan, gender effects, sunlight, timing, agriculture, sleep, work.

3086. International Labour Office. "Environment and the world of work: Report of the Director-General (Part 1)." *International Labour Conference, 77th Session Geneva: Author, 1990,* 106 pages, 1990. Topics discussed include: conservation, pollution, labor markets, wage levels, employment.

3087. Islamov, O. R. et al. "Seasonal Work; Otkhodnichestvo." *Sotsiologicheskie Issledovaniya* 16, no. 1 (1989): 46-49. Topics discussed include: labor markets, supply, demand, USSR.

3088. James, R. et al. "Effects of respirators under heat, work conditions." *American Industrial Hygiene Association Journal* 45 (June 1984): 399-404. Topics discussed include: temperature, physiology, exercise.

3089. Janicek, M. et al. "Effect of work load in a hot environment on the pulse rate of blacksmiths working in hot workshops." *Cas Lek Cesk* 113, no. 9 (28 February 1974): 263-6. Topics discussed include: adaptation, blood pressure, environment, exposure, seasonality, temperature, heat, work-related medicine, work.

3090. Katsuura, T. "Circulatory responses to work at simulated altitudes under different ambient temperatures." *Ann Physiol Anthropol* 4, no. 2 (April 1985): 115-22. Topics discussed include: oxygen consumption.

3091. Kerimova, M. et al. "The importance of vitamin administration, using the preparation Undevit, for enhancing the work capacity of school children during the winter spring period." *Gig Sanit no.* 10 (October 1990): 42-5. Topics discussed include: physiopathology, psychology, Azerbaijan, mental processes, urban areas, population, drug effects, seasonality.

3092. Khaleque, A. "Effects of diurnal and seasonal sleep deficiency on work effort and fatigue of shift workers." *Int Arch Occup Environ Health* 62, no. 8 (1991): 591-3. Topics discussed include: Bangladesh, circadian rhythms, physiology, work tolerance, fatigue, etiology, sleep deprivation.

3093. Knave, B. et al. "Work with video display terminals among office employees. II. Physical exposure factors." *Scand J Work Environ Health* 11, no. 6 (December 1985): 467-74. Topics discussed include: ultraviolet rays, climatic effects, electromagnetic fields, electromagnetics, microclimates, work-related medicine.

3094. Kusuda, K. "Surplus Population and Mobility of Labor." *Developing Economies* 9, no. 2 (June 1971): 154-173. Topics discussed include: Asia, development, economics, demography, labor markets.

3095. Larin, V. "Effect of the seasonal character of agricultural work on the use of medical help by a rural population." *Vrach Delo* 2 (February 1969): 115-7. Topics discussed include: Ukraine, health issues, services.

3096. Lashneva, I. "Evaluation of the efficacy of work during the summer in improving the health of preschool children." *Gig Sanit no.* 2 (February 1990): 39-42. Topics discussed include: USSR, child development, physiology, exercise, seasonality.

3097. Lin'Kov, N. "Seasonal employment and work regimens of women machinery operators." *Gig Sanit no.* 2 (February 1982): 21-4. Topics discussed include: dust, Russia, task performance, agriculture, work.

3098. Lisovskii, V. et al. "Work capacity and the circadian rhythm of certain physiological functions in sailors at different latitudes." *Voen Med Zh no.* 1 (January 1976): 50-2. Topics discussed include: adaptation, arctic regions, seasonality, USSR, circadian rhythms, disabilities, naval medicine, work capacity.

3099. Little, J. I. "Ethnicity, Family Structure, and Seasonal Labor Strategies on Quebec's Appalachian Frontier, 1852-1881." *Journal of Family History* 17, no. 3 (June 1992): 289-302. Topics discussed include: labor markets, economics.

3100. Lovell, M. C. "Least-Squares Seasonally Adjusted Unemployment Data." *Brookings Papers on Economic Activity* no. 1 (1976): 225-37. Topics discussed include: labor markets, wage levels, unemployment levels, vacancy levels, retirement patterns.

3101. Maksimov, A. "Evaluation of the physical work capacity of persons at various altitudes." *Voen Med Zh no.* 11 (November 1977): 66-8. Topics discussed include: adaptation, physical labor, seasonality, timing, disabilities, physical fitness, work capacity.

3102. Maksimova, O. et al. "Characteristics of the work capacity of the machine operators in power installations in different climatic zones." *Gig Tr Prof Zabol* 17, no. 4 (April 1973): 13-6. Topics discussed include: physiology, heart functions, reaction time, seasonality, USSR, electricity, work-related medicine.

3103. Matsui, H. et al. "Seasonal variation of aerobic work capacity in ambient and constant temperature." In: Folinsbee L.J. et al., ed. *Environmental Stress.* New York: Academic Press, 1977: 279-291. Topics discussed include: adaptation, leg blood supply, oxygen consumption, regional blood flow, physical labor.

3104. Medel R., J. et al. *Las temporeras y su vision del trabajo : condiciones de trabajo y participacion social.* 146 pages. Santiago de Chile: Programa Mujer Campesina y Asalariada Agricola, Centro de Estudios de la Mujer, 1989. Topics discussed include: women, agricultural laborers, Chile, seasonality, labor.

3105. Medvedev, V. et al. "The concept of the program 'The influence of global and environmental and climatic changes on human health, working capacity and activity' (problem 6.4 of the State Scientific and Technological Program 'Global environmental and climatic changes')." *Fiziol Cheloveka* 18, no. 5 (September 1992): 5-14. Topics discussed include: efficiency, health issues, monitoring, government, work.

3106. "Migrant and seasonal farmworker programs." U.S. Employment and Training Administration. Dept. of Labor, Employment and Training Administration, Washington, 1977. Topics discussed include: migrants, agriculture, seasonality, labor, economics, assistance.

3107. Miles, D. et al. "Absolute and relative work capacity in women at 758, 586, and 523 torr barometric pressure." *Aviat Space Environ Med* 51, no. 5 (May 1980): 439-44. Topics discussed include: altitude, carbon dioxide physiology, maximal voluntary ventilation, oxygen physiology, atmospheric pressure, physical labor, heart rate, respiration.

3108. Moser, J. W. "Labor Market Transitions: Cyclic, Trend, and Seasonal Effects for Prime-Age Men." *Industrial and Labor Relations Review* 39, no. 2 (January 1986): 251-63. Topics discussed include: labor markets, unemployment levels, vacancy levels, retirement patterns.

3109. Moss, K. N. "Some Effects of High-Air Temperature upon Miner." *Trans Inst Min Eng* 66, no. 5 (1924): 284-314. Topics discussed include: biology, labor.

3110. Mott, F. L. et al. "Seasonal Variations in Labor-Force Activity and Intrahousehold Substitution of Labor in Rural Keyna." *Journal of Developing Areas* 18, no. 4 (July 1984): 449-63. Topics discussed include: rural areas, labor, economics, labor markets, demography.

3111. Nath, S. K. "Estimating the Seasonal Marginal Products of Labour in Agriculture." *Oxford Economic Papers, N S* 26, no. 3 (November 1974): 375-87. Topics discussed include: developing countries, Asia, supply, demand, productivity, labor, capital.

3112. Navakatikian, A. et al. "Effect of the seasonality and strain of work on the basic body functions of workers in agricultural production." *Vrach Delo* no. 7 (July 1981): 100-3. Topics discussed include: physiopathology, cardiovascular systems, hemodynamics, respiration.

3113. Newbold, P. et al. "Seasonal Adjustment of Austrian Labour Force Series." *Empirica* 11, no. 2 (1984): 147-204. Topics discussed include: time series analysis, spectral models, labor markets, unemployment levels, vacancy levels, retirement patterns.

3114. Newman, M. J. "Seasonal Variations in Employment and Unemployment during 1951-75." *Monthly Labor Review* 103, no. 1 (January 1980): 48-52. Topics discussed include: unemployment levels, vacancy levels, retirement patterns.

3115. Odea, J. D. "Possible contribution of low ultraviolet light under the rain forest canopy to the small stature of Pygmies and Negritos." *Homo* 44, no. 3 JA (January 1994): 284-287. Topics discussed include: energy, Australia.

3116. Oshima, H. T. "Seasonality and Underemployment in Monsoon Asia." *Philippine Economic Journal* 10, no. 1 (1971): 63-97. Topics discussed include: unemployment levels, vacancy levels, retirement patterns, development, Asia, agriculture.

3117. Pechinski, J. M. "The effects of a systematic physical conditioning program with and without acclimitization treatments on performance in a hot environment." Ph.D. diss., University of Illinois at Urbana-Champaign, 1974. Topics discussed include: physical education, training, acclimation, heat, physiology.

3118. Pfeffermann, D. "Estimation and Seasonal Adjustment of Population Means Using Data from Repeated Surveys." *Journal of Business and Economic Statistics* 9, no. 2 (April 1991): 163-75. Topics discussed include: econometrics, absenteeism, time series analysis, spectral models, multivariate models, wage levels, fringe benefits, labor, economics.

3119. Pugh, L. "The influence of wind resistance in running and walking and the mechanical efficiency of work against horizontal or vertical forces." *J Physiol Lond* 213, no. 2 (March 1971): 255-76. Topics discussed include: air, biomechanics, gravitation, oxygen consumption, specific gravity, viscosity, physical labor, locomotion, weather.

3120. Quadt, F. "The effect of increased temperatures in the examination room on the efficiency of national servicemen during the EVP (Examination of Qualification and Employment).)." *Wehrpsychologische Untersuchungen* 14, no. 6 (1979): 170-175. Topics discussed include: increased room temperature, concentration, efficiency during testing, temperature effects, enlisted military personnel, concentration, performance.

3121. Rahman, N. et al. "A preliminary survey of filarial infection in a group of labour population at Maijan tea estate in Lakhimpur district, Assam." *J Indian Med Assoc* 56, no. 12 (16 June 1971): 363-6. Topics discussed include: blood microbiology, climatic effects, India, infants, filariasis epidemiology.

3122. Rahman, R. I. "Implications of Seasonality of Rural Labour Use Pattern: Evidences from Two Villages in Bangladesh." *Bangladesh Development Studies* 9, no. 1 (December 1981): 77-96. Topics discussed include: rural areas, labor, economics, unemployment levels, vacancy levels, retirement patterns.

3123. Ramsey, J. D. et al. "Self determined work, rest cycles in hot environments." *International Journal of Production Research* 12, no. 5 (September 1974): 623-631. Topics discussed include: temperature, work, rest, cycles, performance.

3124. Reichert, J. S. "The Migrant Syndrome: Seasonal U.S. Wage Labor and Rural Development in Central Mexico." *Human Organization* 40, no. 1 (March 1981): 56-66. Topics discussed include: rural areas, development, Mexico, labor, wage levels, seasonality, rural development.

3125. Reinders, J. E. "Labor Shortages as a Determinant of the Immigration Climate in South Africa; Het Arbeidstekort als Bepalende Factor van het Immigratie-Kilmaat in Zuid-Africa." *Suid Afrikaanse Tydskrif vir Sosiologie, South African Journal of Sociology* 3 (September 1971): 59-64. Topics discussed include: demography, South Africa, labor shortages, immigration, policy.

3126. Rejda, G. E. et al. "Unemployment Insurance and Full-Cost Experience Rating: The Impact on Seasonal Hiring." *Journal of Risk and Insurance* 57, no. 3 (September 1990): 519-29. Topics discussed include: unemployment insurance.

3127. Rodriguez Labandeira, J. "The Seasonal Unemployment of Farm Workers in Spain during the Restoration; El paro estacional campesino en la Restauracion." *Revista Internacional de Sociologia* 45, no. 2 (April 1987): 297-317. Topics discussed include: unemployment rates, nineteenth century, twentieth century, agriculture, work, Spain, seasonality.

3128. Rourke, B. E. et al. "Seasonality in the Employment of Casual Agricultural Labour in Ghana." *Economic Bulletin of Ghana, Second Series* 3, no. 3 (1973): 3-13. Topics discussed include: employment, unemployment levels, vacancy levels, retirement patterns.

3129. Ryan, J. G. et al. "Labor Market Behavior in Rural Villages in South India: Effects of Season, Sex, and Socioeconomic Status." *Contractual Arrangements, Employment, and Wages in Rural Labor Markets in Asia Economic Growth Center, Yale University series*, 169-83. eds Hans P. Binswanger and R. Rosenzweig. New Haven and London: Yale University Press, 1984. Topics discussed include: wages, fringe benefits, geography, labor markets, rural areas, labor, economics, agriculture.

3130. Rydzewski, L. G. et al. "Seasonal employment falls over past three decades." *Monthly Labor Review* 116 (July 1993): 3-14. Topics discussed include: seasonality, labor, economics, employment levels, industry.

3131. Saltin, B. "Aerobic and anaerobic work capacity at 2300 meters." *Med Thorac* 24, no. 4 (1967): 205-10. Topics discussed include: environment, exposure, physical labor, Mexico, respiratory functions, altitude, oxygen consumption, physical education, training, sports.

3132. Sant, M. "The geography of business cycles: a case study of economic fluctuations in East Anglia, 1951-68." 64 pages. London School of Economics and Political Science Geographical papers, no. 5, London: London School of Economics and Political Science, 1973. Topics discussed include: seasonal unemployment, business cycles.

3133. Sayers, R. R. et al. "A Preliminary Study of the Physiological Effects of High Temperatures and High Humidities in Metal Mines." *U S Public Health Repts Reprint* no. 639 (1921): 116-129. Topics discussed include: temperature, environment, climatic effects, biology, health issues, labor.

3134. Schluter, G. "A simple method for the determination of climatic factors and heat stress for work in hot environment." *Zentralbl Arbeitsmed* 17, no. 12 (December 1967): 391-2. Topics discussed include: environment, exposure, work-related medicine, stress diagnosis.

3135. Schmid, L. "The role of seasonal labor in the economic development of Guatemala." University of Wisconsin, Madison. Land Tenure Center, 1968. Topics discussed include: seasonality, labor, Guatemala, policy.

3136. Schuh, G. E. "Seasonal Farm Labor and U.S. Farm Policy: Discussion." *Seasonal Agricultural Labor Markets in the United States*, 476-81. ed Robert D. Emerson. Ames: Iowa State University Press, 1984. Topics discussed include: policy, labor markets, wage levels, employment.

3137. Scott, C. D. "Agrarian Reform and Seasonal Employment in Coastal Peruvian Agriculture." *Journal of Development Studies* 17, no. 4 (July 1981): 282-306. Topics discussed include: employment, seasonality, agriculture, Peru.

3138. "Seasonal migrants and commuters in Yogyakarta." Universitas Gadjah Mada. Lembaga Studi Pedesaan dan Kawasan. Yogyakarta: Institute of Rural and Regional Studies, Gadjah Mada University, 1977. Topics discussed include: labor mobility, commuters, Indonesia, seasonality, labor.

3139. Senker P.J. *Industrial training in a cold climate: an assessment of Britain's training policies,* The Averbury business school library, Aldershot, England ; Hong Kong: Avebury, 1992. Topics discussed include: work-related training, policy, United Kingdom, employees, industry, education.

3140. Sickert, P. et al. "Effect of the outdoor climate on microclimatic conditions of work sites in buildings." *Z Gesamte Hyg* 34, no. 8 (August 1988): 481-3. Topics discussed include: Germany, heating methods, temperature, ventilation.

3141. Smith, P. "Seasonal Fluctuations in the Motor Vehicle Industry: A Comment." *Journal of Industrial Economics* 21, no. 2 (April 1973): 184-86. Topics discussed include: employment, unemployment levels, vacancy levels, retirement patterns, manufacturing transportation equipment.

3142. Snell, K. D. M. "Agricultural Seasonal Unemployment, the Standard of Living, and Women's Work in the South and East, 1690-1860." *Economic History Review, Second Series* 34, no. 3 (August 1981): 407-37. Topics discussed include: history, Europe.

3143. Solomko, P. "On standards of water consumption in physical work in the desert." *Voen Med Zh* 6 (June 1967): 53-4. Topics discussed include: drinking, desert climates, physical labor, water.

3144. Solon, G. "The Minimum Wage and Teenage Employment: A Reanalysis with Attention to Serial Correlation and Seasonality." *Journal of Human Resources* 20, no. 2 (March 1985): 292-97. Topics discussed include: wages, work hours, unemployment levels, vacancy levels, retirement patterns, labor markets, demography.

3145. Soule, R. G. et al. "Voluntary march rate as a measure of work output in the heat." *Ergonomics* 21, no. 6 (June 1978): 455-462. Topics discussed include: humidity, work output, walking, energy expenditures.

3146. Stone, G. D. et al. "Seasonality, labor scheduling, and agricultural intensification in the Nigerian savanna." *American Anthropologist* 92 (March 1990): 7-23. Topics discussed include: Africa, laborers, Nigeria, seasonality, economics.

3147. Surrey, M. J. C. "The Seasonal Adjustment of Unemployment Statistics-A Note." *Bulletin of the Institute of Economics and Statistics* 34, no. 2 (May 1972): 241-47. Topics discussed include: time series analysis, spectral models, unemployment levels, vacancy levels, retirement patterns.

3148. Sverko, B. et al. "Investigation of attitude towards climatization of work environments." *Arh Hig Rada Toksikol* 30, no. 4 (December 1979): 323-32. Topics discussed include: work-related medicine, Yugoslavia, air conditioning, attitude, occupations.

3149. Tsiupo, G. G. "Temporary loss of work capacity due to hypertension in industrial enterprises in relation to annual factors." *Vrach Delo* 5 (May 1967): 70-2. Topics discussed include: adaptation, disability evaluation, hypertension, work-related medicine, seasonality.

3150. Turnipseed, M. "Seasonal factors, consumer price index: selected series, June 1953-May 1961." United States. Bureau of Labor Statistics. U.S. Dept. of Labor, Bureau of Labor Statistics; for sale by the Superintendent of Documents, U.S. Govt. Print. Off., Washington, 1963. Topics discussed include: consumers prices, seasonality, economics.

3151. Tymanski, S. "Evaluation of Mexican gulf climate from the viewpoint of seamen's work condition oppressiveness." *Rocz Pomor Akad Med* SUPP, no. 10 (1974): 153-9. Topics discussed include: Bahamas, seasonality, ships, naval medicine, tropical climate.

3152. ---. "A research into the degree of exposure of fish cold stores dockers to temperatures inherent in their work." *Bull Inst Marit Trop Med Gdynia* 33, no. 3-4 (1982): 181-8. Topics discussed include: humidity, work-related medicine, seasonality, fisheries.

3153. Van der Hoeven, H. et al. "A Method for Seasonally Adjusted Time Series with Variation in the Seasonal Amplitude." *Journal of Business and Economic Statistics* 4, no. 4 (October 1986): 455-71. Topics discussed include: time series analysis, spectral models, unemployment levels, vacancy levels, retirement patterns, demography.

3154. Verma, B. N. et al. *Seasonal dimensions of surplus labour in agriculture: estimate, impact, and absorption.* 256 pages. New Delhi, India: Uppal Pub. House, 1994.

3155. Vernon, H. M. "Industrial Fatigue in Relation to Atmospheric Conditions." *Physiol Rev* 8 (1928): 130. Topics discussed include: environment, labor, health issues, biology, behavior.

3156. Watanabe, T. et al. "Hemoglobin levels among Japanese farmers: with special reference to climate and work intensity." *Hum Biol* 58, no. 2 (April 1986): 197-208. Topics discussed include: physical labor, Japan, agricultural workers, diseases, epidemiology, anemia epidemiology, hemoglobins analysis.

3157. Werner, J. et al. "Dynamics and topography of human temperature regulation in response to thermal and work load." *Eur J Appl Physiol* 53, no. 4 (1985): 353-8. Topics discussed include: biomechanics, exercise, body regulation, physical labor, heat.

3158. Wilkinson, A. *Big Sugar: Seasons in the Cane Fields of Florida.* New York, NY: Random House, 1990. Topics discussed include: agriculture, seasonality, migrants, labor, industry, government.

3159. Wolters, J. "Persistence and Seasonality in Output and Employmentof the Federal Republic of Germany." *Recherches Economiques de Louvain* 58, no. 3-4 (1992): 421-39. Topics discussed include: wealth, income, wage levels, fringe benefit studies, labor, unemployment levels, macroeconomics, employment levels, wage indexation, production.

3160. Yaglou, C. P. "Temperature, Humidity and Air Movement in Industries. The Effective Temperature Index." *J Indus Hyg* 9 (1927): 297. Topics discussed include: environment, labor, industry, behavior.

3161. ---. "The Thermal Index of Atmospheric Conditions and its Application to Sedentary and to Industrial Life." *J Indus Hyg* 8 (1926): 5. Topics discussed include: temperature, environment, labor, industry, behavior.

Macroeconomics

3162. Anandakrishnan, M. "Climate and Development: Climatic Change and Variability and the Resulting Social, Economic and Technological Implications: Where Do We Go from Here?" *Climate and development: Climatic change and variability and the resulting social, economic and technological implications*, 473-475. eds H. J. Karpe, D. Otten, and S. C. Trinidade. New York, Berlin, London, Tokyo: Springer in cooperation with the United Nations, 1990. Topics discussed include: conservation, pollution, energy.

3163. Bagshaw, M. et al. "Seasonal Variation in Interest Rates." *Review of Economics and Statistics* 59, no. 1 (February 1977): 119-22. Topics discussed include: monetary policy.

3164. Barrett, H. R. et al. "Skin temperature biofeedback for multiple sessions with monetary incentives." *Perceptual and Motor Skills* 65, no. 1 (August 1987): 139-146. Topics discussed include: monetary incentives, biofeedback training, skin.

3165. Barth, J. R. et al. "Seasonal Variation in Interest Rates." *Review of Economics and Statistics* 57, no. 1 (February 1975): 80-83. Topics discussed include: monetary policy, capital and money markets, regulation.

3166. ---. "Seasonal Variation in Interest Rates Revisited." *Review of Economics and Statistics* 59, no. 1 (February 1977): 124-26. Topics discussed include: monetary policy.

3167. Bolch, B. W. et al. "Seasonal Variation in Interest Rates-A Comment." *Review of Economics and Statistics* 59, no. 1 (February 1977): 118-19. Topics discussed include: monetary policy.

3168. Brandt, W. "Climate, Society and the Historical Development." *Climate and development: Climatic change and variability and the resulting social, economic and technological implications*, 22-29. eds H. J. Karpe, D. Otten, and S. C. Trinidade. New York; Berlin; London and Tokyo: Springer in cooperation with the United Nations, 1990. Topics discussed include: conservation, pollution, energy.

3169. Braun, R. A. et al. "Seasonality and equilibrium business cycle theories." 39 pages. Institute for Empirical Macroeconomics. Discussion paper 45, Minneapolis, MN: Institute for Empirical Macroeconomics, Federal Reserve Bank of Minneapolis, 1991. Topics discussed include: seasonal variations, economics, mathematical models, business cycles.

3170. Breuss, F. et al. "Consumer Climate Data in Macroeconomic Consumption Functions." *Empirica* 13, no. 1 (1986): 27-51. Topics discussed include: standards of living, saving behavior, consumer expenditures, consumption.

3171. Chatterjee, S. et al. "A Neoclassical Model of Seasonal Fluctuations." *Journal of Monetary Economics* 29, no. 1 (February 1992): 59-86. Topics discussed include: business cycles, economic cycles, aggregate models, measurement issues, macroeconomics, wealth.

3172. Clark, T. N. "Fiscal Strain: How Different Are Snow Belt and Sun Belt Cities?" *Peterson, Paul E, ed The New Urban Reality*, 253-80. Washington, DC: Brookings Institution, 1985. Topics discussed include: government, finance, economics, policy, sociology.

3173. Clark, T. A. "Interest Rate Seasonals and the Federal Reserve." *Journal of Political Economy* 94, no. 1 (February 1986): 76-125. Topics discussed include: economic history, North America, history of economic policy, monetary policy, central banks.

3174. Cook, T. Q. "The 1983 M1 Seasonal Factor Revisions: An Illustration of Problems That May Arise in Using Seasonally Adjusted Data for Policy Purposes." *Federal Reserve Bank of Richmond Economic Review* 70, no. 2 (March 1984): 22-33. Topics discussed include: monetary policy, central banks.

3175. Darmstadter, J. ed. "Global development and the environment: Perspectives on sustainability." Washington, D.C.: Resources for the Future, 1992. Topics discussed include: economics, development, resources, conservation, environmental management.

3176. English, W. B. et al. "Seasonal Fluctuations and the Life Cycle-Permanent Income Model of Consumption - A Correction." *Journal of Political Economy* 97, no. 4 (August 1989): 988-91. Topics discussed include: consumer behavior, expenditures, consumption, macroeconomics, demand.

3177. Faig, M. "Seasonal Fluctuations and the Demand for Money." *Quarterly Journal of Economics* 104, no. 4 (November 1989): 847-61. Topics discussed include: monetary theory.

3178. Foster, E. "Seasonal Movements of Exchange Rates, Interest Rates and Interest Rate Differentials under the Pre-World War I Gold Standard." Ph.D. diss., SUNY, Albany, 1991. Topics discussed include: interest rates, capital and money markets, regulatory policy, monetary theory.

3179. Ghysels, E. "The economics of seasonality and monetary policy." Produced and distributed by Center for Research in Management, University of California, Berkeley Business School, Berkeley, CA, 1984. Topics discussed include: seasonality, mathematical models, monetary policy, money supply.

3180. Gilbert, R. A. "New Seasonal Factors for the Adjusted Monetary Base." *Federal Reserve Bank of St Louis Review* 67, no. 10 (December 1985): 29-33. Topics discussed include: monetary policy, central banks.

3181. Hein, S. E. et al. "Seasonally Adjusting Money: Procedures, Problems, Proposals." *Federal Reserve Bank of St Louis Review* 65, no. 9 (November 1983): 16-25. Topics discussed include: monetary theory, policy.

3182. Holland, A. S. et al. "The Role of the Federal Reserve as 'Lender of Last Resort' and the Seasonal Fluctuation of Interest Rates." *Journal of Money, Credit, and Banking* 23, no. 4 (November 1991): 659-76. Topics discussed include: economics, history, macroeconomics, growth, cycles, central banks, history, product prices, monetary policy.

3183. Joines, D. H. "Seasonality and the Monetization of Federal Budget Deficits." *Economic Inquiry* 28, no. 3 (July 1990): 413-41. Topics discussed include: budget deficits, monetary policy, central banks.

3184. Jones, J. D. "Seasonal Variation in Interest Rates." *Atlantic Economic Journal* 16, no. 2 (June 1988): 47-58. Topics discussed include: capital and money markets, regulation.

3185. Kamarck, A. M. "Climate and Economic Development." *Finance and Development* 10, no. 2 (June 1973): 2-8. Topics discussed include: development, geography.

3186. Kaufman, H. M. et al. "Seasonal Adjustment of the Money Stock: An Analysis of Alternative Techniques." *Journal of Economics and Business* 29, no. 2 (December 1977): 135-38. Topics discussed include: monetary theory.

3187. ---. "Short-Run Variations in the Money Stock-Seasonal or Cyclical?" *Federal Reserve Bulletin* 60, no. 6 (June 1974): 420-21. Topics discussed include: monetary theory.

3188. Lawler, T. A. "Federal Reserve Policy Strategy and Interest Rate Seasonality: A Note." *Journal of Money, Credit, and Banking* 11, no. 4 (November 1979): 494-99. Topics discussed include: monetary theory, monetary policy, central banks.

3189. Lee, H. S. et al. "Unit Roots and Seasonal Unit Roots in Macroeconomic Time Series - Canadian Evidence." *Economics Letters* 35, no. 3 (March 1991): 273-77. Topics discussed include: wealth, income, economics, cycles.

3190. Liano, K. "Macroeconomic Events and Seasonality of Risk and Returns." *Applied Financial Economics* 2, no. 4 (December 1992): 205-09. Topics discussed include: capital and asset pricing, financial and money markets, capital and money markets, portfolio composition, monetary theory.

3191. Lombra, R. E. et al. "Interest Rate Seasonality and the Specification of Money Demand Functions: A Note." *Review of Economics and Statistics* 57, no. 2 (May 1975): 252-55. Topics discussed include: monetary theory.

3192. MacKrell, N. C. "Seasonal adjustment : applications for Singapore." Monetary Authority of Singapore, Singapore, 1976. Topics discussed include: money supply, Singapore, seasonality, economics.

3193. Mankiw, N. G. et al. "Should the Fed Smooth Interest Rates? The Case of Seasonal Monetary Policy." *Carnegie Rochester Conference Series on Public Policy* 0, no. 34 (March 1991): 41-69. Topics discussed include: capital and money markets, regulatory policy, central banks, interest rates.

3194. Maravall, A. "Effects of Alternative Seasonal Adjustment Procedures on Monetary Policy." *Journal of Econometrics* 14, no. 1 (June 1980): 115-36. Topics discussed include: time series analysis, spectral models, monetary policy.

3195. McCallum, B. T. "Seasonality and Monetary Policy: A Comment." *Carnegie Rochester Conference Series on Public Policy* no. 34 (March 1991): 71-75. Topics discussed include: monetary policy, central banks, capital and money markets, regulatory policy, interest rates, term structures.

3196. Mills, T. C. et al. "Modelling the Seasonal Patterns in UK Macroeconomic Times Series." *Journal of Royal Statistical Society, Series A* 155, no. 1 (1992): 61-75. Topics discussed include: wealth, finance, capital, prices, monetary policy, consumption, saving, production, employment.

3197. Miron, J. A. "Financial Panics, the Seasonality of the Nominal Interest Rate, and the Founding of the Fed." *American Economic Review* 76, no. 1 (March 1986): 125-40. Topics discussed include: cycles, monetary policy, financial, theory, central banks, commercial lending.

3198. Officer, R. R. "Seasonality in Australian Capital Markets: Market Efficiency and Empirical Issues." *Journal of Financial Economics* 2, no. 1 (1975): 29-51. Topics discussed include: economy, finance, monetary policy.

3199. Osborn, D. R. "Seasonal Cointegration: Discussion." *Journal of Econometrics* 55, no. 1-2 (January 1993): 299-303. Topics discussed include: methods, time series analysis, macroeconomics, consumption, saving.

3200. ---. "A Survey of Seasonality in UK Macroeconomic Variables." *International Journal of Forecasting* 6, no. 3 (October 1990): 327-36. Topics discussed include: income, econometrics.

3201. Otto, G. et al. "Seasonal Unit-Root Tests on Canadian Macroeconomic Time Series." *Economics Letters* 34, no. 2 (October 1990): 117-20. Topics discussed include: income, time series analysis, spectral models.

3202. Pierce, D. A. "Data Revisions with Moving Average Seasonal Adjustment Procedures." *Journal of Econometrics* 14, no. 1 (June 1980): 95-114. Topics discussed include: time series analysis, spectral models, monetary policy, central banks.

3203. ---. "Seasonal Adjustment of Daily Data with Special Reference to the US Money Supply." *Economie Appliquee* 32, no. 1 (1979): 49-72. Topics discussed include: time series analysis, spectral models, monetary policy, central banks.

3204. ---. "Seasonal Adjustment of the Monetary Aggregates: Summary of the Federal Reserve's Committee Report." *Journal of Business and Economic Statistics* 1, no. 1 (January 1983): 37-42. Topics discussed include: time series analysis, spectral models, monetary theory.

3205. Pierce, D. A. et al. "Seasonal Adjustment Methods for the Monetary Aggregates." *Federal Reserve Bulletin* 67, no. 12 (December 1981): 875-87. Topics discussed include: monetary policy, central banks.

3206. ---. "Seasonal Adjustment of the Weekly Monetary Aggregates: A Model-based Approach." *Journal of Business and Economic Statistics* 2, no. 3 (July 1984): 260-70. Topics discussed include: econometrics, monetary theory, policy.

3207. Porter Hudak, S. "An Application of the Seasonal Fractionally Differenced Model to the Monetary Aggregates." *Journal of the American Statistical Association* 85, no. 410 (June 1990): 338-44. Topics discussed include: time series analysis, spectral models, monetary theory.

3208. *Seasonal dimensions to rural poverty*, eds Richard Longhurst, Arnold Pacey and Robert Chambers. London; Totowa, NJ: F. Pinter; Allanheld, Osmun, 1981. Topics discussed include: rural areas, poverty, tropics, seasonality, economics, diseases.

3209. Saracoglu, R. et al. "Seasonality and Portfolio Balance under Rational Expectations." *Journal of Monetary Economics* 4, no. 3 (August 1978): 435-58. Topics discussed include: inflation, monetary theory.

3210. Schneeweis, T. et al. "A Note on the Study of Interest Rate Seasonality." *Business Economics* 14, no. 3 (May 1979): 28-32. Topics discussed include: monetary theory, capital and money markets, regulation.

3211. Sealey, C. W., Jr. "Changing Seasonal Movements in Interest Rates and Their Implications for Interest Rate Forecasting." *Business Economics* 12, no. 4 (September 1977): 67-73. Topics discussed include: econometrics, capital and money markets, regulation.

3212. "Seasonal adjustment of the monetary aggregates: report of the Committee of Experts on Seasonal Adjustment Techniques." Board of Governors of the Federal Reserve System (U.S.). Washington, 1981. Topics discussed include: money supply, monetary policy, seasonality, mathematical models.

3213. Sharp, K. P. "Tests of U.S. Short and Long Interest Rate Seasonality." *Review of Economics and Statistics* 70, no. 1 (February 1988): 177-82. Topics discussed include: capital and money markets, regulatory policy, time series analysis, spectral models.

3214. Simpson, T. D. et al. "Recent Revisions in the Money Stock: Benchmark, Seasonal Adjustment, and Calculation of Shift-Adjusted M1-B." *Federal Reserve Bulletin* 67, no. 7 (July 1981): 539-42. Topics discussed include: monetary theory, policy.

3215. Smith, V. K. et al. "Seasonal Variation in Interest Rates: Another Perspective." *Review of Economics and Statistics* 59, no. 1 (February 1977): 122-24. Topics discussed include: monetary theory.

3216. Stokes, H. H. "Effects of Alternative Seasonal Adjustment Procedures on Monetary Policy - Comment." *Journal of Econometrics* 14, no. 1 (June 1980): 137-40. Topics discussed include: time series analysis, spectral models, monetary policy, central banks.

3217. Sumner, M. "Fiscal Policy, Seasonality, and Intertemporal Substitution of Investment Spending in the UK." *Journal of Public Economics* 49, no. 1 (October 1992): 123-34. Topics discussed include: capital, inventories, macroeconomics, fiscal policy, public expenditures, investment, finance, taxation.

3218. Valentine, T. J. "Cyclical and Seasonal Movements in Profits in Australia 1960-1973." *Australian Economic Papers* 15, no. 26 (June 1976): 52-58. Topics discussed include: cycles, business, finance.

3219. Wheelock, D. C. "Seasonal Accommodation and the Financial Crises of the Great Depression: Did the Fed 'Furnish an Elastic Currency?'" *Federal Reserve Bank of St Louis Review* 74, no. 6 (November 1992): 3-18. Topics discussed include: monetary policy, central banks, regulatory policy, capital and money markets, portfolio composition.

3220. Young, P. C. et al. *A systems approach to recursive economic forecasting and seasonal adjustment.* 16 pages. Institute for Empirical Macroeconomics. Discussion paper 8, Minneapolis, MN: Institute for Empirical Macroeconomics, Federal Reserve Bank of Minneapolis, 1989. Topics discussed include: seasonality, economics.

Methodologies

3221. Abeysinghe, T. "Inappropriate Use of Seasonal Dummies in Regression." *Economics Letters* 36, no. 2 (June 1991): 175-79. Topics discussed include: econometrics, single equation analysis, times series analysis, spectral models, statistical approaches, multivariate models.

3222. Abraham, B. "Seasonal Time Series and Transfer Function Modeling." *Journal of Business and Economic Statistics* 3, no. 4 (October 1985): 356-61. Topics discussed include: time series analysis, spectral models.

3223. Akaike, H. "Issues Involved with the Seasonal Adjustment of Economic Time Series - Comment." *Journal of Business and Economic Statistics* 2, no. 4 (October 1984): 321-22. Topics discussed include: time series analysis, spectral models.

3224. Ansley, C. F. et al. "Issues Involved with the Seasonal Adjustment of Economic Time Series - Comment -On Dips in the Spectrum of a Seasonally Adjusted Time Series." *Journal of Business and Economic Statistics* 2, no. 4 (October 1984): 323-24. Topics discussed include: time series analysis, spectral models.

3225. Astier, R. et al. "Analysis of perturbed seasonal time series." (1985). Topics discussed include: operations research, mathematical programming, game theory, economics.

3226. Babich, G. et al. "Short Term Econometric Forecasting and Seasonal Adjustment." *Economic Record* 54, no. 146 (August 1978): 229-36. Topics discussed include: forecasting, models, methodology.

3227. Barkey, P. "Measurement Error and Seasonal Adjustment Error in Economic Time Series." Ph.D. diss., University of Michigan, 1986. Topics discussed include: econometrics, statistical approaches, models.

3228. Barth, J. R. et al. "Cyclical Behavior, Seasonality, and Trend in Economic Time Series." *Nebraska Journal of Economics and Business* 13, no. 1 (December 1974): 48-69. Topics discussed include: time series analysis, spectral models.

3229. Bartolomei, S. M. et al. "A Note on a Comparison of Exponential Smoothing Methods for Forecasting Seasonal Series." *International Journal of Forecasting* 5, no. 1 (1989): 111-116. Topics discussed include: econometric models, methodology, time series analysis, spectral models.

3230. Beaulieu, J. J. et al. "Seasonal unit roots in aggregate U.S. data." *J Econometrics* 55, no. 1,2 (1993): 305-331. Topics discussed include: numerical statistics, stochastic processes.

3231. Bell, W. R. "Estimation and Seasonal Adjustment of Population Means Using Data from Repeated Surveys - Comment." *Journal of Business and Economic Statistics* 9, no. 2 (April 1991): 176-77. Topics discussed include: econometrics, single equation analysis, models with panel data, time series analysis, inferential problems.

3232. Bell, W. "A Note on Overdifferencing and the Equivalence of Seasonal Time Series Models with Monthly Means and Models with (0, 1, 1)12 Seasonal Parts when Theta=1." *Journal of Business and Economic Statistics* 5, no. 3 (July 1987): 383-87. Topics discussed include: time series analysis, spectral models.

3233. Bell, W. R. et al. "Issues Involved with the Seasonal Adjustment of Economic Time Series." *Journal of Business and Economic Statistics* 2, no. 4 (October 1984): 291-320. Topics discussed include: time series analysis, spectral models.

3234. ---. "Reply Issues Involved with the Seasonal Adjustment of Economic Time Series." *Journal of Business and Economic Statistics* 3, no. 1 (January 1985): 95-97. Topics discussed include: time series analysis, spectral models.

3235. Box, G. E. P. et al. "Estimating trend and growth rates in seasonal time series." *J Amer Statist Assoc* 82, no. 397 (1987): 276-282. Topics discussed include: numerical statistics, stochastic processes, time.

3236. ---. *Time Series Analysis: Forecasting and Control*, San Franscisco: Holden-Day, Inc., 1970. Topics discussed include: econometric methodologies.

3237. Brewer, K. R. W. "Seasonal Adjustment of Arima Series." *Economie Appliquee* 32, no. 1 (1979): 7-22. Topics discussed include: time series analysis, spectral models.

3238. Burman, J. P. "Issues Involved with the Seasonal Adjustment of Economic Time Series - Comment." *Journal of Business and Economic Statistics* 2, no. 4 (October 1984): 325-27. Topics discussed include: time series analysis, spectral models.

3239. Burridge, P. et al. "Unobserved-Components Models for Seasonal Adjustment Filters." *Journal of Business and Economic Statistics* 2, no. 4 (October 1984): 350-59. Topics discussed include: time series analysis, spectral models.

3240. Canova, F. "An Alternative Approach to Modeling and Forecasting Seasonal Time Series." *Journal of Business and Economic Statistics* 10, no. 1 (January 1992): 97-108. Topics discussed include: model applications, econometrics, single equation analysis, times series analysis.

3241. ---. "Forecasting Time Series with Common Seasonal Patterns." *Journal of Econometrics* 55, no. 1-2 (January 1993): 173-200. Topics discussed include: methods, single equation analysis, times series analysis, models.

3242. Chan, K. H. et al. "Forecasting of Seasonal and Cyclical Financial Variables: The Wiener-Kolmogorov Method vs. the Box-Jenkins Method." *Advances in financial planning and forecasting Volume 2 A Research*, 103-118. ed Cheng F. Lee. Greenwich, CT and London: JAI Press, 1987. Topics discussed include: specific forecasting, business finance, electrical, gas, communication, information services.

3243. Chatfield, C. et al. "Box-Jenkins Seasonal Forecasting: Problems in a Case-study." *Journal of the Royal Statistical Society* series A 136, no. part 3 (1973): 295-315. Topics discussed include: time series analysis, spectral models, econometric models, methodology.

3244. Cleveland, W. P. et al. "Decomposition of Seasonal Time Series - A Model for the Census X-11 Program." *Journal of the American Statistical Association* 71, no. 355 (September 1976): 581-87. Topics discussed include: time series analysis, spectral models.

3245. ---. "Modeling Seasonal Time Series." *Economie Appliquee* 32, no. 1 (1979): 107-29. Topics discussed include: time series analysis, spectral models.

3246. Collins, D. G. "Measurement of Temperature in Economic Systems Considered as Thermodynamic Models." *Energy Economics* 14, no. 1 (January 1992): 73-78. Topics discussed include: models.

3247. Cromer, T. L. "Harvesting in a seasonal environment." *Math Comput Modelling* 10, no. 6 (1988): 445-450. Topics discussed include: economics, operations research, mathematical programming, game theory.

3248. Cupingood, L. A. et al. "Seasonal Adjustment of Time Series Using One-Sided Filters." *Journal of Business and Economic Statistics* 4, no. 4 (October 1986): 473-84. Topics discussed include: time series analysis, spectral models, transportation.

3249. Dagum, E. B. "The Effects of Asymmetric Filters on Seasonal Factor Revisions." *Journal of the American Statistical Association* 77, no. 380 (December 1982): 732-38. Topics discussed include: time series analysis, spectral models.

3250. Dagum, E. B. et al. "Issues Involved with the Seasonal Adjustment of Economic Time Series - Comment." *Journal of Business and Economic Statistics* 2, no. 4 (October 1984): 328-32. Topics discussed include: time series analysis, spectral models.

3251. den Butter, F. A. G. et al. "The Use of ARIMA Models in Seasonal Adjustment-A Comparative Study of Census X-11, X-11 ARIMA and Burman's Signal Extraction Method." *Empirical Economics* 10, no. 4 (1985): 209-30. Topics discussed include: time series analysis, spectral models.

3252. ---. "Seasonal Adjustment Using Structural Time Series Models: An Application and Comparison with the Census X-11 Method." *Journal of Business and Economic Statistics* 8, no. 4 (October 1990): 385-94. Topics discussed include: time series analysis, spectral models.

3253. Dickey, D. A. et al. "Testing for Unit Roots in Seasonal Time Series." *Journal of the American Statistical Association* 79, no. 386 (June 1984): 355-67. Topics discussed include: time series analysis, spectral models.

3254. Diebold, F. X. "The Effect of Seasonal Adjustment Filters on Tests for a Unit Root: Discussion." *Journal of Econometrics* 55, no. 1-2 (January 1993): 99-103. Topics discussed include: methods, single equation analysis, times series analysis, spectral models.

3255. Fase, M. M. G. "Issues Involved with the Seasonal Adjustment of Economic Time Series - Comment." *Journal of Business and Economic Statistics* 2, no. 4 (October 1984): 333-34. Topics discussed include: time series analysis, spectral models.

3256. Findley, D. F. et al. "New Techniques for Determining if a Time Series Can Be Seasonally Adjusted Reliably." *Regional econometric modeling International Series in Economic Modeling*, 195-228. ed James R. Schmidt and M. Ray Perryman. Norwell; MA: Kluwer Academic, 1986. Topics discussed include: trade relations, time series analysis, spectral models.

3257. Franses, P. H. "Modeling seasonality in bimonthly time series." *Statist Probab Lett* 15, no. 5 (1992): 407-415. Topics discussed include: numerical statistics, stochastic processes, time.

3258. ---. "Seasonality, Non-stationarity and the Forecasting of Monthly Time Series." *International Journal of Forecasting* 7, no. 2 (August 1991): 199-208. Topics discussed include: econometrics, single equation analysis, times series analysis.

3259. ---. "Testing for Seasonality." *Economics Letters* 38, no. 3 (March 1992): 259-62. Topics discussed include: econometrics, single equation analysis, times series analysis.

3260. Gersch, W. et al. "The Prediction of Time Series with Trends and Seasonalities." *Journal of Business and Economic Statistics* 1, no. 3 (July 1983): 253-64. Topics discussed include: time series analysis, spectral models.

3261. Geweke, J. "Forecasting Time Series with Common Seasonal Patterns: Discussion." *Journal of Econometrics* 55, no. 1-2 (January 1993): 201-02. Topics discussed include: methods, single equation analysis, times series analysis, time series analysis, spectral models.

3262. Ghysels, E. "Seasonality and Econometric Models: Editor's Introduction." *Journal of Econometrics* 55, no. 1-2 (January 1993): 1-8. Topics discussed include: methods, single equation analysis, times series analysis, inferential problems.

3263. ---. "A study toward a dynamic theory of seasonality for economic time series." *J Amer Statist Assoc* 83, no. 401 (1988): 168-172. Topics discussed include: numerical statistics, stochastic processes, time.

3264. Ghysels, E. et al. "The Effect of Seasonal Adjustment Filters on Tests for a Unit Root." *Journal of Econometrics* 55, no. 1-2 (January 1993): 57-98. Topics discussed include: methods, single equation analysis, times series analysis, time series analysis, spectral models, statistical approaches.

3265. Goldrian, G. "On the Effects of Seasonal Adjustment Method ASA-II on Economic Time Series." *Jahrbucher fur Nationalokonomie und Statistik* 189, no. 3, 4 (June 1975): 275-85. Topics discussed include: time series analysis, spectral models.

3266. Granger, C. W. J. "Issues Involved with the Seasonal Adjustment of Economic Time Series - Comment." *Journal of Business and Economic Statistics* 2, no. 4 (October 1984): 335-36. Topics discussed include: time series analysis, spectral models.

3267. Hannan, E. J. et al. "The Seasonal Adjustment of Economic Time Series." *International Economic Review* 11, no. 1 (February 1970): 24-52. Topics discussed include: time series analysis, spectral models.

3268. Hansen, L. P. et al. "Seasonality and Approximation Errors in Rational Expectations Models." *Journal of Econometrics* 55, no. 1-2 (January 1993): 21-55. Topics discussed include: estimation procedures, single equation models.

3269. Havenner, A. et al. "A Random Coefficient Approach to Seasonal Adjustment of Economic Time Series." *Journal of Econometrics* 15, no. 2 (February 1981): 177-209. Topics discussed include: time series analysis, spectral models.

3270. Hillmer, S. C. et al. "An ARIMA-Model-Based Approach to Seasonal Adjustment." *Journal of the American Statistical Association* 77, no. 377 (March 1982): 63-70. Topics discussed include: time series analysis, spectral models.

3271. Huot, G. et al. "Analysis of Revisions in the Seasonal Adjustment of Data Using X-11-ARIMA Model-Based Filters." *International Journal of Forecasting* 2, no. 2 (1986): 217-29. Topics discussed include: forecasting, time series analysis, spectral models.

3272. Hylleberg, S. et al. "Seasonal Integration and Cointegration." *Journal of Econometrics* 44, no. 1-2 (April 1990): 215-38. Topics discussed include: time series analysis, spectral models.

3273. Hylleberg, S. ed. "Modelling seasonality." *Advanced Texts in Econometrics*, 476 pages. Oxford; New York; Toronto and Melbourne: Oxford University Press, 1992. Topics discussed include: single equation models, multiple, simultaneous equation models.

3274. Kenny, P. B. et al. "Local Trend Estimation and Seasonal Adjustment of Economic and Social Time Series." *Journal of the Royal Statistical Society*, Series A; 145, no. Part 1, (1982): 1-28. Topics discussed include: time series analysis, spectral models.

3275. Kitagawa, G. et al. "A Smoothness Priors-State Space Modeling of Time Series with Trend and Seasonality." *Journal of the American Statistical Association* 79, no. 386 (June 1984): 378-89. Topics discussed include: time series analysis, spectral models.

3276. Lee, H. S. "Maximum Likelihood Inference on Cointegration and Seasonal Cointegration." *Journal of Econometrics* 54, no. 1-3 (October 1992): 1-47. Topics discussed include: time series analysis, spectral models, statistical approaches, multivariate models.

3277. Maravall, A. "Issues Involved with the Seasonal Adjustment of Economic Time Series - Comment." *Journal of Business and Economic Statistics* 2, no. 4 (October 1984): 337-39. Topics discussed include: time series analysis, spectral models.

3278. ---. "On the dynamic structure of a seasonal component." *J Econom Dynamics Control* 13, no. 1 (1989): 81-91. Topics discussed include: numerical statistics, applications.

3279. Miller, T. W. et al. "Daily Cash Forecasting and Seasonal Resolution: Alternative Models and Techniques for Using the Distribution Approach." *Journal of Financial and Quantitative Analysis* 20, no. 3 (September 1985): 335-51. Topics discussed include: specific forecasting, business finance.

3280. Muscatelli, V. A. et al. "Cointegration and Dynamic Time Series Models." *Journal of Economic Surveys* 6, no. 1 (1992): 1-43. Topics discussed include: spectral models, multivariate models, information theory.

3281. Nourney, M. "Seasonal Adjustment by Frequency Determined Filter Procedures." *Statistical Journal* 2, no. 2 (April 1984): 161-68. Topics discussed include: time series analysis, spectral models.

3282. Oliver, R. M. "Bayesian Forecasting with Stable Seasonal Patterns." *Journal of Business and Economic Statistics* 5, no. 1 (January 1987): 77-85. Topics discussed include: econometric theory, methodology, time series analysis, spectral models.

3283. Osborn, D. R. "The Implications of Periodically Varying Coefficients for Seasonal Time-Series Processes." *Journal of Econometrics* 48, no. 3 (June 1991): 373-84. Topics discussed include: methods, spectral models, multivariate models.

3284. Parzen, E. et al. "An Approach to Modeling Seasonally Stationary Time Series." *Journal of Econometrics* 9, no. 1, 2 (January 1979): 137-53. Topics discussed include: specific forecasting, time series analysis, spectral models, public utilities.

3285. Pauly, R. et al. "Calendar Effects in Structural Time Series Models with Trend and Season." *Empirical Economics* 14, no. 3 (1989): 241-56. Topics discussed include: time series analysis, spectral models.

3286. Perron, P. "The limiting distribution of the least-squares estimator in nearly integrated seasonal models." *Canad J Statist* 20 (1992): 2. Topics discussed include: numerical statistics, stochastic processes.

3287. Pfeffermann, D. "Estimation and Seasonal Adjustment of Population Means Using Data from Repeated Surveys - Reply." *Journal of Business and Economic Statistics* 9, no. 2 (April 1991): 177. Topics discussed include: econometrics, single equation analysis, multivariate models, information theory, inferential problems.

3288. Pierce, D. A. "Issues Involved with the Seasonal Adjustment of Time Series - Comment - Seasonal Adjustment-Models, Assumptions, and Criteria." *Journal of Business and Economic Statistics* 2, no. 4 (October 1984): 340-42. Topics discussed include: time series analysis, spectral models.

3289. Plosser, C. I. "The Analysis of Seasonal Economic Models." *Journal of Econometrics* 10, no. 2 (June 1979): 147-63. Topics discussed include: time series analysis.

3290. ---. "Short-term Forecasting and Seasonal Adjustment." *Journal of the American Statistical Association* 74, no. 365 (March 1979): 15-24. Topics discussed include: econometric theory, methodology, time series analysis, spectral models.

3291. Polasek, W. "Hierarchical Models for Seasonal Time Series." *Bayesian Statistics 2: Proceedings of the Second Valencia International Meeting, September 6, 10, 1983*, 723-31. ed J. M.Bernardo et al. Amsterdam; New York and Oxford, Elsevier Science, North Holland; Valencia, Spain Valencia University Press: Amsterdam; New York and Oxford, Elsevier Science, North Holland; Valencia, Spain: Valencia University Press, 1985. Topics discussed include: time series analysis, spectral models, trade relations.

3292. Raveh, A. et al. "Finding Common Seasonal Patterns among Time Series - An MDS Approach." *Journal of Econometrics* 12, no. 3 (April 1980): 353-63. Topics discussed include: time series analysis, spectral models.

3293. Raynauld, J. et al. "Seasonal BVAR Models: A Search along Some Time Domain Priors." *Journal of Econometrics* 55, no. 1-2 (January 1993): 203-29. Topics discussed include: forecasting, multiple, time series analysis, statistical approaches, multivariate models.

3294. Reinmuth, J. E. et al. "Recursive Models for Forecasting Seasonal Processes." *Journal of Financial and Quantitative Analysis* 9, no. 4 (September 1974): 659-84. Topics discussed include: time series analysis, spectral models, econometric theory, methodology.

3295. Schlicht, E. "Seasonal adjustment in a stochastic model." *Statist Hefte* 25, no. 1 (1984): 1-12. Topics discussed include: numerical statistics, stochastic processes, time, numerical statistics.

3296. Sims, C. A. "Rational Expectations Modeling with Seasonally Adjusted Data." *Journal of Econometrics* 55, no. 1-2 (January 1993): 9-19. Topics discussed include: estimation procedures, times series analysis.

3297. ---. "Comment on 'Issues Involved with the Seasonal Adjustment of Economic Time Series.'." *Journal of Business and Economic Statistics* 3, no. 1 (January 1985): 92-94. Topics discussed include: time series analysis, spectral models.

3298. Stier, W. "Fundamental Problems of Seasonal Adjustment of Economic Time Series." *Jahrbucher fur Nationalokonomie und Statistik* 192, no. 3-4 (November 1977): 315-35. Topics discussed include: time series analysis, spectral models.

3299. ---. "Seasonal Adjustment as a Problem in Filter-Design: On a New Method of Seasonal Adjustment." *Jahrbucher fur Nationalokonomie und Statistik* 195, no. 5 (September 1980): 421-34. Topics discussed include: time series analysis, spectral models.

3300. Terrell, R. D. et al. "Application of an Evolving Seasonal Pattern to an Australian Economic Time Series." *Australian Economic Papers* 9, no. 15 (December 1970): 186-218. Topics discussed include: time series analysis, spectral models.

3301. Thury, G. "The Consequences of Trading Day Variation and Calendar Effects for ARIMA Model Building and Seasonal Adjustment." *Empirica* 13, no. 1 (1986): 3-25. Topics discussed include: time series analysis, spectral models, distribution, retailing.

3302. Thury, G. et al. "Outlier Detection and Seasonal Adjustment." *Empirica* 19, no. 2 (1992): 245-57. Topics discussed include: econometrics, single equation analysis, times series analysis.

3303. Wallis, K. F. "Seasonal Adjustment and Revision of Current Data: Linear Filters for the X-11 Method." *Journal of the Royal Statistical Society, Series A; 145, Part 1,* (1982): 74-85. Topics discussed include: time series analysis, spectral models.

3304. Watson, M. W. "Uncertainty in model-based seasonal adjustment procedures and construction of minimax filters." *J Amer Statist Assoc* 82, no. 398 (1987): 395-408. Topics discussed include: numerical statistics, stochastic processes, time.

3305. Withycombe, R. "Forecasting with Combined Seasonal Indices." *International Journal of Forecasting* 5, no. 4 (1989): 547-52. Topics discussed include: econometric theory, methodology.

3306. Wooldridge, J. M. "A Note on Computing r-Squared and Adjusted r-Squared for Trending and Seasonal Data." *Economics Letters* 36, no. 1 (May 1991): 49-54. Topics discussed include: econometrics, single equation analysis, times series analysis.

3307. Yin, X. R. et al. "Multidimensional fuzzy Markov process and its application to the study of climate impact on economics." *Math Appl* 1, no. 1-2 (1988): 29-37. Topics discussed include: probability theory, stochastic processes, set theory, fuzzy sets.

Author Index

Please note: This index lists citation numbers as found in the select bibliographies (Chapters 3-6).

Subject Index

Please note: This index lists citation numbers as found in the select bibliographies (Chapters 3-6).

Country & Region Index

Please note: This index lists citation numbers as found in the select bibliographies (Chapters 3-6); citations not indexed typically study the country where the work is published, or the United States.

About the Author

PHILIP M. PARKER is an Associate Professor at the European Institute of Business Administration (INSEAD) in Fontainebleau, France, where he teaches graduate courses in multivariate statistics, research methodology, and international marketing. His research interests include evaluating cross-national effects of physical environment on human behavior and the resulting implications for business and economic policy. Dr. Parker has consulted on various projects with international development agencies and has conducted field research and training with private concerns in Europe, Africa, the Middle East, Asia, Latin America, and North America. He has recently published articles in academic and industry journals on various economic and business topics, including the diffusion of innovations, marketing, and industrial organization.

www.ingramcontent.com/pod-product-compliance
Ingram Content Group UK Ltd.
Pitfield, Milton Keynes, MK11 3LW, UK
UKHW021909150926
475560UK00003B/276

9 780313 294006